THE LAST FLIGHT OUT

Chuyến Bay Sau Cùng (original Vietnamese title)

Stories of Bravery, Love, and New Beginnings

by

Nguyễn Văn Ba

For information contact:
Web: https://nguyenvanba.com
Email: mdn425@gmail.com

ISBN: 979-8-9919618-0-6

BVN1ED – November 2024

CONTENTS

Acknowledgment

The family of Nguyễn Văn Ba extends our deepest gratitude to Captain Paul Jacobs and the brave crew of the USS Kirk for their extraordinary humanity and compassion shown on April 29, 1975. Their efforts saved countless lives, including those of our family, during a pivotal moment in history. We are also immensely thankful to Jan K. Herman, whose diligent research brought to light the heroism of that day in his book and documentary, "The Lucky Few: The Fall of Saigon and the Rescue Mission of the USS Kirk."

Special thanks to Rory Kennedy, Moxie Film, Naja Pham Lockwood, and PBS/American Experience for their pivotal roles in producing the 2015 Oscar-nominated film "Last Days in Vietnam." This documentary has been instrumental in sharing our family's story among the many other heartfelt stories during the last few days of April 1975, preserving the memory of those turbulent times.

War and conflict casts long shadows, bringing untold darkness and suffering. Yet, in its aftermath, it also reveals tales of resilience and the human spirit. The original contents of this book are drawn from writings by Nguyễn Văn Ba, written originally in Vietnamese over a period from the late 1980s until 2004 and previously published across various Vietnamese newspapers, primarily in Seattle and other U.S. cities. This translated collection serves not only as a tribute to the 50th-year remembrance (2025) of those pivotal events but also fulfills Nguyễn Văn Ba's heartfelt personal wish for his grandkids and descendants to know their roots and the strength of their heritage and culture.

Amidst recounting the realities of war and conflict from his unique vantage point, it was Nguyễn Văn Ba's hope to see a unified global Vietnamese community — one that shares common ideals and rich traditions, striving together towards reconciliation and brotherhood.

www.nguyenvanba.com

Section 1: The Fall and Escape

Survival and Sacrifice in the Face of Chaos - This section focuses on the dramatic events surrounding the fall of Saigon, the escape from Vietnam, and the immediate aftermath. It captures the intensity, fear, and desperation of those final days, as well as the personal sacrifices made.

Overview: Ch. 1 - "The Last Flight Out"

"The Last Flight Out" (Chuyến Bay Sau Cùng) recounts a dramatic and emotional chapter set during the final days of the Vietnam War, specifically on April 29, 1975. The narrative follows "Nam" (Nguyễn Văn Ba), a South Vietnamese Air Force officer, and his squadron, "Lôi Thanh 237," as they navigate the chaos and danger of the last hours before the fall of Saigon.

Duty and Leadership in Crisis: The story opens at Tân Sơn Nhất Airport, where Nam and his fellow pilots are anxiously waiting for orders. As the situation in Saigon deteriorates, Nam faces a critical decision about whether to flee or stay. Despite the overwhelming fear and uncertainty, Nam remains calm and committed to his duty, waiting for his commanding officer and ensuring his crew is ready for any eventuality.

The Chaos of War's End: As the North Vietnamese forces begin their final assault, the airport is bombarded with artillery, creating an atmosphere of panic. Nam witnesses other aircraft evacuating families of high-ranking officers and realizes the gravity of the situation. Despite the chaos, Nam decides to lead his crew to Vũng Tàu, believing it to be safer.

Sacrifice and Responsibility: Throughout the narrative, Nam is portrayed as a responsible leader who prioritizes the safety of his crew and their families. He carefully plans their departure, even taking the risk of returning to Saigon (hot zone) to rescue his wife and children. His actions reflect his deep sense of responsibility, not just to his family but also to his squadron.

The Personal Cost of War: As Nam and his crew make their final escape, the story emphasizes the personal sacrifices made by those involved. Nam's final flight, which ends with him ditching his Chinook helicopter in the ocean to ensure his crew's safety, symbolizes the ultimate sacrifice—a leader who is willing to risk everything for the lives of others.

Reflection and Legacy: The narrative concludes with Nam reflecting on his journey, the loss of his country, and the sacrifices he made. Despite the chaos and the ultimate collapse of South Vietnam, Nam takes pride in having fulfilled his duty to the very end. The story ends on a poignant note, capturing the bitterness and sense of loss that came with the end of the war.

Overall, "The Last Flight Out" is a powerful account of bravery, leadership, and the personal cost of war, offering a deeply emotional perspective on the final moments of the Vietnam War as experienced by those who lived through it.

Overview: Ch. 2 - "Operation Tống-Lệ-Chân"

"Operation Tống-Lệ-Chân" (Phi Vụ - Tống-Lệ-Chân) tells the harrowing story of a critical military mission during the Vietnam War, focusing on the South Vietnamese Air Force squadron "Lôi Thanh 237." The narrative centers on an operation at Tống-Lệ-Chân, a strategically vital hill near An Lộc, which became a key battleground during the intense fighting of 1972. The North Vietnamese launched a massive offensive aimed at capturing the hill to cut off South Vietnamese defenses and assert control over the region. The hill was defended by a small but determined unit of South Vietnamese Rangers who found themselves isolated and surrounded by enemy forces, making resupply and evacuation missions essential for their survival.

Heroism and Sacrifice in War: The pilots of "Lôi Thanh 237" were tasked with a dangerous mission to deliver reinforcements and evacuate the wounded from Tống-Lệ-Chân, knowing they would be flying into a heavily fortified area. The story vividly depicts the extreme risks they faced, including heavy anti-aircraft fire and the threat of enemy missiles. Despite these overwhelming dangers, the pilots demonstrated extraordinary bravery and a deep sense of duty, fully aware that their chances of survival were slim. Their willingness to undertake such a perilous mission reflects the spirit of sacrifice that defined their service during the war.

Camaraderie and Brotherhood: The narrative emphasizes the strong bonds of brotherhood that developed among the soldiers and pilots involved in this mission. These men were not only united by their shared experiences in battle but also by the deep emotional connections that these experiences forged. Years after the war, these bonds remain unbroken, as evidenced by the reunion of the surviving squad members in California. This gathering underscores the lasting impact of their wartime relationships and the enduring camaraderie that continues to connect them.

The Impact of War on Memory and Legacy: The mission at Tống-Lệ-Chân left a lasting imprint on those who participated in it. The story reflects on how these events have continued to shape the lives and identities of the veterans involved. The sacrifices made, particularly by those who did not survive, are honored and remembered by their comrades. The narrative underscores the importance of preserving these memories as part of the veterans' legacy and as a testament to their courage.

The Tragedy and Futility of War: The story does not shy away from depicting the grim realities of war, including the fear, tension, and tragic losses experienced during the mission. The deaths of crew members, whether from enemy fire or crashes, highlight the immense human cost of the conflict and the tragic futility of war.

Overall, "Operations - Tống-Lệ-Chân" serves as a powerful tribute to the courage, sacrifice, and enduring bonds of the South Vietnamese Air Force. It reflects on the profound and lasting impact of the Vietnam War on those who served and honors the memory of their heroic efforts.

Overview: Ch. 3 - " Footprints of a Refugee"

"Footprints of a Refugee" (Vết Chân Tị Nạn) narrates the emotional and harrowing experiences of Nam, a South Vietnamese Air Force officer, during and after the fall of Saigon in 1975. The story captures Nam's journey from the chaos of the Vietnam War to his eventual escape and life as a refugee in the United States, highlighting the immense challenges faced by him and his family.

The Chaos of War and Displacement: The narrative begins with Nam and his family fleeing the imminent collapse of South Vietnam. Amidst the chaos at Tân Sơn Nhất Airport, Nam manages to reunite with his wife and children, finding temporary relief on a U.S. Navy ship. The scenes depict the panic, confusion, and desperation as thousands try to escape the advancing North Vietnamese forces. Nam's sense of duty and leadership is evident as he navigates his family through the uncertainty, making quick decisions in life-threatening situations.

The Struggles of Refugees at Sea: Nam's experience aboard the ship is filled with tension and hardship. The ship, packed with refugees, sails away from the Vietnamese coast, symbolizing the irreversible break from their homeland. The story details the cramped and uncomfortable conditions on the ship, where refugees, including Nam's family, face a lack of basic necessities like food, clean water, and shelter. The emotional and physical toll of the journey is palpable as Nam struggles to provide for his children, particularly his infant daughter who suffers without proper nutrition.

Resilience and Adaptation: Despite the hardships, Nam's resilience shines through. The narrative follows Nam's journey from the ship to refugee camps in Guam and later to the United States. Each stop on this journey brings new challenges—whether it's the scarcity of resources in the camps or the uncertainty of their future. Nam's resourcefulness and determination to secure a better life for his family are highlighted as he navigates the bureaucratic processes, secures temporary jobs, and eventually transitions to life in America.

Reflection on Loss and New Beginnings: The story is also a reflection on the immense loss experienced by Nam and many like him. As Nam leaves Vietnam, he reflects on the country he fought for, now lost to him forever. His journey is not just a physical one but an emotional and psychological transition from a life of service and battle to one of survival and adaptation in a foreign land. The narrative ends on a note of cautious optimism as Nam and his family settle into their new life, carrying with them the scars of the past but also the hope of a new beginning.

"Footprints of a Refugee" provides a poignant and detailed account of the refugee experience, capturing the trauma, resilience, and enduring hope of those who fled Vietnam during one of the most turbulent times in the country's history.

CHAPTER 1

The Last Flight Out

Original Title: "Chuyến Bay Sau Cùng"

On April 29, 1975, at 4:00 AM, four massive Chinook transport helicopters were parked in parallel in front of the civil aviation terminal of Tan Son Nhat Airport, Vietnam. Nam's crew had taken refuge there the previous day, April 28, because several South Vietnamese pilots who defected and allied with the Viet Cong had commandeered a few aircraft from the Republic of Vietnam Air Force to bomb the airfield, further intensifying the chaos and terror among the people of Saigon.

The fate of Nam's 237th Airborne Battalion was tied to the destiny of South Vietnam. All members of the flight crew were visibly anxious, awaiting Nam's decision, but he hesitated, hoping that the commanding officer of Loi Thanh's squadron would join them before making any final decisions. A few days before, on April 27, the 237th Battalion had been ordered to move from their base in Bien Hoa to Tan Son Nhat, leaving the commanding officer behind to coordinate operations with Brigadier General T., commander of the 3rd Air Division and Colonel T., commander of the 43rd Tactical Air Wing. Meanwhile, Nam was assigned to command the establishment of an air bridge between Bien Hoa and Tan Son Nhat.

At this time, the staff officers of the battalion were notably absent after hearing the news of the bombing of the airfield. The afternoon of April 28 was a time of extreme tension as if they were sitting in a boiling cauldron. Nam noticed several transport planes, including C-130s, C-47s, and C-119s, taking off one by one, carrying the families of senior officers of the Air Force Command to Thailand or Con Son Island. Many of the remaining officers and non-commissioned officers clung to Nam as if he were their last lifeline.

Nguyễn Văn Ba ["Nam"]

After ensuring that the aircraft was in good operational condition and fully refueled, Nam organized four flight crews, consisting of himself, Captain S., Lieutenant Q., Lieutenant K., and other aircrew members; weapons were hidden on board, ready for use if necessary. Nam issued a password for everyone and assigned them to guard duty to prevent any sabotage by enemy commandos in the night. That night, everyone slept under the aircraft, waiting for any new information from the Air Force Command.

Nam had not had a chance to bathe for three days, as after sending his wife and children to stay with family in Phu Lam (a small province in Saigon), he lived in the battalion, strictly confined to the base, like all the other men. Some officers informed Nam that they had seen the families of the Wing Commander, Deputy Wing Commander, and other tactical wing commanders busy evacuating their wives and children out of Vietnam. Occasionally, Nam had asked the commanding officer of Loi Thanh squadron about the evacuation plan, but he only vaguely knew that they were waiting for orders from above.

Nam understood that in this critical situation, the commanders of the divisions, wings, and senior staff officers of the Air Force Command had kept the information secret to avoid causing panic among their subordinates.

Nam advised his crew to sleep on the concrete floor under the aircraft; if the enemy attacked, they would have a better chance of survival and a wider field of vision to counterattack effectively.

In the early evening, Nam noticed many transport aircraft of the Republic of Vietnam Air Force taking off, full of people, heading south. He didn't pay much attention to this. Occasionally, a few foreign civilian aircraft landed and parked next to the Chinooks. The passengers were mostly civilians, well-dressed. Nam thought they were probably the wives of government officials or the families of South Vietnamese elites heading to Thailand.

In these extremely difficult moments, Nam couldn't be sure if the members of his crew shared the same goals, as he suspected that the Viet Cong might have compromised some of them. He silently observed their behavior and encouraged them to wait for orders from their superiors.

After several sleepless nights, coupled with endless worries, Nam eventually fell asleep without realizing it. The sound of a 122mm rocket exploding nearby jolted everyone awake. Nam saw that the southern end of the runway was ablaze, likely from a hit on a fuel depot. It was around 3:30 AM on April 29, 1975. The Viet Cong were concentrating their firepower on the airfield, with explosions echoing continuously. The sound of enemy 130mm artillery, combined with the howling of 122mm rockets, created an eerie, terrifying atmosphere.

The airfield was nearly paralyzed, as Nam didn't hear or see any fighter planes taking off to strike back at the enemy's artillery positions. Nam called all four flight crews together, instructing them to be ready to take off when they saw the first aircraft start its engines. The men were anxious, eagerly waiting for news from the Air Force Command through the Paris frequency.

As an officer standing on the edge of history during a period of national turmoil, Nam remained absolutely committed to military discipline, placing himself under the command of his superiors. Like a knight with a horse (helicopter) at his disposal, he could flee at any moment but chose not to, as he wasn't a deserter fleeing his responsibilities. His duty as a commander didn't allow him to do so. With the commanding officer absent, Nam was the commander of his unit, and the remaining officers and non-commissioned officers saw him as their final hope in this almost hopeless situation.

At around 4:00 AM, Nam received news from the airport security radio that the Viet Cong were attacking and had breached the northwest perimeter of the airfield, not far from where the helicopters were parked.

Nam realized that staying any longer would be futile. He ordered all crews to start their engines and head towards Vung Tau Airfield. He believed that being close to the coast, Vung Tau might be less susceptible to enemy artillery fire. Nam led first, only to discover that Captain S.'s helicopter had an engine malfunction and couldn't start. Nam turned back to pick up the crew, but just then, Captain S.'s aircraft managed to start its engines. All four helicopters headed towards Vung Tau, with the sky still dark. Nam noticed flashes of light around the airfield, likely enemy artillery firing at the airport.

The formation climbed to about 3,000 feet, looking down at Saigon, where many small fires were burning. Glancing back at Tan Son Nhat Airport, he saw that the fuel tanks were engulfed in flames, lighting up the sky. As they approached the Vung Tau Naval Base, the formation was warned to keep their distance as the Navy was firing at aircraft from their ships. Nam instructed his men to turn off the navigation lights to avoid detection from the ground.

Before long, all four helicopters landed safely at the refueling station designated for helicopters. Nam instructed everyone to move the helicopters to a nearby empty field and shut down the engines to listen for any updates from Saigon, giving them time to take care of personal matters. Nam took the opportunity to change out of his flight suit, which he had been wearing for three days. Just as he finished changing and was about to put on his shoes, a sudden "boom-boom" echoed through the air as the Viet Cong began shelling their location again. Everyone quickly started their engines and took off, with Nam ordering them to head towards Can Tho. Once the helicopters were airborne, Nam tried to contact the Paris frequency one last time but received no orders. Nam thought, "That's it then, we are on our own."

A minute later, Nam called the formation to check their positions:

- "Big Eagle 1 here, calling 2, 3, and 4. Report your position."
- "2 here, to your right, one mile away, over."
- "1 copies, 5/5, 3, report your position."
- "4 here."
- "4 copies, 5/5."

It seemed that Big Eagle 3 had separated from the formation during takeoff from Vung Tau.

Nam remained silent in the cockpit, realizing that yet another member of the formation had gone off on their own without a final word of farewell or goodbye. At that moment, Nam didn't have time to be angry or blame anyone for their actions. The aircraft was government property, but at this dire moment, it "belonged" to the men in the cockpit. They could use it however they saw fit, even though Nam was their direct commander. The reality was that at this point, all that was left was the camaraderie of "brothers in arms," as Nam's superiors had all either fled or gone into hiding.

- "Big Eagle 1, this is 2, calling."
- "1 here, 5/5."
- "Requesting permission to land at My Tho Lake to say goodbye to my family, over."
- "Permission granted. Be careful when landing, the area is small. 2 and 4, do you copy?"
- "2 copies, 5/5."
- "4 copies, 5/5."

Flying at a low altitude, Nam noticed that My Tho was completely calm. The market was bustling with shoppers, showing no signs of disorder or war. He ordered the three Chinooks to land, taking up almost the entire lakefront. After shutting down the engines, everyone felt a sense of relief, as it seemed there was no sign of bombing or chaos here, and the locals were probably unaware of what had happened in Saigon.

While Captain S. went to visit his family in the city, Nam discussed with his men whether they should stay there instead of continuing to Can Tho (further south), and everyone agreed. Lieutenant K. was tasked with flying to Dong Tam Airfield to get more fuel for the three helicopters.

At this time, it seemed that most of the crew members were still single, except for Nam and Captain S., who both had families. Nam's wife and three children were still temporarily staying in Phu Lam, so he decided to return to pick them up and take them wherever they could.

At around 9:30 AM on April 29, 1975, after giving instructions to his men to wait for him until around 11:00 AM if possible, Nam and Sergeant C., Sergeant M., and the gunner took off, heading back to Phu Lam.

The weather in southern Vietnam at this time of year was often cloudy, with low cloud cover. Nam knew that the Communists had placed anti-aircraft guns near the outskirts of the city in the Binh Tri area to shoot down planes approaching the airport, so he flew above the clouds to avoid anti-aircraft fire and Sam-7 missiles. About fifteen minutes later, he spotted Tan Son Nhat Airport, still shrouded in smoke and fire. Descending through the clouds, Nam saw his house below. Circling low, he saw his wife, children, and in-laws running out into the yard, looking up. Nam waved and landed in the empty playfield in front of the house.

Sergeant C. opened the rear door of the Chinook and ran out to help

Nam's wife, children, and two adopted nieces on board. By this time, curious onlookers had crowded around to see what was happening. Police cars, sirens blaring, also rushed to the scene, seemingly unaware of what had happened.

Within three minutes, Sergeant C. reported, "All clear, ramp up, ready for takeoff." Nam took off with full power, and the massive aircraft, like a wild stallion, gracefully leaped into the sky, leaving behind a terrifying whirlwind. Later, Nam's mother would tell him that the helicopter's takeoff had caused the collapse of a house behind theirs, and they had to pay to rebuild it.

In a short time, Nam had climbed above the clouds again, just as the red light indicating a malfunction in engine number two came on. Normally, this would require finding a place to land as soon as possible, but at that moment, it was impossible. Nam knew that the area below was Binh Chanh, likely filled with Viet Cong guerrillas advancing toward Saigon. However, Nam was confident in his ability to fly the helicopter to safety. To his left, not far away, Nam saw a Huey helicopter flying towards My Tho. He opened the emergency frequency to request an escort in case of an emergency landing but received no response.

Despite everything, the flight proceeded smoothly, and the helicopter landed safely at My Tho Lake, where the crew was still waiting. They were overjoyed to see Nam return on time.

By 11:00 AM, Nam had given the flight crew a sum of money, both from the battalion's funds, which Lieutenant L. had handed over the day before and from what his wife had brought along. He sent the men off to have lunch and enjoy a final day with friends and family in My Tho.

Nam sat with his wife, feeding their children. The youngest of them was Mina, a daughter just eight months old, while Mika, their son, was three, and Miki, their eldest, was seven. At that moment, the men brought an elderly man, over sixty years old, to Nam. He introduced himself as the father of Brigadier General T., commander of the 3rd Air Division in Bien Hoa. He asked Nam for a ride to Saigon if they were flying back, explaining that the roads between Saigon and My Tho had been cut off by the Viet Cong. Nam explained that they were all fleeing from Saigon and hadn't yet decided where to go in the coming hours. Hearing this, the old man shook his head in disappointment.

"If that's the case, please take me wherever you're going."

"Sure, just wait here," Nam replied.

Nam reviewed the map and discussed with his men that they couldn't stay there overnight due to the insecurity and dangers, as the Communists could attack at any time. It was better to find an island off the coast to spend the night while waiting for news from Saigon.

There were plenty of weapons and ammunition on board, so the crew could hold out if necessary, and everyone seemed to agree with this plan. Nam asked the driver of the old man's Toyota to take him into town to buy provisions. He purchased two 100-kilogram sacks of rice, several kilograms of sugar, a dozen bottles of fish sauce, and some dried snakehead fish, as well as cooking utensils. These supplies were intended to sustain them for a while as they waited.

Lieutenant K. had finished refueling the last helicopter and approached Nam to report the news he had heard on the emergency frequency.

"I heard on the guard frequency that the U.S. Navy is calling all pilots to evacuate to the 7th Fleet. What do you think, Major?"

After a moment of thought, Nam shook his head.

"Why listen to the Americans? We should wait to hear news from Saigon Radio and see what happens."

Lieutenant K. seemed to understand Nam's intentions and didn't argue further, returning to his helicopter to rest.

At around 1:00 PM, everyone was back on board, including Captain S. and his family. Nam suddenly remembered the bottles of Napoleon brandy that Colonel T. had given him a few days before when he moved his family to Saigon. Nam took them out and invited the men to share what might be the final drink of the 237th Airborne Battalion.

As they were enjoying the moment, trying to temporarily forget the heavy burden of their military careers, a jeep arrived, driven by a lieutenant colonel from the infantry, wearing the insignia of the General Staff. The fear and anxiety were evident on his face and in his gestures.

"Where are you planning to go?" he asked.

"I'm waiting for orders from Saigon," Nam replied.

"My God! Don't you know? The Chief of General Staff has already fled, and there's no one left at the General Staff Headquarters. I'm the last one who managed to get out and come down here. If you stay here any longer, the Provincial Governor of My Tho will have his troops hold you to evacuate their families. If you don't believe me, wait, and I'll go get my family to come with you."

"OK, go get your family."

After that, Nam called a secret meeting with the battalion inside the military cemetery in My Tho.

"The colonel's words make sense. If we stay here, we'll all be trapped. What do you guys think? Should we head to the 7th Fleet or fly somewhere else?"

Lieutenant K. quickly responded, "I think we should fly to the 7th Fleet for safety, and then we'll figure out the next steps. What do you think, Major, and the rest of you?"

"If you all agree with Lieutenant K.'s suggestion, we'll go. If anyone has family or wants to stay, feel free. In this difficult time, I won't force anyone to follow me."

A few men chose to stay, as they were uncertain about what their lives would be like if they left. However, most of the crew agreed to leave. Nam shook hands with those staying behind, wishing them good luck. He divided the group into two flight crews: one under his command and the other under Captain S., while the third helicopter remained on the lake shore as a backup.

It was now around 1:00 PM in My Tho. Two Chinooks took off from the lake, heading towards Vung Tau. On the guard frequency, the 7th Fleet had already called and given the position of the U.S. ships moving off the coast of the South China Sea.

Flying at high altitude, Nam looked down at the lush green fields, the winding rivers, and the thatched roofs of the Long Dinh village, where smoke from cooking fires rose in thin wisps. The countryside seemed peaceful and beautiful, with no sign of the turmoil that had engulfed Saigon. The simple, honest villagers below were likely unaware that Nam and his crew were fleeing, about to leave behind their homeland, where they had experienced countless memories from childhood, school days, and military service. At this moment, the war has reached not only the rural outskirts but also the heart of the capital, Saigon, and Tan Son Nhat International Airport. The people of Saigon were now experiencing the full force of bombs and shells from the so-called "liberation forces," who would soon bring with them a regime of poverty, atheism, and backwardness to remake South Vietnam and lead the nation back to medieval times.

Nam was brought back to reality when his copilot, Lieutenant M., exclaimed, "Major, do you see that column of black smoke at 11 o'clock on the sea? That's the checkpoint the Americans gave us."

"OK, let's head towards that. Their fleet shouldn't be far."

Captain S.'s helicopter, faster than Nam's, had already flown ahead. After twenty minutes of flying over the water, the Vietnamese coastline faded into the distance below the horizon. Nam grew anxious, as there were no life vests on board for his wife, children, and crew if there was an accident at sea. If the helicopter malfunctioned, they would all perish in the deep ocean.

Nam strained his eyes, searching for any sign of the fleet, but the fog was too thick, limiting visibility. At that moment, Lieutenant M. pointed towards the 12 o'clock position.

- "Major, do you see that?"
- "Yes, it looks like a ship!"

A wave of relief washed over everyone like a drowning man grabbing a lifeline. Nam instructed Lieutenant M. to keep an eye on the surroundings while he circled the ship, seeking permission to land. However, as it was a patrol boat, it only had a small deck designed for Huey helicopters, and Nam's Chinook was too large to land.

Editor's Note: *This story (chapter 1) was written in the late 1980's. Nguyễn's family did not know the name of the ship, nor had they seen these photos from the US Navy until 2009.*

Nguyễn Văn Ba ("Nam") approaching the USS Kirk.

At first, the sailors on the ship waved their arms, signaling the Chinook not to land nor get close as it was too big to land, and even aimed their high-caliber rifles at the helicopter as a warning to back away. The patrol boat continued to move quickly, not stopping. But Nam persisted, following closely until they eventually gave in and basically reduced speed. With no other choice, the only option to evacuate family and crew was for Nam to hover precisely towards the rear end of the ship, carefully staying away from the many tall antenna masts and equipment on the ship and giving everyone the chance to jump down onto the ship's deck.

Carefully, Nam eased into position, and everyone began to jump from the front starboard (right) side door, with American sailors waiting below to catch them, as a misstep could easily result in broken legs or injury as well as injury to the sailors below. Nam's wife and children were no exception to this perilous descent. Everyone was safely on board within minutes, and Nam turned to Lieutenant M.

"Now it's your turn, but before you jump, check the back of the helicopter to make sure no one is left behind. I'll ditch the helicopter."

Lieutenant M. removed his flak jacket, unbuckled his seatbelt, and stepped out of the cockpit to perform one last check. A few seconds later, he gave a thumbs-up to indicate everything was clear. Nam nodded in thanks, and moments later, he saw Lieutenant M. waving from the ship's deck below. All clear!

At around 2:00 PM on April 29, 1975, Nam flew the Chinook away from the ship and moved a safe distance, about 200 feet away. Several U.S. Navy cameras were locked in on the Chinook, not wanting to miss this rare opportunity, as they knew Nam would have to ditch the helicopter right next to them.

Nguyễn Văn Ba hovered to remove gear and clothing.

The CH-47 was now in a safe position, and Nam pulled the emergency release handle, ejecting the port (left) window out into the sea. Slowly, he began the difficult process of removing his flak jacket, flight suit, and weapons with one hand while flying with the other. Looking back, Nam should have asked Lieutenant M. to take over flying so he could free himself of these cumbersome items before M. jumped out. But now it was too late (a lesson learned, one that Nam hoped never to have to apply again).

Nam was now ready for the final step. The U.S. Navy had already deployed a life raft and was waiting about 200 feet away. Nam gave a thumbs-up to signal that he was ready. The enormous rotor blades spun above, their sharp edges slicing through the air like deadly swords. If Nam made a mistake, the consequences could be fatal, turning him into mincemeat for the sharks below. The truth was, none of the CH-47s in the battalion were equipped for water landings like those practiced at Fort Rucker (Alabama) many years earlier in flight training school, as water would flood the helicopter before the rotor blades stopped spinning.

Nam checked the instruments one last time, and all systems were "green". He gently pushed the cyclic forward and to the right while applying right pedal, causing the Chinook to move forward and tilt to the right. At that moment, Nam quickly jumped out of the left window into the ocean below.

Nguyễn Văn Ba ("Nam") in the water.

He tried to dive deep several times, but the salty water kept pushing him back to the surface. Nam opened his eyes to see the helicopter, now resembling a giant, angry beast, its rotor blades churning violently on the water's surface. The massive machine rocked back and forth, struggling before finally beginning to calm down a bit. Nam attempted to dive again, knowing that the helicopter's proximity posed a significant risk of injury from the still-spinning rotor blades. After a few more seconds, Nam surfaced, and all was quiet. The helicopter hadn't fully sunk yet, still partially floating on the water's surface. Just then, the U.S. Navy's rescue boat arrived and pulled Nam aboard.

Nguyễn Văn Ba's ("Nam") first taste of freedom.

When Nam stepped onto the ship's deck, all he had were his boxer shorts and a short-sleeved shirt (*no money, gold, or identification*). A sympathetic Navy sailor gave him a yellow plastic raincoat to keep warm. The ship's captain then approached Nam, shook his hand, thanked him for his composure in safely evacuating the passengers, and asked to take a commemorative photo.

Nguyễn Văn Ba's ("Nam") and Captain Paul Jacobs.

This was the final flight of Nam's military career, a contribution and sacrifice for his country. Nam had always hoped and dreamed that Vietnam would one day be peaceful and happy, only to be met with a bitter and painful reality. In the end, all he had left from his homeland were his wife, three children, and his immediate clothing.

Reflecting on the men of the 237th Airborne Battalion, Nam took great pride in having fulfilled his duty as their final commander. He hadn't abandoned his unit or deserted his men to save himself. Even in the final moments, Nam had willingly sacrificed himself like a ship's captain, ensuring the safety of others first. Nam didn't like to glorify himself, but this was the reality.

Nam also didn't want to mention the names of the men who had left the formation to flee before him. He believed that out of fear for their own lives and the lives of their families, they had acted out of desperation.

Nam didn't forget to thank Sergeants C. and M., and Lieutenant M., who had flown with him back to Phu Lam to pick up his wife and children at the last moment. He also thanked Captain S., for if it hadn't been for his request to land at My Tho, their fates might have been very different. Nam also thanked Lieutenant K. and all the crew members who had carried out their duties until the very end.

Editor's note: Nguyễn Văn Ba's family wishes to extend our gratitude to Captain Paul Jacobs and the crew of the USS Kirk for their rescue of Nguyễn's family and many other families on April 29, 1975.

USS Kirk – DE 1087

CHAPTER 2

Operation Tống-Lệ-Chân

Original Title: "Phi Vụ - Tống-Lệ-Chân"

At first glance, "Tống-Lệ-Chân" might sound like the name of a beautiful and talented Chinese woman immortalized in history books. However, this name held immense strategic importance during the scorching summer of 1972.

Several North Vietnamese regular divisions, supported by tanks and heavy artillery, advanced from the Cambodian and Lower Laotian borders along the Ho Chi Minh Trail, aiming to capture Bình Long and Phước Long provinces. Their goal was to establish a government they called the "Liberation Front of South Vietnam," gaining international support and putting immense pressure on the Republic of Vietnam's government in Saigon.

Tống-Lệ-Chân was a strategic location, a tactical hill approximately five miles southwest of An Lộc city (northwest of Saigon). Positioned on a high elevation, it was guarded by a battalion of elite South Vietnamese Rangers. This hill was not only the eyes and ears of An Lộc but also of Saigon itself, as it monitored the movement of North Vietnamese troops and war machines heading towards Saigon, potentially threatening Bình Dương, Lái Thiêu, and Biên Hòa, where the III Corps Tactical Zone headquarters was located.

For these tactical and strategic reasons, Tống-Lệ-Chân became a sharp thorn in the side of the North Vietnamese, something they were determined to remove at all costs.

Before the battle, Tống-Lệ-Chân was a serene hill surrounded by vast green forests, with a winding stream at its base, resembling a peaceful landscape painting. If not for the war, it would have remained unnoticed, much like other remote places that suddenly became infamous due to brutal battles—places like Ben Het, Đức Cơ, Đắc Pét, and Đắc Tô. These locations, once unknown to city dwellers, became household names, frequently mentioned in horrifying war reports on the radio, television, and in newspapers, shocking the conscience of the world.

The scars of war had faded into the background of memory, almost forgotten. But this evening, those memories were reignited within me as I unexpectedly met comrades from my old unit after more than twenty long years of separation.

The hero of "Tống-Lệ-Chân," former Captain Lê Văn Cầu, the flight leader of the 237th Chinook Squadron, still had the same slender build and the ever-present open smile that greeted his comrades. We reunited at a gathering in San José, California, organized by a few former members of the 237th Lôi Thanh Squadron, including Nguyễn Văn Tiên, Nguyễn Vũ, and Nguyễn Văn Mai (from Đà Nẵng), along with the wives of Vũ and Mai. It was a great joy and pride for me to meet again with these men, with whom I had once shared life-and-death experiences, including shared rations of rice and pork under the blistering sun in the battlefields of Tây Ninh.

Seeing each other again in this foreign land, we had aged a little, our lives had changed significantly, but the camaraderie remained as warm and strong as ever.

"Tôn" was still as charming as ever, "Ngọc" remained the playboy, "Vũ" was still lively and youthful, "Mai" still fancied himself as Henry Chúc, and "Tiên," having recently arrived, was still adjusting with some apprehension. Senior member "Mai," the former squadron leader from Đà Nẵng, had just retired from IBM and looked as dignified as ever, though perhaps he needed some "freshness" to prevent further hair loss.

We shared drinks, reminiscing from 3 PM until nearly 7 PM, before parting ways, as some of the men had to attend a wedding that evening. We agreed to meet again at the "MINI" club at 9:30 PM.

Senior member Mai took me home to visit his family, while "chú tư Cầu" (a term of endearment for Captain Cầu) went home to fetch his wife and children. Chú tư Cầu was incredibly lucky to have a young and beautiful wife who not only supported him but also gave him two wonderful, well-behaved children. Senior member Mai, possibly saving his energy for his wife later that night, went to rest, leaving me and Cầu to continue drinking and talking late into the night about everything from the sky to the earth, and the thrilling stories of past military operations.

I had known Cầu since the establishment of the first Chinook Squadron, the 237th Lôi Thanh, around 1970-1971. He was one of the young, eager pilots fresh out of school, full of enthusiasm for every mission, while we, though not yet old, were considered "veterans" by comparison.

During the summer of 1972, the battles in the III Corps Tactical Zone grew increasingly fierce. An Lộc city endured thousands of artillery shells day and night, reducing homes and streets to rubble. The residents abandoned their belongings and fled toward Chơn Thành and Saigon.

The outpost at Tống-Lệ-Chân, defended by a battalion of South Vietnamese Rangers, had been isolated for weeks by a North Vietnamese regiment lurking at the base of the hill. Supply and medevac helicopters were under severe threat from SA-7 heat-seeking missiles and anti-aircraft guns, making daytime landings nearly impossible. C-130 transport planes attempted to drop supplies and medicine by parachute, but most of the cargo was blown off course into enemy-controlled areas.

The Viet Cong regularly bombarded the hill with artillery, turning it into a barren landscape. Yet, the courageous brown-bereted South Vietnamese Rangers dug deep into the earth like venomous snakes, valiantly repelled wave after wave of human-wave assaults by the Viet Cong. The outpost commander, Lieutenant Colonel Ngôn, requested air support to bring in reinforcements and evacuate the wounded so he could continue the fight.

The General Staff of the South Vietnamese Armed Forces ordered the III Corps Air Division Commander to assign the giant Chinook helicopters of the 237th Squadron to carry out this suicidal mission. Lôi Thanh 01, the squadron leader of the 237th, received the secret order and convened a closed meeting with a group of experienced combat pilots to select volunteers. After Lôi Thanh 01 explained the mission, the room fell silent, heavy with tension and a sense of foreboding. No one spoke a word, adding or subtracting from the instructions.

I looked at all the pilots and aircrew, my eyes lowering as my mind went numb, thinking to myself, "They're about to make us sacrificial lambs." In truth, none of us were afraid of dying or wanted to avoid dangerous missions, but the frustration was unbearable. It seemed that our commanders either didn't understand the technical capabilities of the Chinook or were afraid of responsibility, leading to this grave mistake.

It wasn't a matter of political, psychological, or tactical considerations; the mission could have been accomplished with smaller helicopters like the U-1H, each carrying ten soldiers at a time and landing in different locations over five attempts. The U-1H was designed for such surprise tactics deep in enemy territory, with the ability to maneuver quickly and land troops and supplies within a few seconds, minimizing exposure to enemy fire.

In contrast, a massive Chinook CH-47, carrying 50 soldiers, landing on a hilltop outpost with no cover, surrounded by North Vietnamese regulars entrenched in a maze of trenches armed with advanced Soviet and Czech anti-aircraft guns, 82mm mortars, and 122mm artillery with pinpoint accuracy, was a recipe for disaster. Even if the enemy was blind, deaf, and asleep, the slow approach and departure of the Chinook would take at least 200 seconds, far too long to avoid being shot down.

To compare, an F-5 jet could dive, drop its bombs, and pull up in a matter of seconds, yet could still be easily shot down by SA-7s or anti-aircraft fire. The lumbering Chinook, twice or three times the size of a jet, with thousands of liters of JP-4 fuel on board, could be a massive target, turning into a blazing torch at the slightest spark, killing not only the crew but also the 50 soldiers inside.

In this narrative, I don't intend to criticize any individual commander, as orders came from the highest level. I merely highlight the mistakes in the tactical use of mechanical warfare. But in the military, orders must be followed. If one survives without injury, it's considered a blessing from the gods above.

After a tense minute of silence, Lôi Thanh 01 stated that if no one volunteered, he would have to assign someone to this critical mission. Time seemed to slow down. I asked what the primary objective of the mission was, though I was the Operations Officer of the squadron at the time. Lôi Thanh 01 refused to divulge the details, maintaining the secrecy of the landing zone. When he finished speaking, I saw Captain Lê Văn Cầu, the leader of Flight 1, raise his hand to volunteer, followed by Captain Huỳnh Bá Hùng, the leader of Flight 2.

We still needed a flight engineer, a gunner, and an escort. Sergeant Nguyễn Văn Tranh and Corporal Nguyễn Văn Hoàng also volunteered. I recall that several other officers and non-commissioned officers volunteered as well, but in the end, the crew selected for the mission consisted of:

- Aircraft Commander: Captain Cầu
- Co-Pilot: Captain Hùng
- Flight Engineer: Sergeant Tranh
- Gunner: Corporal Hoàng
- Escort (unfortunately, I don't remember his name).

The meeting ended around 4 PM. To maintain the mission's secrecy, the squadron was ordered to stay on base until the mission was completed. I took the volunteer crew members to the mess hall in the NCO's quarters on the airbase for dinner.

At 7 PM on December 26, 1972, the aircraft took off from Biên Hòa Airfield, maintaining radio contact with C&C (Command and Control center) on the FM frequency as they headed for An Lộc to receive a briefing and orders from the friendly unit.

As it was late December, the Christmas season, the sun set earlier than usual, and the last golden rays of sunlight faded behind the airfield, leaving me alone to bid farewell to the crew.

In the last world war, the Japanese kamikaze pilots were honored before their final missions with a ceremonial sake toast from a senior officer before flying off to their deaths. But who was there to send off our men? Where was the honor guard to salute them? I knew they didn't need such pomp and circumstance, but with iron determination, they were ready to sacrifice their lives for their motherland, willingly accepting a mission to land in what could only be described as "hell on earth."

The aircraft disappeared into the twilight, heading northwest. I quietly drove back to the Operations Room to monitor the mission, praying for the crew's safe return.

The responsibility of the mission was shared among the crew: Captain Cầu commanded the flight, handling the aircraft and its systems, while Captain Hùng managed radio communications with C&C and the friendly unit, checked flight maps, and navigated coordinates. The flight engineer, gunner, and escort ensured the aircraft's security, monitored enemy SA-7 positions, and used the two heavy machine guns mounted on either side of the aircraft to suppress enemy fire.

The clock showed a little after 8 PM when the aircraft reached the designated area. Outside, there was only darkness, with the dense forest enveloping the hills. An Lộc was completely dark, with no streetlights standing or burning to indicate the location of the city. Tống-Lệ-Chân Hill was not far away, and the aircraft approached what they thought was An Lộc, guided by a flashing light from below.

A minute later, the sky was filled with the relentless tracers of anti-aircraft fire, like fireworks on New Year's Eve. The crew realized they had veered slightly toward Tống-Lệ-Chân. Captain Cầu calmly pulled the control stick to the right, steering the aircraft out of the enemy's line of fire. At the same time, Sergeant Tranh reported seeing lights at the 1 o'clock position, not far away. After careful observation and communication with friendly forces, Captain Hùng gave the thumbs-up, signaling Captain Cầu to prepare for landing south of An Lộc.

After ensuring the aircraft was in good condition, the crew was escorted to the Forward Operations Command Center for a mission briefing. The intelligence officer informed them that North Vietnamese regulars, armed with various anti-aircraft guns, were entrenched around the base of the hill. It was also likely that they were equipped with SA-7 heat-seeking missiles to counter South Vietnamese Air Force aircraft.

The mission was straightforward: transport 50 Rangers to reinforce and replace those who had been in the outpost too long, and evacuate some of the wounded. If this mission were at any other location, it wouldn't have been significant or worth mentioning.

The rendezvous at the landing zone was set for midnight. Lieutenant Colonel Ngôn, the Ranger Battalion commander, had to ensure the signal lights were placed in trenches to avoid detection by the Viet Cong.

The crew discussed the landing technique and divided responsibilities. The challenge was landing with 50 fully armed soldiers on board, making the aircraft too heavy to maneuver quickly to avoid anti-aircraft fire. Flying low and fast, from the base of the hill straight into the small landing zone surrounded by enemy trenches, was impossible, especially at night. This task was more suited for smaller helicopters like the U-1H or H-34. The only viable option was a 360-degree landing approach. Both Cầu and Hùng agreed, knowing it was dangerous but provided a better chance of survival than a "hop" landing at night.

At 11:45 PM, the sky was pitch black. The 50 Rangers were seated in the aircraft, and Captain Cầu started the engines. The roar of the twin turbines, combined with the spinning rotors, echoed through the still night, like the roar of an angry forest. Captain Hùng checked the communication frequencies with C&C and the friendly units one last time, while Sergeant Tranh reported that the aircraft was in good condition and ready for takeoff.

A little before midnight, C&C gave the order to take off, climbing to an altitude of about 2,000 feet, heading southwest towards Tống-Lệ-Chân. Captain Cầu ordered Corporal Hoàng to load the two heavy machine guns, ready for action, and turned off the external navigation lights. The aircraft, like a lone knight from ancient times, charged into the battlefield.

Captain Hùng pointed ahead, where the landing zone lights were blinking from the Tống-Lệ-Chân outpost, now visible in front of them.

"OK! Pitch down!"

The entire aircraft seemed to plunge towards the ground, spinning sharply in a 360-degree turn to the left towards the lights below. The rapid descent caused a stomach-churning sensation, as the aircraft dropped from a high altitude.

"Boom… Boom!" Large explosions flashed below as enemy artillery began shelling the landing zone. Hell on earth had awakened! Tracer fire from anti-aircraft guns at the base of the hill crisscrossed the night sky, red streaks converging on the hovering aircraft. The Viet Cong fired flares to illuminate the area, making it easier to target the aircraft.

In the artificial light, the aircraft, now exposed and isolated, was riddled with bullets from the fierce enemy guns, like an eagle struck in a vital spot, yet still defiantly landed on top of the enemy.

Captain Hùng lost communication with C&C, as the radio transmitter was hit and caught fire. The hydraulic system failed, leaving the control stick locked and unresponsive. Captain Cầu, feeling numb, instinctively tried to control the aircraft, guiding it down from an altitude of over 30 feet to the landing zone.

The aircraft hit the ground hard, bouncing back up and twisting like a dying dinosaur. The rotor blades smashed into the outpost's bunkers, shattering into pieces. Flames and smoke billowed from the rear, while enemy mortars and rockets continued to pound the area, creating a thunderous roar as dust and debris filled the air.

After a few seconds, the aircraft finally came to a stop, accepting its fate on this foreign hilltop. The fire grew more intense as the surviving Rangers rushed forward to escape. Captain Cầu felt a sharp pain in his right heel but struggled to crawl out through the left window of the aircraft. In the right seat, Captain Hùng was trapped, his safety harness jammed, pinning him to the seat. Burdened with a flak jacket and ammunition, he was crushed under the weight of the escaping Rangers. The fire, fueled by thousands of liters of JP-4, burned hotter, nearly overwhelming him with heat and smoke.

At that moment, the last Ranger to climb out was hit by shrapnel in the head and fell backward, grabbing the seat cushion behind Captain Hùng for support. The cushion came loose, creating a gap that allowed Hùng to wriggle free and escape through the hole in the cockpit ceiling. Corporal Hoàng, miraculously protected, had already managed to escape before the aircraft exploded.

The survivors crawled to a nearby trench as artillery fire continued to scream through the air, creating eerie, haunting sounds. The burning aircraft illuminated the night sky, possibly providing the Viet Cong below a reason to celebrate their victory, dancing around the souls they had just killed.

Lieutenant Colonel Ngôn ordered the Rangers to carry Captain Cầu to safety and tend to his wound. They checked the crew roster but could not find Sergeant Tranh. The next morning, as the situation calmed, they searched the wreckage and found the remnants of Sergeant Tranh, identified by his dog tags and the equipment he carried. He was likely killed by anti-aircraft fire before the aircraft landed. Captain Hùng carefully collected his remains, intending to bring them back for burial.

Three days later, the crew was informed that a U-1H helicopter would be sent to extract them. They stayed up all night waiting, only to be told that the rescue mission had been canceled due to security breaches. Time dragged on forever, with restless, haunted sleep as death hovered close, blurring the line between life and death.

The brave brown-bereted Rangers fought on, isolated and enduring relentless artillery fire. The eerie sounds of drums and loudspeakers echoed from the slopes below, as the Viet Cong taunted them from their trenches. The Rangers clung to their weapons, ready to rise up and crush any attempt to turn this place into a mass grave.

Around 9 PM on the fifth night, Lieutenant Colonel Ngôn informed the crew that the Air Force Operations Command had issued a secret order for their extraction at midnight on New Year's Eve. All evacuation plans were meticulously prepared. Corporal Hoàng would carry the wounded Captain Cầu, while Captain Hùng would cradle the remains of Sergeant Tranh as they ran to the helicopter.

It was now 11:55 PM on December 31, 1972, just five minutes before the start of a new year. While people in other countries were celebrating with champagne, toasts, warm embraces, parties, dances, and songs, here, in a remote outpost in the middle of the jungle, the consequences of two opposing ideologies were playing out.

Why, my fellow Vietnamese, why?

The sound of the approaching helicopter grew louder from the northeast, like an angel descending from the heavens to rescue those in peril. Anti-aircraft fire crackled like New Year's fireworks, and rockets exploded around the landing zone as the Viet Cong celebrated their victory.

With calm composure and a defiant smile, Lieutenant Phát, also known as "Phát Strong," who had faced death many times, skillfully brought the helicopter into the landing zone. His co-pilot, Lieutenant Bằng, a former Border Ranger and a hero who always placed his duty to the nation above his own life, was by his side.

Artillery fire continued to rain down as the helicopter landed in the small clearing, its rotors stirring up dust and debris. Corporal Hoàng struggled to carry the injured Captain Cầu aboard, while Captain Hùng clung tightly to Sergeant Tranh's remains, determined to bring him home. The crew boarded the helicopter, and Lieutenant Bằng gave the thumbs-up to Lieutenant Phát, signaling readiness for takeoff. Like a warhorse leaping into battle, the helicopter took off, disappearing into the thick night sky.

Goodbye, Tống-Lệ-Chân, and farewell to the brave brown-bereted Rangers. As the radio broadcast from Saigon announced the New Year with the ringing of church bells, the crew—Cầu, Hùng, Hoàng, and the spirit of Sergeant Tranh—along with the U-1H rescue team, Phát and Bằng, became the heroes of the South Vietnamese Air Force and the Republic of Vietnam Armed Forces.

Fate scattered these men across the globe, but the spirit of "Tống-Lệ-Chân," forged over twenty-three years ago, remains a testament to their brotherhood and selfless sacrifice for their comrades. I pray for them and wish them happiness and good fortune in the years to come.

To Sergeant Tranh, I offer my heartfelt prayers for your eternal peace. Rest in peace, my friend.

CHAPTER 3

Footprints of a Refugee

Original Title: "Vết Chân Tị Nạn"

After setting foot on the U.S. Navy ship of the Seventh Fleet (USS Kirk), Nam reunited with his wife and children. Everyone was safe, which gave him peace of mind. Sitting huddled in a raincoat to keep warm, because after being rescued by the ship, he was left with only a wet short sleeve shirt and boxer shorts soaked in seawater, someone brought him a pack of cigarettes, saying it was sent by a Vietnamese Brigadier General.

Thanks to the stimulating effect of the cigarette, Nam became fully alert again. It was then that he started to pay attention and walk around the ship to observe. The distant shore of Vietnam was now just a faint and distant blur on the horizon. He tried to look towards it, but even he did not know why.

Nam encountered a few Air Force officers, such as Lieutenant Colonel L., commander of the 251st Squadron. Nam did not know when he had arrived, but he noticed the Lieutenant Colonel was well-dressed and seemed very relaxed, cheerfully chatting with those around him as if nothing had happened.

The sun had set on the other side of the water, and the sea began to sparkle, reflecting the stars that appeared earlier than usual. Nam's wife had put the children to sleep in one place while he remained huddled in his shorts, wearing a yellow nylon raincoat like a firefighter. Nam found a quiet spot on the ship's deck. Resting his chin on the railing, he gazed out at the ocean. The tropical night sea breeze made him feel very comfortable, but his mind was filled with scattered, disordered thoughts, coming and going suddenly. Despite himself, his eyes were drawn towards the shoreline, searching for the familiar electric lights of his homeland, where he had grown up and matured. Though he was not crying, he suddenly felt a stinging sensation in his eyes, and something warm seemed to trickle down his cheeks. For more than thirteen years, as a combat soldier across four military regions, Nam's sole duty had been to fly on missions. His life was very simple, trusting only in the present moment without worrying about what might happen tomorrow. Hence, tears rarely came to Nam as easily as they did now. He could not determine whether these tears were for his motherland, Vietnam, or from the shameful fate of a defeated soldier like him.

In the distance, flickering lights appeared on the sea's surface, possibly from fishing boats gradually approaching the anchored ship. However, each time they got close, the U.S. ship would move further away. Later, Nam learned that the U.S. Navy feared Viet Cong commandos, so they only rescued people during the day.

The cigarette pack from the afternoon was already more than half gone. Now, it was late at night, and the cold was starting to set in. Nam returned to lie next to his children on the ship's floor, hoping to get some sleep that night. Nam's wife, perhaps uncomfortable with the unfamiliar place and filled with worries, woke up very early and urged Nam and the children to wake up.

"Wake up, dear. It looks like they're preparing to transfer us to another ship."

The sun had already risen above the water on the eastern shore. Thanks to the cool sea breeze and fresh air, along with a few nights of sleeplessness, Nam had a sound sleep as if he had taken a strong tonic, making him feel completely refreshed and alert. Everyone around was busy packing their belongings into suitcases, lining up to descend the rope ladder to board another boat.

"Pack up our things so we can go with them."

"What is there to pack? We threw everything into the sea yesterday afternoon."

"That's fine. I'll carry Mika, you hold Mina, and Miki can go with adopted sister Mai and her older brother Châu to the ladder."

They called it a ladder out of formality, but it was actually a makeshift rope ladder lowered from the ship for emergencies or rescue operations. The ladder was narrow, just wide enough for one person to climb. Any mistake or slip could easily result in a fall into the sea.

After carefully instructing his wife on how to climb, Châu, Mai, and their eldest son, Miki, quickly climbed down the ladder first. Nam had a bit of difficulty because his second son, a chubby three-year-old, clung tightly to his chest, so Nam had to be very cautious, stepping down each rung carefully. Following Nam was his wife, holding their youngest daughter, Mina, an eight-month-old who clung tightly to her mother as they carefully descended. Eventually, Nam's family, like the others, safely boarded the second ship.

It was now noon on April 30, 1975. Nam noticed a few Vietnamese generals sitting on the deck below in the ship's hold. These officers were still in full military uniform, complete with ranks on their shoulders and caps, surrounded by many of their guards. Nam overheard some unflattering comments from a few soldiers standing nearby about the behavior of one of the generals, who had a reputation for less-than-commendable deeds.

Here, everyone was provided with military rations. The biggest hardship was for their youngest, Mina, who was still bottle-feeding. However, finding milk on the ship was impossible, so her mother had to mix coffee creamer with sugar and cold water for her to drink. At first, Mina resisted, but when she got too hungry, she had to drink it. The few cans of Guigo milk they had saved for her had all sunk into the sea when they jumped onto the ship.

The most selfish and despicable act Nam remembered was from a fellow Air Force member who also had a small child. His wife and he were lucky to have brought along more than half a dozen milk bottles. Nam's wife begged them to lend one so she could mix the creamer for their daughter, as using a cup was messy and spilled everywhere. But the couple heartlessly refused, leaving the baby girl with no choice but to drink from the cup. Fortunately, with the help of heaven and earth, Mina only suffered mild diarrhea from the cream and sugar combination drink and did not fall seriously ill.

More and more people crowded onto the ship as many fishing boats fled from the coast. Many people had brought radios and were able to listen to Saigon Radio, which reported that the Viet Cong had already entered the city and that the Republic of Vietnam's government, led by President Dương Văn Minh, had surrendered.

Nam observed that many helicopters from the Republic of Vietnam Air Force as well as those from the American company Air America were continuously flying out to the fleet. After landing, the pilots had to help push the helicopters off the deck to make room for others to land.

It seemed that Nam's ship had been ordered to rescue refugees, so it kept circling around without moving further away. As night fell, more and more people were fleeing by boat. The rope ladders used to bring people aboard had been pulled up as the sun began to set, casting a grayish hue over the western coastline. Each time a boat approached, the ship would use its fire hoses to spray water, forcing them to stay away.

The dim lantern lights on the boats flickered on and off with the waves, resembling fireflies in the night. The cries for help from women and the wails of children mixed with the sound of waves crashing against the ship created a terrifying and eerie atmosphere.

Occasionally, some helicopters that took off late circled above, unable to find a landing spot on the ship, so they made ditch landings at sea. Others, suffering from low fuel, plunged headfirst into the water like moths drawn to a flame. Nam had no idea whether those unfortunate pilots survived or perished, as in the darkness, he did not see any life rafts being deployed to rescue them.

When people are faced with desperation and extreme fear, they often become reckless and make daring decisions. Nam still remembered an incident in early April 1975. He was ordered to deliver emergency supplies to the Phan Rang Air Base. After dropping off the supplies, he brought the aircraft to the fuel depot to refuel before returning to Biên Hòa. As soon as the ramp was lowered, four soldiers guarding the fuel depot, armed with M16 rifles, jumped onto the aircraft. With bullets already chambered, they pointed their guns directly at the crew and ordered:

"If you don't take us to Saigon, we'll shoot you."

It should be noted that Khánh Hòa province and the city of Nha Trang had already fallen to the North Vietnamese Army, leaving only Phan Rang and the airbase as the last defensive outpost. No one knew how much longer the base could hold out before it would be forced to make a tactical withdrawal, as President Nguyễn Văn Thiệu had ordered during the evacuation of Pleiku.

Nam unbuckled his seatbelt and stepped out of the cockpit.

"What do you want?"

"We want to go to Saigon."

Nam looked them in the eyes and observed their gestures, noticing the overwhelming fear and anxiety etched on their faces.

"Where's your commanding officer?"

"He's already fled!"

Nam thought that refusing them would not help and might make the situation worse.

"Okay, you can go, but we will have to get down to refuel first, or we won't make it to Saigon."

The situation calmed down as the four men no longer pointed their guns at the crew. Two of them jumped down to operate the fuel pump while the other two remained on the aircraft to keep watch. About ten minutes later, the aircraft was ready to take off, carrying four deserting soldiers. As soon as they passed over Cà Ná at an altitude of about 5,000 feet, Nam radioed Biên Hòa to report the incident. When the aircraft landed, military police were already waiting to greet the four men.

Deep down, Nam did not harbor resentment towards those soldiers because he was certain they had no intention of hijacking the aircraft or harming the crew. They were simply terrified of being abandoned. Just like their direct superiors who had fled first, leaving them behind, when the aircraft arrived, it was like a life-saving buoy that they had to grab at any cost to survive. This desperation led them to take such a reckless action.

Nam had also heard about some of his fellow pilots flying F-5s to Thailand who, due to fuel shortages, were forced to land on highways or had to squeeze together two people on one seat in the cockpit. There were also stories of L-19 reconnaissance planes landing on aircraft carriers and similar ditching incidents involving HU-1 helicopters at night, which were understandable.

After two days and nights, the ship had rescued a large number of people, mostly coastal residents and fishermen of various ages. The ship then began to leave its designated position, heading further out into international waters. More than a day later, the refugees were informed over the loudspeaker that they would be transferred to a larger, faster ship with a capacity of over 4,000 people. Upon hearing this, everyone cheered enthusiastically as if it were Tet (Lunar New Year). Nam's greatest hope was that they would have proper bottled milk for his daughter and water for bathing because for many days since being rescued from the sea, Nam had only been able to rinse his mouth and bathe in the sea breeze. His body was on the verge of fermenting!

It wasn't long before the refugee ship docked alongside a large cargo ship in Subic Bay, Philippines. A makeshift bridge was set up to connect the two ships. The refugees began to lead each other onto the larger ship, and Nam and his wife hurriedly carried their children, following the crowd, hoping to find a decent spot below deck for the children to lie down. He had no idea how long this journey would last or where it would take them.

Everyone crowded together, descending into the ship's hold. Nam led his wife and children down the stairs as quickly as possible. Standing on the deck, his hopes faded as he saw the rusty, damp interior of the ship. The large hold was dimly lit by only a few weak light bulbs. People sat in long rows on relatively clean, dry spots. Nam felt like he was in a prison or in a scene from a movie where white people captured African slaves in the past to transport them for sale. The musty smell was overwhelming, and the scene of people jostling for space and arguing loudly was disheartening. Nam felt disappointed and dejected. The baby still didn't have milk, and many people nearby were moaning in sickness. He turned to his wife and whispered:

"It's too dangerous down here, easy to catch diseases. Let's climb back up; the air is better, only worry if it rains."

Nam's wife immediately agreed: "Let's go up then; we can use a towel to cover up if it rains."

The whole family carried their belongings and climbed back up the stairs to the deck. The deck was also packed with people, and it was difficult to move. Fortunately, Nam found an empty corner just big enough for his family to sit down. There, Nam reunited with Sergeant M. and Sergeant C., whom he had lost contact with after jumping onto the ship. Since both were single and without families, they had moved quickly. They had hung hammocks nearby, giving Nam someone to talk to and pass the time with.

"M., is there a place to bathe around here?" "Hey! The Lieutenant Colonel doesn't know? They're bathing out there like crazy."

He pointed to the back of the ship.

"Okay, I need to take a shower right away."

Making his way to the back as directed, Nam saw about a dozen hoses pouring water down onto people scrubbing themselves below.

Men, women, and children of all ages were bathing together, fully clothed, seemingly very comfortable. Without hesitation, Nam joined in, tilting his head back to sip some water and rinse his mouth. Immediately, he spat it out in discomfort, muttering to himself:

"My God, what kind of water is this? So salty!"

A young man nearby, who had probably made the same mistake, laughed loudly: "Don't you know? Sea water is great for rinsing your mouth, no need for toothpaste."

Since birth, Nam had never been on a ship like this.

Nam still remembered that in 1969, the Air Force Command had sent him to the United States for training, learning to fly a new type of aircraft that the U.S. military was about to transfer to Vietnam. While at Fort Rucker, the training school had organized a "deep-sea fishing" trip in Miami, Florida, for allied officers. Even on a fairly large ship, Nam had gotten seasick and vomited almost to the point of death, so he had disliked sea voyages ever since, even for pleasure. Thus, his experience with survival at sea was next to none.

At first, Nam assumed that the seawater was filtered into freshwater for bathing or washing, but it turned out to be 100% pure seawater. He chuckled to himself and thought, "Bathing in seawater is refreshing and disinfectant. People even pay to go to Vũng Tàu or Nha Trang for a sea bath. Now that it's free, what am I complaining about?!" But the difference was that after a seawater bath, people could rinse off with freshwater, whereas after Nam bathed, he soon turned into a pickled radish. His hair, back, and armpits were all covered in glistening white salt, and even his private parts turned into a salty "hot dog."

Despite the blazing tropical sun, the fast-moving ship and the fresh sea air made Nam feel very comfortable. Large fish, whose names he didn't know, followed the ship's wake, creating a beautiful sight. Surrounded by nothing but the blue of the ocean, Nam felt as if he were flying, lost in some mysterious, boundless universe. This gentle, endearing sensation often came to him during very early flights.

Before 1969, when Nam was still with the 215th Squadron stationed in Nha Trang, he often had to go on missions to Pleiku, Ban Mê Thuột, or Qui Nhơn, requiring the crew to take off very early to provide support to allied forces. The weather in central Vietnam, in the second military region, often had thick white clouds hanging low in the early morning, with cold, misty droplets forming heavy cloud clusters, covering the entire forested area ahead. The air outside was still cold, and the sky had not yet fully brightened.

The cockpit windows were closed tightly. The sound of the jet engine mounted at the rear of the aircraft and the whirling rotor blades had quieted down. Nam felt as light as a butterfly, silently gliding over the still-sleeping forests and nameless mountains and winding streams as slender as a woman's body. The scenery was pristine and magnificent as if preparing to welcome the birth of a new day. In those transcendent moments, Nam often felt as if he were transcending this world, becoming a free-flying bird, seeking new and wondrous discoveries in nature.

While still lost in his reverie, Nam was startled by a loud noise behind him.

"Why don't you go get some canned food? They're handing it out over there."

It turned out his wife had been waiting for him to bring food back and had come to remind him.

"Okay, I'll go get it."

A pathetic and cunning scene was unfolding before his eyes, as a group of people who had gone to get the food were unevenly distributing it among the refugees. Arguments and disputes ensued because the person who went to get the food had hidden all the good, edible items for their family and friends, only sharing the less desirable items that few people wanted. Fortunately, no one had starved to death yet.

Nam felt disheartened by the behavior of some people.

After many years of colonization, first by the Chinese and then by the French, followed by the influence of the Americans under the guise of being allies, many Vietnamese people had lost the inherent virtues of their nation, learning instead to be deceitful and cunning. Proverbs like "A drop of water reflects a whole river. People of the same country should love each other" or "The healthy leaf covers the torn leaf" seemed, at times, to be mere words used to comfort those who had fallen on hard times.

Some people could sell their conscience for just a few hundred đồng, becoming paid slanderers for others, or for a few dozen đồng, writing defamatory articles or attacking those with differing views. They would even go so far as to desecrate the dead. Nam felt disgusted when he heard people boastfully claim that they had studied at France or American schools, not Vietnamese ones, or that they had worked as servants for the French or Americans to appear more cultured and progressive.

But when the country was in turmoil and needed them, these people proudly bragged that they had foreseen the situation and fled. Once in the West, in France or America, they rejoiced as if they had returned to their native homeland or their ancestral land. Poor motherland Vietnam, so kind and virtuous, yet burdened with such pretentiousness!

After more than two days adrift at sea, everyone was talking, not knowing where the ship would take them. Nam was also completely in the dark, unable to determine their direction. He told himself, "Whatever, wherever it goes..." He only hoped they would soon reach land so he could find fresh water to bathe, as the itching and discomfort had become unbearable. His hair was burnt, full of sea salt, and stood up like bamboo roots. However, he was fortunate that neither his wife nor his children had fallen ill.

After lunch, Nam was sitting with his eyes closed, leaning against the steel wall, trying to take a nap. Suddenly, Sergeant C. woke him up.

"Lieutenant Colonel, we're almost there."

Still half-asleep, Nam asked again to make sure he had heard correctly.

"Almost where? I don't see anything."

Sergeant C. pointed to some birds flying around the back of the ship.

"See those birds? It's a sign we're nearing land."

Indeed, for the past few days, there hadn't been any birds, just blue water and white clouds surrounding them. Besides, birds never fly far from land. Nam stood up excitedly and walked to the front of the ship, where a group of people had already spotted the faint outline of land ahead. They were pointing and discussing whether it was an island or the mainland.

A few hours later, everyone could clearly see that it was the mainland, inhabited, with many boats coming and going in the area. However, no one could determine which country's waters they were in. Not long after, the loudspeaker announced that the ship was about to dock at Guam and requested everyone to prepare their belongings and maintain order while completing immigration procedures upon disembarkation.

It was now around noon on May 7, 1975. After filling out the necessary paperwork with the help of immigration officers, the refugees were guided to temporary accommodations in tents that had been set up on a large open field near the dock. The International Red Cross provided them with blankets, clothes, and food for the children. After more than a week of living on sugar water and creamer, Nam's youngest daughter was now able to drink fresh milk and wear clean, brand-new clothes. She looked visibly happier. The next priority was to bathe and rinse their mouths.

Not far from there, Nam noticed a few bathing facilities built with wooden planks, clearly marked for men and women, with some children playing and bathing there. The first cool drops of water from the showerhead above cascaded down his head and neck, and the sensation was incredibly refreshing! Nam closed his eyes to fully enjoy the coolness, which flowed from the top of his head to his toes, giving him a light, euphoric feeling, like someone who had just found something precious they had long lost. Nam's two sons also delighted in playing with the water. Nam no longer felt like a "pickled radish," as his hair and skin returned to normal, no longer stiff and salty like they had been just a few hours earlier.

That evening, Nam's family dressed in clean, though slightly oversized, clothes, even after selecting the "small size." The pants and sleeves had to be rolled up to fit. Nam's wife was very skilled at tailoring, but without a needle and thread, they had to wear the clothes as they were. The dinner included hot, delicious food, fresh fruit, and soft drinks at the makeshift military club, making up for the days spent at sea eating purple beets and dried rice. They were fortunate to get a bit of bacon or tuna, as the more desirable food had already been hoarded by those who had gone to collect it first.

Returning to the tent at dusk, the sea breeze was refreshingly cool, and with a warm blanket, Nam quickly fell asleep, not knowing when. It wasn't until the sound of engines and people calling to each other woke him up that he realized it was around seven in the morning. The sun was still hidden behind the mountains, and his children were still sound asleep. Nam quietly stepped out of the tent to observe his surroundings, as there had been many urgent tasks the previous afternoon, leaving him no opportunity to do so.

Standing by the roadside, lost in thought like a country bumpkin in the city, where everything seemed new and noteworthy, a large military-style bus suddenly pulled up in front of him. As the door opened, the soldier driving the bus waved for everyone to get on, signaling that it was time to move to another camp. Nam quickly returned to the tent, urging his family to gather their things and board the bus.

About an hour later, the bus dropped them off at Guam Airport, where many of their compatriots were already gathered in groups inside a hangar. Nam's wife, slightly puzzled, remarked:

"They might be taking us somewhere else, not here?" "I don't know either; wherever we go, we're basically homeless now."

Indeed, at this point, Nam no longer had anything to hold on to, no lingering attachments or regrets. After thirteen years of fighting and sacrificing for his motherland, all Nam had left were a pair of shorts and a shirt. He felt sorrow for his wife and children, who had endured so much hardship because their husband and father were soldiers. But Nam found comfort in being luckier than many others; he had his family by his side, easing the loneliness during this difficult and tumultuous period.

As noon approached, everyone seemed anxious, waiting. Nam kept an ear out for news and learned that they were likely going to Wake Island for temporary shelter, though nothing was certain. In this situation, everyone felt like leaves blown by the wind, carried wherever the wind took them, with no way to predict their destination. The sound of a plane's engine grew louder in the distance, prompting everyone to stand up and look around.

Indeed, a large C-130 transport plane was approaching for landing. Without needing to be told, everyone was ready with their luggage in hand. About ten minutes later, Nam's group was called to board the plane. As a military cargo plane used for resupply and troop transport, it only had two rows of canvas seats along the sides of the aircraft. Those who boarded first got to sit in the seats, while the rest had to sit on the floor. However, it was still more comfortable, clean, and pleasant-smelling than the cargo ship that had taken them to Guam a few days earlier.

After the flight from Guam, Nam was dropped off at Wake Island's military airfield, where they were temporarily housed in solid, brick-built houses that Nam learned were formerly used by U.S. military families. The families were grouped two or three to a house, depending on their size, with clean and well-maintained facilities, including bathrooms, meeting American standards.

Nam's family shared a house with a large fishing family from Quảng Ngãi and an infantry officer and his son, whose family was still stuck on Guam. Nam chose the living room for its spaciousness and coolness. These houses had been unoccupied for a long time, so there was no furniture; they simply swept and cleaned the tiled floors to sleep on.

In the first few days, life was comfortable, with nothing much to do, so after meals, they often went to the nearby beach to catch fish. The fish were abundant and bold, swimming in schools close to the shore. Some were as big as a calf's thigh, and the smallest were the size of a wrist. They didn't know the names of the fish, but grilled or stewed, they tasted delicious.

Nam and Trần Văn L. "L. Cowboy," the squadron leader of the 223rd Squadron in Biên Hòa, used army-issued mosquito nets as makeshift fishing nets. Nam's wife and Mrs. L. gathered firewood to build a stove on the beach, where they cooked the fish. Meanwhile, L. kept watch for the American military police, who probably prohibited fishing and cooking on the island for sanitation reasons. In the evenings, the group strolled to watch outdoor movies.

At this location, Nam reunited with Colonel T., the tactical air wing commander, Cao Văn T, P., "Coconut," Sử Ngọc C., Thái Văn Â., and others. While many were contemplating their future life and careers in the U.S., P., T., and some others were determined to return to Vietnam on the "Việt Nam Thương Tín" ship. Many advised them to stay and figure things out later, but they were resolute, as most had family still trapped in Vietnam and needed to go back.

A week later, the camp was overcrowded with over 4,000 people. The communal dining areas were limited, with only three or four locations, leading to long lines of people waiting to eat, sometimes stretching hundreds of meters. On some occasions, people had to wait two or three hours to get a meal, especially on days when fresh fruit was served. Often, after lunch, people would rest for a bit and then start lining up for dinner.

The days here dragged on, each one monotonous like the last. Occasionally, immigration officials would call people in to complete paperwork, or the health department would call for vaccinations, or there would be clothing and cigarette distributions from the military. As this was a purely strategic military base, there were no civilian shops like in town. One afternoon, while Nam was wandering along the road, Sử Ngọc C. called out to him from behind.

"Hey, Nam! Want to have a beer with me?" "What are you talking about? Where would we get beer?" "I have a brother who has some U.S. dollars. He asked the Filipinos to buy it from the club." "Okay, let's go, but let's keep it out of sight from the MPs."

Sử Ngọc C. had been fortunate to send his wife and children ahead, so now he appeared relaxed and carefree, like a single man. The three of them headed to the beach, sharing two cans of Budweiser each. Nam chose a flat rock to sit on comfortably, savoring the bitter taste of the beer. As they drank and smoked, the gentle sea breeze added to the sense of relaxation. Nam felt as if he were enjoying a vacation at Quán

Số Năm in Nha Trang or a thatched hut in Vũng Tàu.

Nam's family stayed on the island for about twenty days, just long enough for his adopted nephew to fall ill. The doctor ordered the child to be sent to the mainland for treatment in a hospital, so the entire family completed the paperwork to leave the island and followed the child to Camp Pendleton in San Diego, California. After spending a night in Hawaii for a plane transfer, everyone got to bathe and change into cleaner clothes before heading to the civilian airport to catch their flight.

Everyone's faces were now filled with hope, no longer anxious and worried like during the previous waits. People seemed more polite and well-mannered, without the selfish, greedy behavior Nam had witnessed earlier. The old saying from their ancestors held true: "Wealth begets manners; poverty begets thieves." People's attitudes can change depending on their environment and current circumstances.

About half an hour later, the group boarded a Boeing 747 civilian aircraft. The flight attendants greeted them warmly at the door, just as they had on Nam's previous trips to study in the U.S. in 1963 and 1969 on Boeing 707s. The plane had long since left Honolulu Airport, flying at an altitude of 33,000 feet, with thick white clouds below resembling snow and the vast blue sky above. The hum of the jet engines created a constant, uninterrupted sound as the plane sped through the air.

Nam had truly left Vietnam, leaving behind the devastation of war and a small, impoverished nation filled with fear and pain. People of the same blood and ancestry had eagerly taken up foreign weapons to kill each other, all in the name of being the frontline against communism or imperialism, as dictated by the superpowers of Russia and America.

Nam and other young men had been born into the wrong generation, dedicating more than half their lives to protect and defend their beloved motherland. The departure today was filled with sorrow and regret, but it was a necessary, justified decision with no other choice.

The sun's rays no longer shone brightly through the small windows on either side of the plane, likely because it was nearing evening. The plane had crossed several time zones, flying westward against the sun, so daylight passed quickly. A flight attendant informed Nam that the local time was 5:30 p.m., and they were expected to land in San Diego at around 7:00 p.m., meaning they had about an hour and a half left. Nam closed his eyes to shorten the wait.

A gentle tap on Nam's shoulder woke him up. "Look, it seems we're flying over land now, not water anymore." Through the plane's window, Nam could see faint outlines of green forests below, visible through the drifting white clouds.

"It looks like we're flying along the U.S. coastline. We should be landing soon."

A light, but worrisome, thought crossed Nam's mind. Years before, he had come to this country as a student, planning to stay for a few years before returning, with everything well taken care of by the government. Everyone was cheerful and carefree. But now, everything was different. "I left with the promise of return, but now there's no return...," Nam thought. With no money or possessions, what would life be like in this unfamiliar land?

Nam thought that the U.S. government would take everyone to some remote mountainous area to clear the land, farm, and start a new life, similar to the new economic zones in Vietnam under the Ngô Đình Diệm government, which resettled refugees from the north to the south in 1954. He comforted himself, "What's there to fear? If others can do it, so can I." Besides, Nam had always believed in his own hands and abilities.

The plane had come to a complete stop at the terminal of San Diego airport. After all the regular passengers disembarked, Nam and the group of refugees exited to collect their luggage. The group was then directed to board military buses to be taken to Camp Pendleton refugee camp. This was a U.S. Marine Corps training base, located in a remote, mountainous area far north from the city.

By now, it was already getting dark outside. The bus carried the refugees across wide highways and onto narrow, winding roads through sparsely populated towns. Looking through the bus windows, Nam saw houses scattered on either side of the road, with lights glowing warmly and softly. Nam tried to keep his eyes fixed on those lights, imagining and picturing a scene of family happiness, where a father sat drinking tea and reading the newspaper, a mother sewed, and children gathered around to listen to her stories.

It was like the image of a happy family from a third-grade reading book that he still remembered from his school days in the village. A simple, ordinary life that everyone should have been able to enjoy. But, tragically, how many families in Vietnam could claim such peaceful, happy evenings? How many innocent children who had just learned to say "father" were suddenly left with a mourning band on their heads? The girl from the neighboring village who had just finished her wedding celebration a few days ago received bad news a few days later, and the love letter she was writing to send to her husband at the front line ended up as a farewell letter.

Suddenly, Nam felt a lump in his throat and turned away to push those unpleasant images out of his mind. The bus was passing through a dark forest, with tall trees on either side, making it impossible to see the surroundings. A few minutes later, bright lights appeared ahead, and the female soldier driving the bus announced that they were almost at the camp.

The refugees were then directed to collect blankets and other necessities before settling into the camp.

It was the evening of May 30, 1975, and the group was officially moved into Camp Six. The first few days were quite busy with paperwork, updating missing documents, getting social security cards, taking photos, registering with volunteer agencies to find sponsors, and so on. During this time, Nam also reconnected with two of his sisters, who had arrived earlier, and with some old friends. Within about a week, everything had settled down temporarily.

The military tent was located next to a forest behind the camp. On some nights, Nam could hear the distant howling of dogs, which some said were wolves from the nearby mountains. Nam had never heard a wolf howl before, so he couldn't be sure. One day, he asked a soldier who was setting up tents, and the soldier warned him about the dangerous rattlesnakes in the bushes behind the camp. These stories made the women quite fearful, and they would wrap themselves tightly in blankets every night.

The weather here was warm, even hot during the day, but quite cold at night, so everyone had to huddle together to keep warm. Nam didn't know who had invented the "elephant eggs," but they were extremely effective at keeping warm. These were carefully guarded, and when someone received their release papers, they would pass them on to their closest family member. Sometimes, there would be arguments when someone accidentally took someone else's egg. Nam's wife was particularly careful, writing her name on their egg to prevent any mix-ups.

The "elephant egg" was a brilliant invention of the refugee community. Made from large or small nylon water bottles, these were smuggled out of the kitchen by "undercover agents," who had secretly brought them back to the tents. After cleaning them thoroughly, the bottles, which might have been used to store soy sauce or other condiments, could be filled with hot water before bed, making them warm and cozy. It was a simple but ingenious way to stay warm. If anyone fears the cost of heating, they should try this method, but be careful to seal the bottles tightly, or they might end up with burns in uncomfortable places.

The United States was different. Here, there was no need to wait in line for hours to get a meal, as there were plenty of spacious dining halls with ample food and drink. Clothing and shoes were available in large bins for everyone to freely choose what they liked. The only downside for smokers was that there were no free cigarettes like on Wake Island. Nam's wife had two dollars in her pocket, a gift from a distant relative while they were still on the ship, which she guarded like gold, while the Vietnamese money they had was useless.

In his free time, Nam often wandered up to the volunteer agencies to gather accurate information because there were so many rumors circulating in the camp. For example, Nam heard about an Army lieutenant colonel who had lost a leg and was sponsored by an individual who then gave him a gun and assigned him to guard a tent on a farm at the edge of the forest, bringing him back to the camp every evening. This story spread like wildfire, sending shivers down everyone's spines. People also debated whether church sponsorship was better than individual sponsorship or whether one parish was better than another. As we know, idleness breeds gossip, and stories often grow more dramatic and frightening with each retelling.

One day, Nam ventured out of the camp gate to explore and make contact with the outside world. Although the camp was vast, it was isolated from society by the surrounding mountains and forests. Those who had been released from the camp would sometimes return to visit relatives or friends still inside. One day, Nam ran into Trần Z., a former observation pilot who had driven in to visit family. Nam was amazed at how sophisticated he looked, driving himself in a nice car, dressed impeccably, and clearly enjoying life. He was living the dream everyone else aspired to at that time. Nam looked at him with hope for the future, wishing for the same fortune.

Trần Z. told Nam he had secured a job at a gas station in a small district near San Diego, earning three dollars an hour. He had just bought the car for two hundred dollars, paying in installments to his sponsor. He seemed happy and content.

Back at the tent, Nam shared the story with everyone, and they listened intently, enjoying every word. The hope and joy were evident on the faces of those around him, and Nam was sure that everyone was hoping for the same good fortune as Trần Z. now had.

During his wanderings, Nam also ran into Lt. Col. Lưu Văn T., a former helicopter pilot who had worked at the Air Force Headquarters in charge of helicopters before the fall of the country. Nam had only heard of him and knew little about him until he accompanied the delegation of exceptional soldiers to Cambodia, led by Brigadier General Trần Bá D. When he reported to headquarters to complete the paperwork for the trip, Nam learned more about him through his request for Nam to buy deluxe tennis clothes. However, he hadn't given Nam the money in advance. Nam, being a field soldier, didn't have extra money to ingratiate himself with anyone. However, upon returning, Nam gave him a good-quality t-shirt as a gift to avoid any ill feelings. The colonel accepted it but didn't seem pleased.

Now, by chance, they met again, and Nam learned that the colonel held an important position, as he was sponsored by an American pastor who headed the Lutheran relief agency in the camp. The agency needed translators to help refugees with their paperwork and find sponsors, and the colonel was considered the head of the translation team, so just a recommendation from him would secure Nam a job.

After a conversation filled with disagreements and arguments, the colonel insisted on a condition: Nam had to split his wages with him. Nam wasn't sure if the colonel had exploited others before, but now, in this time of chaos and hardship, he still treated his comrades this way. Nam thought that a man like him, who valued money above all else, was unworthy of respect but agreed to let the colonel exploit him.

While temporarily working at the relief agency, Nam learned more accurate details about the refugee situation. It wasn't as bleak as the rumors had made it out to be. Nam also contacted Lt. Col. Lại Như S., who was living in Marysville, Washington, at the time. He encouraged Nam to bring his family up there, saying, "Good land attracts birds," and it would be nice to be close to friends.

In the years before, when Nam visited the U.S., he had always been in the southern states, so he wasn't familiar with the climate and scenery in the Pacific Northwest. However, out of curiosity and with Lt. Col. S.'s encouragement, he decided to make a trip north to see for himself.

After several weeks of working at the relief agency, Nam received his first paycheck, a $300 check. After cashing it, he took out half, $150, and handed it over to Col. Lưu. Nam thought to himself that the colonel's name should be changed to "Lưu Văn Mạnh" (somewhat derogatory) to better suit his character, especially when he smiled and eagerly took the money. Despite his frustration, Nam kept silent, waiting for the day he could leave the camp. About two weeks later, Nam's family completed the sponsorship paperwork and was officially sponsored by a Lutheran Church in Bellevue, Washington (25 minutes east of Seattle).

Before officially leaving the camp, Nam introduced Trần Văn L. as his replacement at the relief agency and took the opportunity to say goodbye to everyone. His final paycheck from the agency was $150, and this time, Nam was determined not to let the colonel exploit him again. He kept the money and went straight back to the tent to prepare for his departure.

That evening, after taking a shower, Nam's wife told him that the colonel had come by to collect his "share." Nam knew he would likely return the next morning, so he prepared to board the bus early to avoid seeing that "butt face" again. Sure enough, just as Nam had anticipated, as his family boarded the bus, he caught a glimpse of the colonel's face and his scraggly beard outside the window, trying to get on the bus but being turned away by the driver. As the bus began to move, the colonel was still standing there, muttering as if he were cursing. Nam looked at him with disdain and waved goodbye.

Today was July 22, 1975, and Nam was on his way to the San Diego airport, traveling the same road he had taken over a month and a half ago, only this time it was daylight instead of night.

This morning was beautiful, and Nam's family was dressed smartly. Everyone had on warm sweaters, and Nam's wife had made sure he wore a tie to look formal. Nam had been told that many representatives from the church would be there to greet them at Sea-Tac airport.

The gates of Camp Pendleton had completely disappeared behind the hills. The bus was now crossing a grassy field with scattered trees on either side. Occasionally, a white rabbit would dart across the road. Nam leaned back against the seat, inhaling the fresh air of the natural wilderness.

In front of him, the early morning sunbeams seemed to be welcoming the arrival of a new, pristine day. "Goodbye, Camp Pendleton!"

Section 2: Resettlement and Adaptation

Struggles and Resilience in a New World - This section covers the journey of adapting to life in the United States after fleeing Vietnam. It delves into the challenges of resettlement, the struggle to keep cultural identity, and the personal and familial struggles to rebuild a life from scratch.

Overview: Ch. 4 - "Changes in Life"

"Changes in Life" (Đổi Đời) is set in the aftermath of the Vietnam War, focusing on the life of Nam, a former South Vietnamese Air Force officer, and his family as they navigate the challenges of resettling in the United States. The narrative begins on July 22, 1975, as Nam and his family arrive in Washington state, marking the start of a new chapter in their lives after a prolonged period of uncertainty and displacement in various refugee camps.

Anxiety and Hope of New Beginnings: As Nam and his family disembark from the plane at SeaTac Airport, they are greeted by members of a local church who have agreed to sponsor them. The warmth and hospitality of their American hosts provide a stark contrast to the anxiety and uncertainty that have dominated Nam's life since the fall of Saigon. The family is welcomed into their new home and begins to adapt to the unfamiliar surroundings and the culture of their host country.

Cultural Adjustment and Everyday Challenges: Nam and his family face many challenges adjusting to American life. Grocery shopping, navigating the school system, and learning English are significant hurdles. Nam's wife struggles with unfamiliar supermarkets, often relying on others for help. Nam himself encounters humorous yet embarrassing situations due to language barriers.

Struggle for Employment and Financial Stability: The story outlines Nam's efforts to find work in his new country. He begins at a factory assembling electronic components, far from his previous military career. Despite low pay and monotony, Nam perseveres to support his family. The narrative also highlights the economic struggles faced by many Vietnamese refugees, adapting to menial jobs in a new environment.

Resilience and Adaptation: As Nam continues to work and adjust, he enrolls in vocational training to improve his prospects. Despite the initial difficulties, including financial constraints and the challenge of balancing work, study, and family responsibilities, Nam's determination and resilience shine through. He eventually graduates with a certification in electronics, opening up new opportunities for him and improving his family's situation.

Reflection on Loss and New Identity: Throughout the story, Nam reflects on the immense changes in his life—from a respected military officer to a refugee struggling to make a new life in a foreign land. The narrative highlights the emotional and psychological toll of these changes, as Nam grapples with the loss of his homeland and the need to forge a new identity in the United States.

"Changes in Life" is a poignant tale of resilience, adaptation, and the quest for a better life in the face of overwhelming odds, capturing the profound challenges and small victories that define the refugee experience.

Overview: Ch. 5 - "The Village of Somber Bridges"

"The Village of Somber Bridges" (Xóm Cầu Nỗi) is a deeply emotional narrative set in the backdrop of Vietnam's turbulent history, focusing on the life of Nguyễn Văn Châu, a man whose life was drastically altered by the long conflict between the Communist forces and the Nationalist government, culminating in the events of April 30, 1975.

The Simplicity of Rural Life Before the War: Before the war, Châu was a simple farmer in the "Cầu Nỗi" village, living a life similar to that of his ancestors. His days were spent working in the fields, following the tradition passed down through generations. The story paints a picture of a peaceful rural life, deeply connected to the land and traditional values.

The Impact of War and Change: However, the war brought significant changes to Châu's life. The long conflict between the Communist and Nationalist forces, and the eventual fall of Saigon, marked a turning point in his life. The narrative highlights how these historical events disrupted the traditional way of life, forcing Châu to leave his village and join the military service under President Ngô Đình Diệm's conscription policies.

Love and Separation: Amidst the chaos of war, Châu develops a romantic relationship with Năm Lài, a beautiful girl from a neighboring village. Their love story is a touching aspect of the narrative, filled with moments of tender affection and youthful romance. However, the impending war and Châu's conscription cast a shadow over their relationship, leading to a painful separation that deeply affects both of them.

The Struggles of Military Life: Châu's transition from a farmer to a soldier is depicted with great detail. His time in the Quang Trung training camp, the hardships of military life, and the challenges he faces in adapting to this new reality are all explored. The story reflects on the broader experience of many young men who were forced into military service during this period, leaving behind their families and loved ones.

The Loss of Innocence and the Reality of War: As Châu's story unfolds, it becomes clear that the war has stripped him of his innocence. The narrative conveys the harsh realities of war, including the destruction of lives, families, and communities. Châu's journey is marked by loss, not only of his peaceful life and his love, but also of the traditional values and simple joys that once defined his existence.

"The Village of Somber Bridges" is a poignant exploration of the impact of war on rural Vietnamese life, capturing the sorrow, loss, and resilience of those who lived through one of the most challenging periods in the country's history. The story is a reflection on the enduring effects of war, as well as the profound changes it brings to individuals and communities.

Overview: Ch. 6 - "Sorrowful Past"

"Sorrowful Past" (Dĩ Vãng Buồn) is a deeply reflective narrative that captures the life and experiences of Nam, a man born during a time of upheaval and change in Vietnam. The story delves into his childhood memories, shaped by the socio-political landscape of his village and the broader impacts of French colonial rule, the Japanese occupation, and the subsequent return of the French.

The Innocence of Childhood Amidst War and Turmoil: Nam's early years are depicted against the backdrop of a peaceful village life, surrounded by fruit trees and a close-knit family. However, this tranquility is disrupted by the harsh realities of war and occupation. Nam recalls how the idyllic life in his village was marred by the return of French forces, who imposed brutal measures on the local population, leading to frequent evacuations and a constant state of fear. These experiences shape Nam's early understanding of the world, blending the innocence of childhood with the harshness of war.

The Impact of Colonialism and Conflict on Identity: As Nam grows older, he witnesses the effects of colonialism and conflict on his community. The story highlights the stark divisions between the colonizers and the local population, with Nam internalizing a sense of national pride and resistance against foreign domination. This period is marked by significant upheaval, as Nam's family is forced to navigate the dangers posed by both the occupying forces and local conflicts. The narrative captures the tension between the desire for a peaceful life and the reality of living under oppressive regimes.

Struggle and Resilience During Times of Change: Nam's family endures numerous hardships, including displacement, economic struggles, and the constant threat of violence. The story details the resilience of Nam's parents, who work tirelessly to provide for their family despite the instability. Nam himself takes on responsibilities beyond his years, contributing to the family's survival while also trying to pursue his education. The narrative reflects on the broader challenges faced by many Vietnamese families during this period, highlighting their determination to overcome adversity.

Reflection on Lost Traditions and Cultural Shifts: As the story progresses, Nam reflects on the loss of traditional values and the changes in his community brought about by modernization and war. He recalls the simple pleasures of village life, now replaced by the fast-paced, materialistic culture of the cities. This shift is portrayed as both inevitable and bittersweet, as Nam mourns the disappearance of the world he once knew.

"Sorrowful Past" is a poignant exploration of the complexities of memory, identity, and the lasting impact of historical events on personal lives, offering a rich portrayal of a man's journey through a rapidly changing world.

Overview: Ch. 7 - "Teacher Bảo"

"Teacher Bảo" (Thầy Giáo Bảo) tells the poignant story of Bảo, a literature teacher in Saigon, who gradually becomes disillusioned with life in post-war Vietnam. The story opens with Bảo feeling frustrated after a local retired teachers' association meeting, where he was questioned about why he returned to Vietnam despite the communist regime still being in power. This question triggers a reflection on his past and the current challenges he faces.

Passion for Teaching and the Impact of War: Before 1975, Bảo was a passionate literature teacher who loved classical Vietnamese poetry. He was known for his unique teaching style, captivating students, who affectionately called him "Tú Uyên" after a famous literary figure. However, his life took a sharp turn when he was drafted into the Army of the Republic of Vietnam during the war. After returning to teaching post-service, Bảo found it difficult to recapture the joy and fulfillment he once found in his profession due to the profound impact the war had on him.

Disillusionment with the New Regime: The post-war period brought significant changes that alienated Bảo. The communist takeover led to forced personal history declarations and mandatory political re-education sessions. Bảo was humiliated by a former thug, now a local official, during one such session, highlighting the dramatic shift in power dynamics. The new ideological standards imposed on him stripped away the passion and creativity that once defined his teaching style, leading to a growing sense of disillusionment.

Isolation and Loss of Purpose: Bảo's growing frustration led to feelings of isolation in a society that no longer valued his ideals. The oppressive political climate eroded his love for literature, causing him to question his place in a world that had changed so drastically. The story explores the emotional toll of living under a dictatorship, where personal freedoms are suppressed and former passions fade away.

The Struggles of Intellectuals in a Totalitarian State: Bảo's experience reflects the broader struggles faced by intellectuals in post-war Vietnam, where the state's demands for ideological conformity stifle creativity and intellectual freedom. The story portrays the tension between individual expression and the oppressive demands of the state, a common theme in literature about life under totalitarian regimes.

"Teacher Bảo" is a moving narrative that delves into the personal and professional turmoil experienced by those who lived through Vietnam's post-war transition. The story highlights themes of disillusionment, loss of identity, and the harsh realities faced by individuals trying to adapt to a society that has undergone profound and painful changes.

CHAPTER 4

Changes in Life

Original Title: "Đổi Đời"

Today is July 22, 1975. It's an extremely important moment for Nam and his family because it truly marks the beginning of a new life in formal American society, rather than wandering from one refugee camp to another, full of anxiety and uncertainty as before.

The airplane began its final approach to land at Sea-Tac Airport in Seattle, Washington State. Nam's wife was busy adjusting the children's clothing, making sure everything was in order, and she didn't forget to straighten her husband's tie. The sunlight outside was beautiful, casting moving shadows through the windows. Everyone waited quietly in anticipation. The plane came to a complete stop at the passenger gate, and a gentle voice from the flight attendant over the loudspeaker said, "Welcome to Washington State..." Nam's family followed the people ahead of them and stepped out of the plane.

"Hello! Are you Mr. Nguyen!?"

Nam quickly halted, right in front of a man about forty years old, with a kind and open smile, standing with a woman who was probably his wife.

"Yes, I am."

A firm, warm handshake filled with affection made Nam feel comfortable, like meeting an old friend. The man introduced himself as Richard Hanner, and the woman as Mrs. Darlean Hanner, whom Nam had spoken with several times over the phone while working as an interpreter at Camp Pendleton. They were members of a church, here to welcome Nam's family. After a brief exchange, they guided the family to the waiting area to retrieve their luggage. Here, Nam also met several other church members who had come to greet them, such as Reverend and Mrs. Baseler, Reverend and Mrs. Joe Grandi, and Mr. and Mrs. Gorman Colling. Nam noticed that everyone was cheerful and eager as if they were welcoming a close relative. Nam wasn't sure if this was the first time the church had sponsored a refugee family because everyone seemed so enthusiastic and dedicated to helping.

Seattle, USA – late July, 1975.

After collecting their luggage, which was nothing more than two bags of old clothes obtained from the camp, packed in two cardboard boxes, Nam's family split into two groups to head to their temporary residence.

The convoy left Sea-Tac International Airport, driving through seemingly endless highways. About twenty minutes later, they stopped at a roadside restaurant where everyone got out to have coffee and lunch. Nam's children loved the hamburgers and Coca-Cola, and they seemed very comfortable. The sun was now directly overhead, and Nam's wife, feeling warm, took off her thick coat. Nam also fumbled with his tie, which had become an uncomfortable nuisance around his neck.

This was Nam's first time setting foot in the far north of the United States, and he had assumed that the weather here was always cold year-round, with snow never melting on the mountain peaks.

Nam had learned this lesson back in 1965, after completing his helicopter pilot training on the CH-34, while in route back to Vietnam, the Boeing 707 had landed at Anchorage Airport, Alaska, to refuel and pick up more passengers before crossing the Pacific Ocean to Vietnam. Sitting comfortably inside the warm airplane, he looked outside at the bright blue sky and thought that it must be warm outside, so he lazily left his coat behind, not wanting the extra hassle. But as soon as he stepped out onto the stairway to descend, his ears felt numb from the cold. There was no wind, but the cold was as biting as ice. Nam and his friends had to quickly run into the waiting room to escape the bitter cold of the far north. Since then, like a bird hit by an arrow, Nam had feared any tree branch that looked like a bow, and that's why his family was dressed more heavily than necessary and not in season.

After finishing lunch, the convoy continued on its way. Nam quietly watched the scenery around him. Everything here was so different; the trees were abundant and lush, unlike in California, where the sun scorched nearly year-round, and the fields were a uniform yellow, stretching to the horizon with only a few sparse clusters of trees standing like the last remaining teeth in the gums of old people. It seemed the car was passing through a farm, and Nam noticed some cornfields to the right of the road. The corn was growing tall, almost ready to bear fruit. This reminded Nam of the story of the colonel tending a garden that had been rumored in Camp Pendleton. He reassured himself, "It probably won't be like that..." because the church sponsorship involved many people helping, so it wouldn't be as bad as he had imagined. Moreover, before accepting sponsorship in Bellevue, Nam had asked for advice from Mrs. Beryl E. Cox, who assured him that this area was the most beautiful and affluent in the state, so he felt somewhat reassured. About ten minutes later, the car entered a residential area with well-maintained and beautiful streets. Mr. Hanner, who was driving, turned to Nam and said,

"We're almost at your temporary residence."

Nam nodded gratefully. Finally, the car stopped in front of a charming white house with a smooth, green lawn and various flowers blooming in the garden leading to the front door.

Nam's wife and children had already gotten out of the car and followed Mr. and Mrs. Colling inside, while Nam lugged the luggage behind them. The family was assigned to the basement level of the

Colling's house. Nam and his wife slept in one room, while the children slept in the family room outside, which was relatively well-furnished. Mrs. Luala Colling mentioned that Nam's family would stay there for about a week until their apartment was ready for them to move in. This was also the first time Nam had lived with an American family, so he carefully instructed his wife and children to observe the customs and habits of the host family to avoid any mistakes.

On the second day, the church members took turns helping Nam's family with the necessary paperwork, such as applying for social security benefits. There, after answering a few questions, they wrote Nam a check for five hundred dollars and issued some food stamps. For a moment, Nam was stunned, not understanding why they would give him such a large sum of money. He handed the check to the church representative, who explained that it was his money and that the government would send the same amount every month until he found a job. In truth, Nam had no idea that the social services would provide such assistance because he had always thought that he would have to find a way to support his family on his own. After leaving the social security office, Nam's wife wanted to visit the Chinese market to buy some Asian food. At that time, there were no Vietnamese grocery stores, only a Chinese market called "Wahsan," where many Asians shopped, and occasionally, one might be lucky enough to meet a fellow Vietnamese. Nam and his wife were so naïve that they thought food stamps couldn't be saved for long, so they spent all of them at once on rice, Thai fish sauce, and various snacks. They couldn't finish eating it all, so they ended up giving some of it to the church members. While staying with the Collings, Mrs. Colling taught Nam's wife how to use common household appliances, such as the dryer, washing machine, and dishwasher, while Nam's wife showed her how to cook Vietnamese dishes and mix fish sauce. Initially, their children couldn't stand the smell of fish sauce and would cover their noses and ears, sticking out their tongues, but once the sauce was mixed with lime and garlic, the whole family sat down to eat and praised the delicious flavor. To this day, the Colling family still uses fish sauce instead of salt and pepper as they used to.

A few days later, the church contacted a Vietnamese family living nearby with their sponsor, so one evening after dinner, Mr. and Mrs. Colling took Nam to visit them and introduce him to the first Vietnamese person he met on this foreign land, Mr. Vu Duc V. After a brief conversation, Mr. V. revealed that he was also a former Air Force officer who had been transferred to work for Prime Minister Nguyen Cao Ky. He was a writer in the Air Force, so he intended to return to school to study journalism at Bellevue Community College. Not long after, he launched the first Vietnamese newspaper in Washington State, called "Dat Moi" (New Land), providing a valuable source of mental nourishment for the Vietnamese community in Seattle and surrounding areas, thanks to his extraordinary efforts.

In the following days, church members taught Nam's wife how to recognize different denominations of money and coins, as well as how to shop and understand the necessary items in supermarkets. It wasn't as simple as shopping in the markets back home, where meat, fish, and vegetables were all pre-packaged in plastic bags, with English labels that were hard to understand. Nam's wife had the habit of buying food based on visual and tactile cues, so she sometimes confused one thing with another, especially since essential items couldn't be easily described with hand gestures.

Since arriving, she had only been able to engage in everyday social exchanges, mostly smiling politely and "going mute," or speaking English with hand gestures. Nam had a young adopted niece who quickly learned English, and she told him that English was easy, giving examples like saying "ếch xào lăn" (stir-fried frog – "excellent") to praise someone or "xe đạp" (bicycle – "shut-up!") to tell someone to be quiet, implying that English was just like Vietnamese and not difficult at all.

Due to the language barrier, Nam's wife once bought pork to cook in coconut milk as a New Year dish with pickled vegetables, but she mistakenly bought lamb instead. After marinating and carefully cooking it, when she served it at the table and Nam took a bite, he was startled by the strong gammy smell of lamb. Unaccustomed to the taste, he had to set down his chopsticks, but his wife thought he was being picky, so she tried it herself. After barely chewing it, she quickly spat it out, agreeing with Nam. In the end, their eldest son, curious about the new taste, devoured it all within a day or two. The loosely translated Vietnamese saying "better to be stabbed than to be mimicked" certainly applied, as Nam found himself embarrassed at least a few times due to his poor pronunciation when speaking English.

The story Nam is about to tell isn't much different from the story of the minister. It goes like this: During the Nguyen dynasty in Vietnam, King Khai Dinh summoned a high-ranking official responsible for bridges and roads to ask about the construction of a bridge being built by the French Governor - General Paul Doumer. After reporting on the progress of the project, the king asked for the name of the person who initiated the bridge project. The official turned red and hesitated, asking the king for permission to speak, explaining that the name was not appropriate. After the king insisted, he reluctantly stammered out, "Your Majesty, his name is (Doumer)," leaving the court stunned by the strange and vulgar-sounding name. **Editor's Note**: *"Doumer" as pronounced in Vietnamese is loosely equivalent to "mother F'kr." Hence the awkwardness and slight humor of various words.*

Nam's story wasn't quite as bad because, after all, he had already been to this country twice, though it was more for training than anything else. In reality, Nam had only attended English classes at Lackland Air Force Base (San Antonio, TX) for about six months, while the other years were spent studying aircraft mechanics, weather, and flying techniques on various aircraft types for the South Vietnamese Air Force. One day, when Nam's wife was hosting a meal for church guests, and everyone had arrived, Nam invited them to sit at the table. Instead of saying, "Please sit down," Nam mistakenly said, "Please take a shit." making everyone awkwardly look at each other, nearly blushing. Realizing his mistake, Nam quickly corrected himself by saying, "Ladies and gentlemen, please sit down."

It was a true mishap for someone learning to speak English and pronouncing words incorrectly, especially for older folks like Nam, whose tongue had already "stiffened" with age. So, before speaking, one must twist their tongue and curl their lips many times to avoid making the same mistake Nam made when he first arrived.

About a week later, Nam's family moved into an apartment nearby that the church had rented for them. The first month's rent had already been paid by the church, and the apartment was relatively comfortable, with two floors, three clean rooms upstairs, and a living room and dining room downstairs. There was a TV, enough furniture, dishes, and utensils to get by, though everything was old, it was much appreciated. Now, everything was relatively stable, and they could settle down. Nam began studying to take his driver's license test. Mr. Colling gave Nam more than an hour of driving lessons to get used to the controls and city streets. Eventually, Nam got his license, but he still didn't have enough money to buy a car.

The day after they moved into the apartment, Mr. Anderson from the church took Nam for a job interview at Sunstrand, where they offered him a job as an electronic assembler. The company agreed to hire him temporarily at $2.32 an hour, but he had to work the night shift before he could be moved to the day shift.

This was Nam's first job in a private company. Since leaving school, Nam had immediately joined the military, where everything from A to Z was taken care of by the government. He considered his unit as his family, and his daily work as household chores, so he had grown accustomed to the military lifestyle and discipline. Now, working for a private company, Nam didn't know what to expect.

So after the interview, whatever the company said, Nam accepted without hesitation, even the low wage. In reality, Nam didn't know how to negotiate or understand his rights. These were the disadvantages of not having someone experienced to advise him. For the church, it seemed that as soon as Nam got a job, it relieved them of a burden, so they never pointed out any potential benefits or drawbacks. Nam didn't mean to imply that they were bad people; most of the church members who helped his family were kind and dedicated. But that was the nature of things—if it didn't involve their interests or beliefs, they wouldn't bother or get involved.

Since it was the summer of 1975 and schools were on break, Nam's eldest son, Miki, hadn't started school yet. The church ladies either taught him at home or took him to learn English with his mother at Bellevue Community College. Each week, Nam's wife was taken to shop at American supermarkets to buy groceries for the week, with the church usually covering the cost. Nam's wife saved the food stamps to buy Asian groceries at "Wahsan" in Seattle's Chinatown.

Nam still didn't have a car, so for any trips beyond walking distance, he had to rely on others, which was quite frustrating. So, he often stayed home watching TV to avoid bothering others. The TV only had two channels that worked, but you had to bang on the box several times to get the picture to stabilize so you could watch "Gilligan's Island" on channel 11 and "Mr. Rogers" on channel 9. The other channels were just black and white stripes. After two weeks, the picture was completely lost, and Nam had to ask Mr. Hanner to take it to the shop for repairs, costing twenty-seven dollars. Even after the repair, the black-and-white picture wasn't much better than before.

However, thanks to the TV, Nam's family gradually learned to understand a bit of English.

Every morning, they would take turns picking up dust under the carpet because they didn't have a vacuum cleaner. Sometimes, when it got too dirty, Nam would go to the apartment office to borrow a vacuum. The house didn't have a single nail, hammer, or anything useful for repairs, which made it difficult when something needed fixing. It wasn't that they were too stingy to buy the necessary tools, but they didn't know where to buy them or which store to go to. Cars roared by on the street, but no one walked, so they couldn't just walk to the store.

Nam worked the night shift, leaving around two in the afternoon with someone giving him a ride. He returned late at night with someone else picking him up. The strict schedule and reliance on others for transportation were extremely frustrating, but he endured it.

Nam's job at the company was monotonous, just assembling small electronic parts. Most of the workers around him were women, with only the supervisor, George Bubar, a Vietnam War veteran named Gary, who worked as a welder, and Nam as the only men. The work was repetitive, the same tasks day in and day out. A few weeks later, while distributing paychecks, the supervisor told Nam that he understood his situation and knew that the wages weren't enough to support a wife and three children. He introduced Nam to an electronics instructor at Lake Washington Vocational School to take night classes.

He took Nam to meet Mr. Charles Ayling in the next room to discuss going to school. After their conversation, Mr. Ayling promised to help Nam with transportation to attend evening classes since Nam didn't have a car yet. The supervisor also promised to move Nam to the day shift so he could attend night school.

The following week, Nam started working the day shift. For the first few days, he had to ride his bicycle to work because the church hadn't arranged for someone to give him a ride yet. The distance wasn't too far, about ten kilometers, but the road had many hills, making the ride exhausting. Since there was no bus route, walking or biking was the only option.

During this time, Nam befriended a man from the Montagnard tribe in Kontum, Vietnam, who had served in the Special Forces with the U.S. Army. His family had come over around the same time as Nam's. He got a job digging ditches and building roads nearby, but he didn't have money to buy a car, so he also had to bike to work like Nam. But his situation was a bit tougher because, early in the morning, he had to ride past several houses with dogs. The dogs would bark and chase him, biting his legs and pulling at his pants, so he had to pedal for his life. Every morning, he would tightly roll up his pants and pedal as fast as he could past those houses. In the evening, he did the same to avoid getting bitten by the dogs. Occasionally, he would complain that his sponsor wouldn't let him apply for social security benefits or food stamps and forced him to work to support his family. He also didn't have the opportunity to go to school or learn any skills.

Every evening after work, after dinner, Mr. Charles Ayling would pick Nam up around seven to go to school, and he would bring him back home around ten o'clock. He explained that studying in this field would lead to better pay and easier work, not as physically demanding as other jobs. A few times, he took Nam to his shop to show him all the electronic measuring devices. At the time, Nam was clueless and didn't understand anything, but he admired Mr. Ayling's knowledge in this field.

A week later, Nam bought a car and began driving himself to school, no longer relying on Mr. Ayling for rides or the church members for transportation. Nam saved up about seven hundred dollars from his job and social security. Nam told Mr. Hanner that he wanted to buy a car for transportation and asked him to help find a used car in the church.

About a week later, Mr. Hanner called Nam to tell him about a used 1965 red Plymouth Valiant with a good engine. The owner was asking $650.00. After a test drive, Nam agreed to buy it. When signing the title transfer, Mr. Hanner said the church would cover $250, and Nam would have to pay the remaining $400. Once everything was finalized, Nam drove the car home to show his wife and children, and everyone was overjoyed.

Ba & Nho in front of their first car in the US ('65 Plymouth Valiant)

Nam's wife, besides attending English classes at BCC, also had church members teaching her conversational skills, so her English improved a bit. She could greet and ask simple questions, no longer just smiling or using hand gestures. The church members came to visit often, and Nam didn't know all their names. His adopted niece came up with a way to rename them with Vietnamese names that were easier to remember, such as Mr. and Mrs. "Lăn Chao," Mr. and Mrs. "Y Địch," and Mr. and Mrs. "Bắc Mận." It worked because whenever someone mentioned these names, they immediately knew who was being referred to.

Because there were so many people coming and going, helping and visiting, Nam's wife assumed that every American who visited was a church member. One day, the neighbor next door, noticing that Nam's wife didn't have a car, offered to take her shopping. Nam's wife assumed it was a church member as usual, taking her to buy groceries for the week. The kind neighbor even pointed out what was good, what was on sale, and what could be stored for a long time. Nam's wife filled the cart, and when they reached the checkout, the total came to nearly forty dollars. The cashier spoke a string of English words, but Nam's wife stood there, staring blankly, not understanding a word. The neighbor, feeling awkward, pulled out her wallet and paid for everything. That evening, when Nam came home from work, the neighbor told him what had happened. Nam explained that his wife had mistaken her for a church member, which made the neighbor laugh heartily. She apologized for the misunderstanding and offered to let Nam's wife keep the groceries, but Nam insisted on paying her back in full.

During the early days of settling in, Nam's wife had to endure many sudden changes, from the external material conditions to the internal emotional strain. The language barrier and the completely different social environment often left her looking more sad than happy.

Nam's younger sister, who was a nurse and had worked at the Christian hospital in Saigon, had come to the U.S. with the hospital delegation, working in a small town in California. She occasionally called to talk to Nam's wife, and the two sisters would lament, feeling homesick, and end up crying together on the phone for hours.

Nam's children were initially shy and hesitant around the local children, but soon they blended in easily as if they had known each other for a long time. They spoke Vietnamese mixed with a few English words they had just learned, but somehow the American kids seemed to understand. One day, Nam called his second son, Mika, to come in for dinner while he was playing with the neighborhood kids. Nam listened as Mika told his American friends, "My daddy kêu me đi ăn cơm.." ("my daddy called me in to eat dinner") The American kids understood immediately and nodded "OK."

As for Nam, since setting foot here, he had only stayed home for about a week before starting work, so he didn't have much free time to think or ponder. During the day, he worked, and at night, he studied his trade. Any spare time was spent reviewing his lessons because his English proficiency was limited, so he had to read and re-read the technical terms in his textbooks multiple times to fully understand them, which took up a lot of time.

When Nam received his first paycheck, the social security benefits were immediately cut in half, so he had to save carefully just to cover the rent and car insurance for the following month.

In reality, everyone likes "free" money, but Nam was fed up with the social services constantly calling him in for interviews and paperwork, making it difficult for him in various ways. Even when he had a bit of extra money, he didn't dare deposit it in the bank because anything over five hundred dollars required additional paperwork. When using food stamps to buy groceries, the people around him would stare, looking down on him. Nam's wife felt extremely uncomfortable. Poor her, she bore the brunt of these bitter experiences. Whenever they used food stamps, Nam would sneak out to the car, leaving her to handle the payment. When it came to medical or dental care, they had to ask around to find doctors who accepted Medical coupons, which were usually taken by doctors who had few patients or were new to the field, reluctantly accepting them for treatment because Nam's job was temporary, and the company didn't offer insurance for his family like it did for long-term employees.

A few weeks later, school resumed after the summer break. The oldest child started attending Hillaire Elementary School nearby, taking the school bus, which conveniently stopped right in front of the apartment, so there wasn't much to worry about. The second child, Mika, was taken to kindergarten daily by church members, leaving only the youngest daughter, Mina, at home with her mother. Daily life settled into a routine. Initially, Nam only remembered appointments without writing them down, but within a few weeks, there were too many things to remember, so he had to keep a schedule like a timetable, checking it daily to see if there were any appointments or tasks to complete. Life began to feel mechanical, with set times for waking up and sleeping, unlike back home, where the days were more flexible. Only the weekends remained free, allowing them to do as they pleased.

Nearly six months passed, and Nam became acquainted with a few Vietnamese families living in nearby apartments, where the rent was relatively cheaper. Nam decided to move to a new place. The church members helped Nam's family with the move. The new location was closer to the market and school, reducing the need to rely on others. Work at the company began to slow down. For the past few weeks, Nam noticed that every Friday, someone would leave with their toolbox, looking sad. Upon inquiry, he learned that the company was laying people off.

About two weeks later, the supervisor called Nam and asked to speak with him privately upstairs, in the cafeteria where employees had lunch. After sitting down, the supervisor looked at Nam with a sad expression and quietly informed him that, unfortunately, the company couldn't keep him any longer due to the current situation. He handed Nam a thick stack of papers with the names, phone numbers, and addresses of other companies to apply to, along with a letter of recommendation praising Nam's work. He also encouraged Nam to continue his studies in the trade and assured him that if the company's situation improved, he would be called back if he was still interested.

That evening, Nam came home earlier than usual, carrying his toolbox. His wife was surprised and asked,

"Oh! Why are you home so early today?"

Nam set the toolbox aside and slowly sat down,

"I got laid off today because there's no more work."

His wife picked up the toolbox to put it away and calmly said,

"Well, take a few days off, then slowly start looking for another job."

Nam sat quietly, staring out the window. Normally, Friday evenings were a time to be happy, but today was different. He didn't feel like cleaning or preparing the car for a Saturday or Sunday outing as he usually did since buying the car. This was the first time in his life that he had been laid off, so he wasn't sure if it was because he wasn't doing a good job or if the company simply didn't have enough work.

At that moment, Nam felt that being laid off was something shameful, but in reality, it was quite normal in advanced societies like America or Europe.

Being laid off can sometimes be a lucky opportunity because once people have a job, they often don't bother looking for better or higher-paying opportunities elsewhere.

That evening, Nam called Mr. Anderson, the church representative in charge of his job, thinking that he was someone with a lot of influence and connections with many companies in the area, so finding a low-wage job like Nam's should be relatively easy for him to arrange.

After the dull weekend days passed, on the following Monday evening, Nam received a phone call from him, informing him that there was someone interested in interviewing him for a job. After getting the name, address, and time for the interview, he told Nam that the job would be as a custodian at a school in the Bellevue district. Upon hearing this, Nam had no idea what kind of job it was. This was a completely new term for him, given his limited English at the time, and there was no dictionary at home to check what it meant. Nam discussed with his wife, saying, "I'm so small; how can I do a 'custodian' job?" His wife didn't know any better either, but she encouraged him, saying, "Just go to the interview and see how it goes."

The entire night, Nam was restless and couldn't sleep, constantly thinking about the custodian job. He remembered back in the days when he was a student at Cao Thắng school, sometimes in the afternoons, he would visit the Phan Đình Phùng stadium to practice sports. During that time, there was a famous bodybuilder named Costo Nguyễn Công An, who was an idol among the youth. Occasionally, he would perform to promote physical fitness among the young students.

Nam kept thinking that they had mistakenly called the wrong person, but he comforted himself, thinking, "Never mind, just go to the interview and see what happens; there's nothing to lose."

Two weeks later, Nam picked out the best outfit from his old clothes, polished his shoes to a shine, combed his hair neatly, and drove to the meeting point ten minutes early.

After greeting the receptionist, Nam sat and waited for about five minutes before a man in his fifties, wearing a proper vest and tie, came out, shook hands with him, and then guided him to his car to drive together to a nearby school, saying it was to introduce Nam to the night shift team leader there. Although a bit surprised, Nam remained silent to see what would happen.

Not long after, the car stopped in front of the school's flagpole. The man led Nam around to the back of the school, and as they entered the basketball court, they met the elderly night shift leader, who was struggling to push a huge trash bin outside.

After asking a few cursory questions, the man introduced Nam to the night shift leader and told him to show Nam the tasks he would be doing. Meanwhile, he himself had some business to attend to in the principal's office and would return in half an hour to take Nam back.

Nam felt like he was in a dream, moving from one surprise to another. Yet by then, he had started to get a clearer picture of what was happening, though still not entirely.

The elderly team leader was a black man who seemed to have uneven legs, making his gait slightly unsteady. His face and dark complexion made it hard for Nam to see the wrinkles on his forehead, so he couldn't guess his age. However, based on his appearance, Nam estimated that he must have been over fifty. With a husky voice, like someone with a drinking habit, he didn't bother with pleasantries or asking Nam's name and age. He simply gestured for Nam to follow him, leading him to each classroom, showing him how to sweep the desks, mop the floors, wipe the boards, take out the trash, etc.

Now, Nam fully understood what a custodian job was. To this day, whenever he recalls the word "custodian," Nam can't help but laugh at his own ignorance of the language. He thought the Americans were playing with words—what we would call a school janitor back home, they refer to with more civilized terms like "Building Engineer," "Building Maintenance," "Housekeeper," etc.

After returning to the management office of the Bellevue district, the supervisor asked Nam about his work experience. After Nam explained his military background and mentioned that he was a very clean person who always kept his house meticulously tidy, the supervisor nodded in agreement. Finally, the supervisor told Nam that they agreed to hire him on a temporary basis ("work on call") and that he would be paid $4.32 per hour.

In truth, when Nam was shown the job duties, he felt very discouraged. However, when he heard the relatively good wage, he felt somewhat more relieved. After all, in just a short time, less than a week, he had suddenly received a pay raise from $2.32 an hour to $4.32 an hour, much more than he had expected. When he was working at Sunstrand, he had dreamed of getting a job like Gary's, where he earned $3.00 an hour as a welder. But now, although the job was different, Nam would be earning much more than that, so he immediately accepted the offer.

On the way home, Nam's thoughts were mixed between the work he would be doing and the attractive wage. However, it seemed he hadn't had enough time to erase the memories, the past that was still lurking within him. Nam wanted to forget and bury it in some corner of his life, but sometimes it would spring back like an uncontrollable spring, causing storms to rise in his mind for a few brief moments.

Today was Nam's first day as a school janitor. After instructing his wife to pack some rice and a few dishes in a lunchbox for dinner, he carefully chose an old set of clothes to wear and put on a black cap. Standing in front of the mirror, Nam felt like he could blend in with the elderly black team leader who had shown him the job during the interview the previous week.

Nam's wife handed him the food box as he got into the car. Glancing at her face and the half-smile, Nam suddenly remembered a song from before, dedicated to a heroic red beret soldier named Đương, who had gloriously sacrificed himself for the country: "You won't die, my love... You'll come back..."

It was similar to when Nam had gone on distant military assignments,

76

and his wife would worry about all sorts of things, reminding him of every little detail as she handed him his bag and saw him off to the car. Nam felt like a soldier going off to war, just as in the verses of "Chinh Phụ Ngâm..." With a black flight suit, a gun at his side, and a black cap adorned with two white willow branches proudly on its brim.

Now, dressed in black with a black cap, but in two completely different circumstances, instead of sitting in a cockpit soaring through the skies to the battlefield, Nam clutched a large feather duster, standing under the school's roof, sweeping back and forth on the wooden floor, as if trying to erase the traces of his past.

At exactly 2:30 in the afternoon, Nam got out of the car and walked to the back of the school to report to the elderly night shift team leader for work instructions. Since it was the first day and he wasn't yet familiar with the tasks, Nam just followed the leader to help push desks, arrange necessary items in the gym, and collect trash from the bins in the classrooms.

Whatever the leader asked him to do, Nam did without adding or questioning anything. In those early days, he felt like a girl who had missed her chance and accepted a marriage out of desperation without truly loving or valuing it. He hadn't fully accepted or integrated into the work he was doing.

Nam often avoided places where parents or teachers frequently gathered, perhaps because of his deep sense of discouragement from the sudden change in his life, leading to thoughts filled with self-pity and inferiority.

In fact, Nam wasn't overreacting to the situation, but he did feel a sense of inferiority at that time. He might have been one of the first Vietnamese people to work as a school janitor, while most of the Vietnamese people he knew had jobs as teaching assistants, social workers, store clerks, or warehouse workers. They might have been paid less, but their jobs were more respectable.

Meanwhile, the people around Nam who did similar jobs were mostly elderly, with gloomy, worn-out faces, dressed in dirty, smelly clothes that were very unpleasant. Nam thought that if there had been a few more Vietnamese people working with him, he might have felt less embarrassed and would likely have had a different outlook on the situation.

Because it was a temporary job, Nam didn't work at any particular school consistently. He would move around wherever they needed extra help, or he would stay home if there was no need for additional staff that day. On average, he worked three nights a week.

During this time, Nam became close friends with Mr. Lại Như X. and Mr. Lại Như S. Mr. X. was working as a teaching assistant, while Mr. S. was working at the Pay-n-Save store. Nam still remembers that Mr. X. was the first person to own a color television, a 25-inch set he had just bought to watch the 1976 Olympic Games. Every Saturday evening, Nam's family would go to his house in Kirkland, where they would cook, drink, and watch TV until late at night.

Sometimes, they would drive up to Marysville to catch crabs. In those early years, the sea crabs were abundant. All they had to do was buy chicken wings and necks, put them in a trap, lower it into the water, and after five minutes, dozens of large crabs would be caught, each one about the length of a hand. They would bring the crabs back, fill the bathtub to wash them, then boil or salt them for a feast with red wine.

One day, Nam discussed with Mr. X. about learning a new trade to change careers. Mr. X. was very enthusiastic and agreed immediately. He said that they should learn something practical and short-term so that they could earn money to support their families. Nam told him that he had been studying electronics for more than six months at a vocational school in Kirkland, but it wasn't going well. He wanted to apply to a professional school to become a fully qualified electronics technician.

After a short time following the programs at the community colleges, Nam wasn't very satisfied because it took too long and included too many unnecessary subjects.

During a visit to friends in Tacoma, Nam met Mr. Hồng Thiên V., another former Air Force officer. Mr. V. told him about a well-known private technical school in Seattle that specialized in this field. It had experienced instructors, modern laboratories, and well-equipped training facilities. However, the tuition was quite expensive, about four hundred dollars a month. But when students graduated, they would surely have the knowledge and practical experience to find a job.

Not wanting to miss this opportunity, a few days later, Nam and Mr. X. went to check out the school and learn more about the vocational program. After being shown around by a school representative and observing students practicing in the labs, Nam felt that this was exactly what he had been looking for. The representative also mentioned that the school would guide and assist with financial aid applications for students in need.

About a week later, Nam and Mr. X. were called in to take a math test and other related subjects to assess their analytical skills in electronics.

The test wasn't too difficult, and both Nam and Mr. X. passed and were accepted into the program. During this time, Mr. X. received financial support from the social program (CITA) to cover his tuition fees, while Nam was denied because he already had a job at the school and was enrolled in a local vocational program. Feeling disappointed, Nam realized that he wouldn't be able to afford the high tuition fees.

On Sunday, when Nam's family attended church, they were fortunate enough to hear a sermon from the pastor about teaching people "how to fish" instead of giving them fish. The sermon was long, but Nam remembered the gist of it: "Give them a fish, and they eat for a day; give them two fish, and they eat for two days... But if you teach them how to fish, they can feed themselves for life."

Not wanting to miss this rare opportunity, after the service ended, Nam stayed behind to meet the pastor. He explained his desire to learn a trade and asked the church for financial help to pay for tuition, using the words from the pastor's sermon: "I just want to be taught how to fish, not to extend my hand for fish..." The pastor agreed but said he needed to consult with the church council to approve the funding. A few days later, the pastor called Nam to inform him that the church council had agreed to sponsor his tuition.

From that point on, Nam attended school full-time and worked at the school in the evenings. He had become accustomed to the work, so he was able to complete tasks quickly, leaving time to work on his daily school assignments. Despite the hard work and lack of rest, Nam and his wife were determined to rebuild their lives and set an example for their children.

If Nam had given up and remained stagnant thirty years later, he would likely have ended up like the old black team leader, who only knew how to bend his back and take out the trash.

Everything was hectic and challenging, with more and more assignments. Nam stayed up many nights studying for exams because he couldn't endure the pressure for long. So he requested to reduce his working hours to four hours a night to have more time for studying.

Normally, when he worked eight hours, he had to clean all twelve classrooms and the gym. Now, working only four hours, he was supposed to clean six classrooms and half the gym. But because Nam worked quickly, he managed to save some time for studying. Usually, after finishing his duties, he was rarely disturbed, but today, Nam was assigned to a new school.

After completing his tasks, the team leader wanted to give Nam more work, but Nam told him that he had already finished his assigned duties and wouldn't do any extra work. The team leader seemed disgruntled, but Nam simply walked to his car and drove straight home. The next morning, Nam called the school supervisor to explain the situation and expressed his desire to quit. The supervisor promised to assign him to another school, but Nam firmly refused.

This was the first time Nam had voluntarily quit a job out of dissatisfaction. A few days later, the church contacted him, saying they needed someone to clean up after services, offering a monthly salary of three hundred dollars, and they didn't require specific working hours. Nam eagerly accepted.

From then on, after school, Nam would take his whole family to help with the cleaning. His eight-year-old son would collect trash from the small bins and put it in trash bags, his wife would dust and clean the furniture, and Nam would vacuum the carpets and mop the floors. Everyone knew their tasks well, so it only took about fifteen to twenty minutes to finish. They would lock up and head home without anyone watching or instructing them, as long as the church remained tidy.

Time passed by, with Nam attending school during the day and his wife staying home to take care of their youngest daughter, Mina, who had just turned one, and keeping an eye on the older two while also studying English. Occasionally, some neighbors would bring their children over for Nam's wife to babysit for a few hours during the day, bringing in a little extra money each month. Most of the clothes and household items they used were donated by church members. Although they were old, Nam's wife had a knack for mending and repurposing them, so they didn't look too shabby.

The Montagnard family from Kontum, living across from their apartment, seemed unable to cope with the outdoor work during the cold winter days, so they had quit their job a few months ago. Now, they had switched to lawn care and gardening for a contracting company. One day, the husband proudly told Nam that he had bought an old Vega car, even though he didn't have a driver's license yet. He was so excited that he bought it just to look at, planning to learn how to drive later. Every evening after work, he would carefully clean and polish it, starting the engine and moving it back and forth in the driveway. A few weeks later, he came to tell Nam that he had finally gotten his driver's license. To celebrate this momentous occasion, Nam bought a gallon of red wine and spent the entire afternoon drinking with him.

At that time, there were no associations or political organizations yet, just the church occasionally organizing gatherings for newly arrived

Vietnamese people in the area to connect and support each other emotionally. It seemed that every family had someone who had lost or been separated from loved ones, so these meetings were a chance to share the pain and sorrow.

Musician Lê Quang A. would often play the piano and sing songs like the national anthem and "Vietnam, Vietnam, My Country." Everyone would sing together, their voices choked with emotion, tears streaming down their faces, and hands tightly clasped, holding back their overwhelming feelings. The homeland and nation were represented through the songs, the gentle faces of the elderly mothers, and the elegant áo dài dresses that carried the soul of the Vietnamese people.

After more than a year of intense study, Nam and Mr. Lại Như X. dressed more neatly and carefully than usual. It was a very important day, not only for them personally but for their families as well. Nam felt a youthful energy in his spirit, like a schoolboy eagerly waiting to see his name on the results board after an exam.

There were fifteen students in total who passed the graduation requirements. Nam graduated with honors and received his diploma from the hands of the school director. He felt a sense of pride and joy. Nam and Mr. X. had officially become certified electronics technicians, like fishermen who had finally gotten their fishing rods in hand. It wouldn't be long before they caught their first fish. Nam saw a bright light of hope shining before him after years of struggling through a difficult transition in life.

The following week, Nam eagerly began job hunting, carrying his new diploma with him. He applied everywhere he could, but most of the responses he received were polite rejections, stating that he lacked practical experience and would have to wait for another opportunity. Nam began to doubt the excitement and confidence he had felt when receiving his diploma and the encouragement from his instructors while he was in school. He had gotten a taste of the bitter reality of job hunting.

He returned to Sunstrand, where he had worked before, and called his former supervisor for support. Two days later, the company called him in for an interview. Nam dressed in a suit and tie, looking like a professional technician, and met with Gary, the supervisor of the thermoelectric switch manufacturing department for NASA's space program. After reviewing Nam's file, Gary said that while Nam didn't have much practical experience in electronics, he was willing to give Nam a job monitoring computer systems and recording temperature signals in reports. However, he would have to work in this role for a year before being transferred to another department.

After spending several weeks driving around from one company to another without success, Nam had no other choice but to accept the offer, even though it wasn't exactly what he wanted. So he agreed to Gary's proposal.

The night shift work was quite easy, just sitting and monitoring the computer systems and recording the temperature changes of the thermoelectric switches according to set standards. Many nights, Nam found it boring and struggled to stay awake, but he still had to diligently record the temperatures in the reports.

After about six months, Nam met with Gary to request a raise and a transfer to another department. Gary wasn't pleased and insisted that Nam complete a full year in the role. Nam then took his complaint to the HR department, but it didn't lead to any significant results.

Around that time, Nam got in touch with Mrs. English, whom he had known from his previous job at Sunstrand, but who was now working at the "Teltone" electronics company in Kirkland. She informed him that the company was looking for electronics technicians, so Nam quickly went to apply for a job. A few days later, Nam was called in for an interview and was offered a new job. He returned to his current job to inform them that he would be leaving in two weeks. Gary tried to persuade Nam to stay, offering a raise and a transfer to another department, but Nam told him it was too late since he had already accepted the new position.

Mr. Lại Như X. had also left his teaching assistant job and started working at the GTE telephone company in Kirkland. He mentioned that he was very satisfied with his new job.

Teltone specialized in manufacturing electronic components for telephone companies. The work was very busy, with three different shifts: day, night, and weekend.

Nam was assigned to the weekend shift, working from Friday to Sunday, twelve hours a day. He also worked overtime on other days, so his total earnings were quite reasonable. The job aligned with his interests, and he had many opportunities to use the advanced electronic tools and equipment to repair faulty components.

As the old saying goes, "You can't thrive without a stable home," or "You need a house to live and a grave to die in." A few months later, thanks to frugality and saving, Nam's family had managed to save up a little money, so they decided to look for a house to buy. At that time, toward the end of 1977, houses were still relatively cheap. However, with Nam's low income, which totaled just over a thousand dollars before taxes, and with no significant assets, buying everything in cash meant he didn't have a good credit history to prove to the bank that he was a reliable person who had never defaulted on a loan or gone bankrupt, which would make it easier to get a loan.

This was completely opposite to the customs in Vietnam, where those who were in debt were considered poor, unreliable, and untrustworthy. But in Western society, it was different—more practical. If you borrowed money and paid it back on time, it showed you were responsible and had good credit, making it easier to borrow money from the bank than those who only used cash.

Nam still remembers that the government of the Republic of Vietnam used to urge people to tighten their belts and save money during economic crises. In contrast, in the United States, when the economy was shaky, economists often encouraged people to spend more, and banks would send out tempting letters offering loans for easy spending, which could be repaid gradually, even though the interest rates were exorbitant.

Nam didn't fully understand the cycles and principles of economics, but he vividly recalled a simple example his high school economics teacher gave to explain the interrelated nature of trade and the law of supply and demand. The teacher said: If a cyclo driver, Mr. Hai, hoards all the money he earns and doesn't spend it on things like eating noodles or drinking coffee, then Mr. Ba, the noodle shop owner, won't have money to buy new clothes, and Mrs. Tư, the fabric seller, won't have money to go to the theater. Consequently, Mr. Năm, the theater owner, won't have extra money to ride a cyclo for a leisurely outing. Thus, Mr. Hai, the cyclo driver, ends up without customers because he has disrupted the flow of money in the supply and demand chain. Nam mentioned this example just for fun; it wasn't as simple as that.

Returning to the house-buying issue, Nam had to make several trips to the bank before they finally agreed to lend him money to buy a small three-bedroom house in Bellevue, priced at $37,500.00. In early 1978, Nam's family finally moved into their first real home, where they truly felt like it was their own, no longer having to rent apartments. Even though the house belonged to the bank and they had to make monthly mortgage payments, the joy and pride of owning it were immense. This feeling of ownership was something even Americans dream of—"The American Dream." The best part was the freedom to paint, renovate, or do whatever they wanted without anyone scolding or restricting them, as long as they followed the rules.

Nam's children were growing up; Miki was now ten, and Mika was six. In just three years, they had changed schools three or four times. The youngest daughter, Mina, was four and attended preschool for a few hours each day. Nam's wife was making significant progress in learning English. However, with the children still young, she stayed home to care for them. Nam believed that raising children was more important than anything else, so his wife didn't rush into getting a job. Even though they were financially stretched after buying the house to supplement their income, Nam's wife took in sewing work and babysat the neighborhood children, earning a little extra each month.

Teltone Inc., where Nam had been working for several months now, was also experiencing a slowdown, so there was no overtime. Staying home from Monday to Thursday was boring, so Nam went out looking for a part-time job. Thanks to his existing experience, he was quickly hired to work in the research department for electronic invention projects at a small company near his home. They agreed to hire him for three days a week, from Monday to Wednesday, leaving Thursday as his day off.

Nam's job was to test and identify the causes of deviations from expected standards through signals, reporting the findings to the project manager. The work was highly technical and sometimes felt as easy as playing video games, but there were moments when figuring out the cause of a malfunction nearly broke his brain.

In those early years of resettlement, Nam and his wife worked extremely hard, doing everything they could. He believed that this wasn't unique to his family; it seemed like all Vietnamese immigrants were doing the same. Everyone was striving to rise above, working hard, saving money, and surprising the locals with how quickly they had managed to establish themselves in the middle class of this new society.

Many Vietnamese children excelled in academics, and many Vietnamese-owned businesses started to open. The Vietnamese immigrants during this period were like trees that had been uprooted and transplanted to a new and unfamiliar land. Now, they were beginning to absorb the nutrients of this new soil, sprouting branches, leaves, and flowers and blooming beautifully once again.

Although life was becoming easier economically and materially, Nam still felt a lingering sadness. Occasionally, he would receive letters from his parents in Saigon, sent via Europe because there were no postal connections between the U.S. and Vietnam at that time. Nam would carefully examine the familiar handwriting, written in blue ink by his mother, on the old, yellowed, musty paper that was as thick as rice paper and enclosed in an envelope marked "Par Avion." Holding the letter in his hand, Nam felt a lump in his throat.

His parents and homeland were suffering such extreme poverty! Even the letter paper his mother had sent was a coarse type that Nam had seen over forty years ago when he was still a student at the village school. Over time, that type of paper had become so outdated that no one used it anymore, disappearing from the Vietnamese market. Nam was shocked to realize that the so-called "brilliant leaders of humanity," the self-proclaimed "apex of human intelligence," had led the Vietnamese people backward in history in such a bitter and tragic way.

His parents were elderly and could no longer work, so they had to gradually sell off their belongings to survive. Nam's two younger brothers, former officers in the Republic of Vietnam Armed Forces, had been sent to reeducation camps, leaving their wives and children in poverty, struggling to get by. Nam's family had to stretch their finances to send money back to help their relatives survive.

Due to the deep political division between the U.S. and Vietnam at the time, no one could have imagined a day would come when they could return to visit their homeland. Even within the Vietnamese expatriate community, there were varying political opinions. Some held extremist views, advocating for the downfall of the communist regime by refusing to send money to their relatives or encouraging people not to visit their homeland, believing that such actions would help prolong the communist regime's rule.

At the same time, there were also many who held onto the nostalgic sentiments of their national identity. They did not support the communist leadership, but they still wished for their parents, relatives, and friends to have enough food and clothing and for their homeland to prosper.

Their reasoning was based on the principle that human needs develop according to the stages of daily social life. The communist regime's strategy was to "rule the people by their stomachs." If everyone was lacking food and clothing, starving, then where would they find the time to think about or demand higher ideals?

Our ancestors used to say, "You can't preach a hungry man into heaven." Only when people's bellies are full can they think clearly and demand other rights. Has there ever been a case where someone, on the brink of starvation, stood up to demand "human rights"? Or is it usually prominent figures in society, those who are well-off and have a surplus of resources, who step forward to press the government for human rights for all citizens? These are the observations of Vietnamese political views at that time. Nam did not conclude whether they were right or wrong; he simply presented these real-life examples for us to consider.

After many years away from home, Nam focused on working hard to secure a future for his children and to support his elderly parents back in Vietnam. He avoided engaging with political parties or organizations in his new country. However, he regularly sent money to support social causes, such as rescuing boat people or helping fellow soldiers in need back home through charitable, religious, or Vietnamese community associations.

It wasn't until around 1990, when the first wave of "H.O." program participants arrived in the U.S., that Nam had the opportunity to reunite with old friends and comrades who had fought alongside him from the 17th parallel to the southern tip of Vietnam, sharing the dangers and the ups and downs of military life. How could Nam forget friends like Huỳnh Văn B., Thái Văn A., Nguyễn Văn C., Phạm Văn T., Nguyễn Văn T., and many others. Some had sacrificed their lives for the cause, while others were still struggling to build a new life. Their names were ingrained in him, like historical markers in his life.

During this period, the international communist regime was gradually disintegrating after the fall of the Berlin Wall. The communist leadership in Hanoi began to feel uneasy, fearing that the people might rise up following the global trend. They started to loosen their iron grip on the people, allowing a bit more freedom. Many Vietnamese expatriates had the opportunity to return to visit their homeland.

They brought with them a new breeze of life, thoughts of freedom, and the prosperity of the West. They helped their families with warm, caring gestures, honoring their parents, siblings, and neighbors. This had a profound effect on local communist officials, who silently began to reflect on their own miserable, harsh circumstances.

Years earlier, Nam's mother had written letters expressing her longing for her children and grandchildren, but she always advised them not to rush back too soon, fearing the deceitful tactics of the Vietnamese communists that might cause trouble for Nam. It wasn't until 1992, when his father fell seriously ill and was bedridden that Nam decided to return to Vietnam to visit his parents. Nam's wife also wanted to accompany him to visit her mother and siblings in Nha Trang.

A four-engine jet took off from Seattle-Tacoma International Airport, operated by Thai Airways, carrying Nam and his wife back to their homeland after seventeen long years of separation, dating back to April 29, 1975, at 10 a.m. As he sat in the cockpit of the CH-47 transport helicopter, Nam could only vaguely see the figures of his parents standing at the front door, hastily sending off their daughter-in-law and grandchildren onto the ship, without a single word of farewell or a message for Nam. He knew that his parents' hopes had died that day.

Today, he was returning to mend and conclude a prolonged internal conflict to address the lingering doubts and unresolved questions that had plagued him for nearly two decades.

Nam wanted to see and meet the people on the other side of the battle lines, people he had never met but who had played a role in pushing his generation into the darkest period of their military careers. Nam did not harbor hatred towards them despite being a defeated soldier who was forced into exile. He was still proud of being a former soldier of the Republic of Vietnam. In any war, there are winners and losers, but Nam took pride in having fulfilled his duty as a man during wartime.

Nam's perspective on fighting was simple and unwavering: it was for the people, for a free, just, and prosperous Vietnam. He did not believe in the ideologies imposed by foreign powers, used to incite their allies to fight each other while they sat back and reaped the benefits.

After spending a night in Thailand to change planes, Nam was now truly flying over Vietnamese airspace. Looking down through the window, he saw vast fields, winding rivers, and clusters of poor, dilapidated houses scattered in the middle of the vast waterlogged fields, seemingly forgotten and isolated by the country's turbulent history. His homeland, the southern region with its simple, honest people, content with a bowl of water spinach soup or a plate of boiled bitter greens.

After thirteen years of military life, Nam had seen much, understood much, and cried much for the bitter fate of his people. How could he forget the times when he landed troops along the canals of the Plain of Reeds, where elderly mothers, their hair white as snow, leaned on their walking sticks to kneel at their front doors? Nam's hands trembled on the controls, for she could have been his mother or the mother of his friends. But why did Nam have to witness such painful scenes? Nam was neither a demon nor a monster.

Nam had the warm, genuine heart of a Vietnamese person, and he carried the blood of the Hồng Lạc lineage. Today's young people's fight was not based on any foreign ideology but was deeply rooted in the nation's spirit, just as their ancestors had fought to build and protect the country for thousands of years.

The plane began its descent, passing through white clouds. Nam guessed they were flying over the Gò Đen or Bến Lức areas, where the houses and streets along the national highway leading to Saigon were more crowded than before. Someone mentioned that after the war, many northern families who visited relatives in the south were so impressed by the prosperity here that they decided to stay and make a new life in the south.

Similarly, many cadres who considered themselves victors quickly brought their families to the south to seize the spoils, so the men could add a few gold teeth and the women could get fuller chests to enjoy life. The plane's wheels touched down on the runway, and Tan Son Nhat Airport had completely changed. Tall grass grew wild and yellow on both sides, untended, while the rows of workshops and a few empty aircraft hangars stood forlorn, weathered by time.

The Vietnam Airlines bus dropped passengers off in front of the civil aviation terminal to complete immigration procedures. Nam was deeply moved, seeing the old scenes again after seventeen years of absence. Before him, it was at this very place on the night of April 28, 1975, that all four flight crews slept under the plane, waiting in the most anxious moments.

In a gentle feeling that suddenly came over him, Nam heard a whisper in the wind, as if the trees, the tarmac, the workshops, and the old paint were welcoming him back, filling his heart with warmth and joy, recalling old memories.

Nam's wife took care of the immigration procedures and collected the luggage. She shook her head in frustration at the basic lack of courtesy shown by the customs officers at Vietnam's international airport. Nam felt it too—their faces were always stern and stony, as if they harbored ancient grudges from thousands of years ago. Nam thought to himself, "Why can't they have the warm, friendly smiles typical of Vietnamese people? Could it be that the communist regime trains customs officers to greet international tourists with such cold, icy faces?"

Nam advised his wife to remain calm and quickly complete the "first" procedure so they could leave without having to look at those "beloved" faces any longer. Thanks to a magic wand extending a big circle of friendship, more than half an hour later, Nam and his wife were out of that stifling, suffocating place.

His mother, siblings, and nieces had been waiting outside the gate for who knows how long. Their faces and eyes suddenly lit up with excitement. After seventeen long years of waiting, his mother's hair had turned completely white, and her forehead was furrowed with wrinkles like waves, folding away the endless years of longing for her children.

Nam's mother was a Vietnamese mother, patiently sacrificing and enduring the storms of life to care for her children. Since Nam could remember, he had never once seen his mother enjoy herself. Because the family was poor, she always saved the best for her children, giving them the best food and clothes, carefully saving every penny to ensure they could compete in life. Her children's future was her future.

The day her son passed the baccalaureate and entered university was the day his mother cried tears of joy. But then the war came, and he had to leave school to join the military. With his military uniform and academy cap, he went abroad for further studies. Though sad to part, his mother was overjoyed, as if celebrating, "My son has become a man." Nam embraced his mother, saying, "Mother, I'm back." She was so choked up she couldn't speak a word and kissed him like a three-year-old child.

No matter how old or wise a child becomes, they will always be the innocent little ones in their mother's eyes! His siblings and nieces told him that because their father was ill, he couldn't come out, but he had been waiting and reminding them for days since hearing that Nam was returning.

After loading their luggage into his brother's car, Nam finally had a chance to observe his surroundings, trying to recall anything familiar from the past. The road from Tan Son Nhat Airport, passing through the Phi Long Gate and Lăng Cha Cả, then along Phú Thọ racetrack, and finally turning right toward Phú Lâm, was as clear in his memory as if he had just driven it yesterday. But today, Nam couldn't recognize anything at all! Everywhere he looked, there were people and houses crowding the streets. Even the road turning into his childhood home, where he had lived from childhood to adulthood, was unrecognizable.

Seventeen years had passed, yet it felt like only yesterday. The images and words of the past remained faithfully stored within him. But now, Nam felt like he was standing as a man who had lived nearly two decades ago. He remained steadfast, but the outside world of Vietnam had changed, swept along by time like a rushing river, even the every day words people used were entirely unfamiliar to him.

The car stopped in front of the gate. Nam's father, though ill for a long time, mustered the strength to come to the door to greet his returning son. He looked frail and much thinner than before. Nam embraced him to lead him into the house, saying, "Father, we're here to visit you." Though weak and struggling to breathe, his father's face lit up with joy.

Nam remembered that when he was young, he sometimes accompanied his mother to visit his father in Cholon. Every morning, father and son would go out for coffee and noodles at the shop in front of the alley. His father would pour a little milk coffee for Nam to taste, but not being used to it, Nam would close his eyes and say it tasted like burnt rice water, making his father laugh heartily.

A few days later, with some free time, Nam tried to reconnect with a few old friends from his school days. Most were lost or had disappeared; he only managed to meet two, Nguyễn Văn N. and Trần Văn N. Nguyễn Văn N. was now an elected official in the district council, while N. was the head of the electricity department in Lâm Đồng. However, after April 1975, disillusioned by the changes, he quit his job and became a recluse, living a vegetarian monk life ever since. Nam was deeply moved by the reunion after nearly thirty years of separation. Everyone had assumed Nam had died in the war.

N. took Nam on a tour of Saigon, revisiting the boulevards he used to travel daily to school, though they were now entirely different. Even the street names were strange to him.

Nam and N. revisited their old school, the place where he had spent seven long years studying day after day. He still remembered every brick with carved names, the potholes worn by rain, and the tall tamarind trees providing shade in front of the school gate, which also served as a shelter from the rain and sun for the bicycle repairman under the tree.

Nam's school had once been a dignified, beautiful, old building, like a modest, high-born lady. But now, it had completely changed, perhaps thanks to the "blessings" of the party and government. The school had been transformed into a rundown, sleazy place. The high walls and iron bars had been torn down, replaced by numerous small stalls selling all sorts of junk, creating an unsightly and backward appearance.

Saigon no longer had the distinct charm that once earned it the title of "Pearl of the Far East." Saigon had died along with the impoverishment of the people, environmental pollution, and the social degradation of the state-controlled economy. In the past, one had to travel all the way to the Ngã Ba Chuồng Chó to visit a brothel, but now, these women work right on the main streets of the city center.

Stopping by a café, Nam was surprised to see so many young men, aged around twenty to thirty, sitting idly, drinking coffee, and chatting away for hours. Nam wondered about their jobs, and N. explained that some were the children of high-ranking cadres who spent money like water because their parents made money so easily. Many didn't need to study or work, spending all day flirting with girls, moving from one music venue to another, while others met up to discuss shady business deals or scams. Only a small minority were honest customers, simply there to enjoy coffee for a moment of relaxation.

Most young people of working age were unemployed or didn't have stable, long-term jobs. Their future was so uncertain that they became disillusioned, spending everything they earned on a carefree, extravagant lifestyle, leaving tomorrow's worries for later.

The Saigon of the past, with its long, shady Duy Tân Street, its romantic rendezvous at the Phú Lâm cape, its gentle, fragrant love stories, its young men newly in love... Saigon, filled with Nam's cherished memories, was no more. It had been renamed like a violated woman, stripped of her purity by traitorous ideologies.

Saigon was now like the virtuous Thúy Kiều, forced to live under the dim yellow streetlights, luring passersby into nightclubs that stayed open until dawn, entertaining the wealthy. Today's Saigon was a façade of false prosperity, hiding its age with layers of makeup.

Two days remain before Nam's leave from work ends, and he has to return to his routine life. But this was also the day his father's illness worsened, and he had to be hospitalized due to asthma. His siblings and nieces took turns watching over him day and night. The hospital was overwhelmed with patients, yet there were very few doctors and nurses, so they worked tirelessly and devotedly. Family members often helped care for their relatives, performing other necessary tasks. The hospital seemed to lack everything, even the most basic sanitation and maintenance.

An elderly patient lying next to Nam's father, after hearing that Nam was a former soldier of the Republic of Vietnam, confided in him about the injustices he had suffered under the communist regime after 1975. He told Nam that after years of hard work, he and his wife had built a successful printing business in Saigon, their business thriving and growing.

But then, suddenly, the southern regime was forcibly dismantled. Feeling regret for the years of effort he had put into building his career, he decided to stay, believing that the regime, after all, was run by fellow Vietnamese, so it couldn't be that bad. Moreover, his work had never harmed anyone or created enmity.

After a moment of despondent silence, the man sighed deeply and said, "I never imagined they would be so ruthless. They stole the fruits of my and my wife's labor, the sweat and tears of our entire lives..." After taking over the South, they accused him and his wife of being exploitative bourgeoisie and forced them to join a state-run cooperative. Even the multi-story building where his family lived was taken over by communist officials. Gradually, he and his wife were sent to the new economic zones.

Unable to adapt to life in the swampy farmlands, they returned to Saigon, living on the streets, scavenging for plastic bags and scrap metal from garbage heaps to survive.

Sitting silently, listening to the man's story, Nam felt a burning sensation in his chest, an overwhelming sense of frustration at the injustices and mistreatment his people were enduring post-1975.

The next morning, Nam and his wife went to the hospital to say goodbye to his father before returning to the United States. Though still unwell, his father mustered the strength to sit up, hold Nam's hand, and advise him to take care of his health and wish them a safe journey. Nam had no idea that this would be the last time he would hear his father's voice or feel his father's hand. Over a year later, his father passed away at home, as was his wish, to die at home as he had lived. Nam was sure that his father's soul was at peace because, before he died, he had reunited with all his children after so many years of separation.

Today, it's near the end of autumn in 1995. The Pacific Northwest is not very cold, but it often rains heavily all day. The Boeing plant where Nam and his wife work has been on strike for over eight weeks, so they have no work to do and often spend their time watching TV, waking up late, going out for meals, and wandering around to pass the time.

Nam's children are now grown. The eldest, Miki, has graduated from university and has been working for three or four years. The second son, Mika, has also finished his studies and is looking for a job. The youngest daughter, Mina, who was still bottle-feeding when they first arrived in America, is now twenty-one and in her fourth year at the University of Washington. Although all three are old enough to be independent, none of them have dared to leave the nest yet.

About a year earlier, the eldest son, Miki, had boldly quit his job and moved to California, dreaming of turning San Jose into a green valley. He teamed up with his mother's younger brother to start an electronics company. His uncle provided the capital and created the product, while Miki sought out the market for distribution. Both worked without pay, and after nearly five years of hard work, they faced intense competition, requiring significant capital investment with slow returns.

After assessing their financial situation and researching the market, the two decided to temporarily halt their venture. Miki packed his bags and flew back home. With his professional experience in the market, he quickly found a job with a small company in Bellevue about a month later.

After two decades of building a life in a foreign land, thanks to the blessings of heaven and the help of above, Nam's family had moved five times. The first two moves were from one apartment complex to another, and the last three were from one house to another, each time to better suit their needs and work. Both Nam and his wife had been fortunate to work for Boeing for over ten years, and Nam was already planning for the day when he would retire to tend a garden, not too far in the future.

It's now 9 a.m. on November 29, 1995, and Nam recalls that at this very hour, twenty years and seven months ago, he was standing on the shore of Mỹ Tho Lake, filled with anxiety and worry, as if it had just happened yesterday.

Holding a cup of hot coffee that his wife had brewed, steam still swirling like graceful fairies dancing in the air, Nam instinctively ran his hand through his thinning hair. The outside world was still and silent; time seemed to move slowly as a few yellow leaves fell, swaying in the breeze. Nam sat quietly, savoring the peace and tranquility of the autumn passing softly by his window.

Ba Van Nguyen & Nho Tran Nguyen (wife)

Today, Nam felt he was finally living in relative comfort, with his inner self at ease, the wounds and uncertainties mended. The long and arduous Vietnam War had truly ended within him. He had had the opportunity to return and see again the black bà ba tunics, the checked scarves of the gentle rural people like the crops and the sweet potato fields along the Hậu River, the tilted conical hats, the pure white áo dài of the young school girls just beginning their journey in life.

His homeland, though poor, remained beautiful and heroic, steadfast and resilient against the passage of time, regardless of who came and went, despite the morning sun and evening rain, the oppression of the regime, or the curses of those who sold their souls and denied their roots.

His homeland, where the graves of his ancestors still stood, where the spirits of the departed sought refuge, where his elderly mother, with hair as white as snow, continued to endure the hardships of life with unwavering loyalty to her nation.

That was his homeland—a Vietnam that will endure forever.

CHAPTER 5

The Village of Somber Bridges

Original Title: "Xóm Cầu Nỗi"

If there hadn't been over twenty years of struggle between the Communists and the Nationalists, and if April 30, 1975, hadn't happened, the life of Ut Chau would still have been that of a farmer. Daily, he would wear his black shorts, go shirtless, lead his buffalo, and carry a hoe to the fields, following in the footsteps of his ancestors like thousands of other farmers in the "Cầu Nỗi" hamlet.

But as the old saying goes: "No one is rich for three generations, nor is anyone poor for three generations."

Indeed, the life and fate of Nguyễn Văn Châu differed greatly from those of his great-grandfather, grandfather, and father. Chau didn't know much about his great-grandfather, but when he was about six or seven years old, his grandfather often told him stories about his great-grandfather, which sounded like tales from a distant past.

Chau's great-grandfather was a tenant farmer for Mr. Hội Đồng Bền. A tenant farmer means someone who rents land from Mr. Hội Đồng to cultivate rice every year. He worked three mẫu (about 1.08 hectares) of land, which was enough to support a family with one wife and seven children. Chau's grandfather was the eldest of those seven children.

Mr. Hội Đồng Bền was a wealthy man, with wealth passed down from his ancestors. It was said that one of his ancestors held a high-ranking position during the Nguyễn Lord's rule, which earned him vast lands and rewards from the Nguyễn dynasty.

Mr. Hội Đồng Bền was perhaps destined from a previous life because he was born into wealth and never had to worry about working hard. Pampered by his parents like a noble child, Cậu Hai Bền (a respectful term for the second son) was sent to Saigon to study in a French school. When he obtained the "Thành Chung" diploma, equivalent to a lower secondary school diploma, his parents called him back to the village to marry and help them manage their land, gardens, and houses. His new wife was a young and beautiful high school student from Long An province, the daughter of Mr. Cả Lân from a neighboring village.

The newlywed couple was well-matched, and their families were highly compatible. In the southern style, their land stretched as far as the eye could see, with no end in sight.

The wedding was grand, with hundreds of tenant farmers from both sides bringing chickens, ducks, and bananas, gathering together to help with the slaughter of pigs and cows to feast the officials, local authorities, and neighbors for three days and nights.

At that time, Cậu Hai Bền and his wife were seen as highly educated individuals by the villagers. It was said that sometimes when they quarreled, they would argue in French, which sounded like popcorn popping.

The French colonial authorities stationed in Cần Đước district heard of Mr. Hội Đồng Bền's wealth and education and summoned him to present himself before the French officials. Afterward, the district chief conferred upon him the title of "Hội Đồng," which essentially meant a council member, a title that allowed him to attend weddings and funerals and sit in the front rows, but without any real power. For this reason, the villagers no longer called him "Cậu Hai" but instead used the more formal and bureaucratic title of "Ông Hội Đồng."

Although his parents were wealthy and influential in the village, they treated the tenant farmers and servants with great kindness. When they became too old to manage their affairs, they passed everything on to Mr. Hội Đồng Bền, their only son.

Mr. Hội Đồng, like his parents, was generous and kind to those around him. However, his wife was very strict and harsh with the tenant farmers working on their land. According to Chau's grandfather, after every harvest, the tenant farmers had to dry their rice and then carry it to Mr. Hội Đồng's house to pay their rent. Mrs. Hội Đồng herself would bring out a bowl of water and handpick the rice from the tenant farmers. If any grains floated to the surface, she would reject the entire load and make them take it back to remove all the bad grains before bringing it again. In years of poor harvests, she didn't care and still demanded the same quantity as in previous years. If anyone pleaded with her, explaining their family's difficulties, she would make them pay extra the following year to make up for the shortfall.

Chau's great-grandfather fell into this unfortunate situation. After a flood during the Year of the Dragon, his grandfather had to eat porridge instead of rice because there wasn't enough rice that year.

That was the story of Chau's great-grandfather, a tale from a few generations ago. By the time Chau came of age, Mr. and Mrs. Hội Đồng had long since passed away, leaving only the echoes of their reputation among the villagers. Chau's parents now worked on communal land, which belonged to the government, not private individuals like before. After each harvest, they only had to pay taxes, making life much easier.

Chau's family included his parents, two older brothers, and an older sister. They rented four and a half mẫu (about 1.62 hectares) of communal land. Everyone worked together, and after paying the taxes, there was still enough rice left to feed the family and manage their household needs.

Their house was by the canal, so after the harvest, the siblings would take turns fishing, catching shrimp, draining ponds, and selling their catch to earn extra money to help their parents.

Chau's sister was married to a farmer in the upper hamlet, and his eldest brother, who was married but childless, still lived with the family. Chau was the youngest in the family.

After finishing elementary school in the village, Chau had to drop out to help with farming.

Around 1962-1963, President Ngô Đình Diệm's general mobilization order was issued, and Ut Chau had just reached the age to perform his military service.

Thus, he had to leave his family, bidding farewell to his parents and siblings, to report to the Quang Trung training camp to fulfill his military duty, like all the other young men across the country.

This was a significant turning point in the life of Nguyễn Văn Châu, a farmer. It marked a new chapter, a bold line that bridged the gap between his past and future.

War is destructive, lacking morality. However, war also offers opportunities to demonstrate the caring and compassion of humanity. We don't want to say that war makes humanity progress because it drives competition for survival. The modern world today clearly demonstrates this.

Thanks to the war and President Diệm's general mobilization order, the life of Ut Chau, a farmer with dirt on his hands and feet, changed. A man who had spent his life in the acidic fields, up to his belly in mud, today finally had the chance to say goodbye to his pair of buffaloes and the calf that had just turned one year old. Every morning, he used to lead them to the fields to plow and then bring them back home to feed them straw at night.

But the saddest part for Ut Chau was having to say goodbye to the girl next door, whom he had just started dating half a year ago.

Miss Năm Lài, a young woman from the "Dừa" hamlet, was just eighteen years old. She was quite beautiful, with tan skin. When she went to the market, she would let her long hair flow down her back, but when she went to plant rice, she would tie it up neatly. Ut Chau was most enchanted when Miss Năm wore her black áo bà ba (a traditional Vietnamese outfit), hugging her small waist and rounded hips, with a small, tight waist and a prominent chest.

Their love story was both naive and romantic. It all started more than half a year ago when Ut Chau was herding his buffalo to the field to pull the dried straw back home to store for use as fuel and animal feed during the dry season. Miss Năm Lài was at the village pond, using a water bucket to draw water when the rope holding the bucket snapped. She was struggling to figure out how to retrieve the bucket from the deep pond.

By chance, Ut Chau passed by with his buffalo, noticing her predicament. She glanced at him, perhaps wanting to ask for help but feeling too shy. Chau had admired Miss Năm Lài for a long time but hadn't had the chance to get to know her. Being a shy young man, he often wanted to speak but didn't dare to. But today, seeing that no one was around, just Miss Năm alone at the pond, he became bolder. He mustered the courage to ask:

"What are you doing here all by yourself, Miss Năm?"

Seeming a bit shy, she bent down, clasping her hands between her knees, and stammered:

"I... over there..."

Chau stepped closer and saw the coconut leaf bucket floating on the water.

"Oh! Your rope broke. Let me get it for you!"

Chau quickly removed his black áo bà ba, tying it around his head like a turban to keep it from getting wet. He carefully waded into the pond, but because the water was deep and the pond's edge was steep, after only a few steps, the ground gave way, and he slipped right into the water like a car without brakes. Covered in mud from head to toe, he tried to act as if nothing had happened and quietly swam out to retrieve the coconut leaf bucket for her.

Watching from the bank, Miss Năm Lài felt a mix of pity and amusement, seeing him covered in mud like a clown. She didn't dare laugh out loud, so she quickly turned away. After composing herself, she asked:

"Are you okay down there?"

"I'm fine! But I can't climb back up because the ground is too slippery, and there's nothing to hold onto. Could you throw me a rope and help pull me up?"

Miss Năm quickly retrieved a rope:

"Hold on, I'll throw you the rope. Just grab it, and I'll pull you up."

Chau began to climb up with her help, but after just a few steps, the rope slipped out of her hands, and he fell back into the water. Frustrated, he called out:

"Come on, hold on tighter!"

"I'm holding on as tight as I can, but it's too heavy, and I can't hold on. Maybe I should tie the rope around my waist for better leverage."

"Do whatever you think is best!"

After tying the rope tightly around her small, slender waist, she threw the other end to Chau again. This time, it held better, but the outcome was quite different.

On the bank, Miss Năm screamed as her entire body slowly slid down the steep bank, with no force able to stop it. Moments later, her petite figure lay fully in Chau's arms. Embarrassed and flustered, she pushed him away.

"You're so strange!"

Chau, feeling pleased with himself, laughed and joked:

"I didn't do anything!"

Miss Năm, still a bit angry, gave him a sideways glance:

"You didn't do anything? You were holding me!"

"Well, you fell right into my arms. If I hadn't caught you, you would've rolled into the water!"

Trying to make amends, Chau offered:

"Let me get you some water to wash off that mud!"

Miss Năm Lài seemed satisfied and no longer complained. She splashed water from the coconut leaf bucket that Chau had filled, washing her face. Then she asked:

"Do I still have mud on me?"

Noticing that she kept speaking in the third person, Chau reminded her of his name:

"My name is Châu, but everyone in the hamlet calls me Ut Châu. Don't you know?"

After a long sigh, she said:

"Everyone knows that!"

"So, call me Châu or Ut. It sounds strange when you keep referring to me indirectly!"

After the incident at the pond, over a week passed, but Ut Chau found himself missing something. Sometimes, when he herded the buffalo to the upper hamlet, he would make a point to pass by the village pond at least once a day, looking at the spot where he had once held Miss Năm in his arms due to that accidental mishap.

He missed most of all her blushing, pink cheeks, which looked so beautiful when she was embarrassed, and the way her chest had inadvertently pressed against his. He had wanted to hold her longer, but she had pushed him away afterward. Similarly, Miss Năm Lài's thoughts were not very different from those of Chau. At eighteen, just entering the prime of her youth, she had many dreams.

Those fleeting, subtle thoughts of early youth—despite having caught the attention of several young men in the village—hadn't yet led her to choose anyone. But after the one-in-a-thousand-years accident, she found herself often thinking about it. She started to feel a bit of affection for the young man from the lower hamlet.

Today, the weather was quite pleasant. The sun wasn't too hot, and a gentle breeze rustled through the dry rice stalks, which had turned pale white. Chau grabbed his fishing rod and went to dig worms to use as bait, planning to fish for rô mè (a type of fish) at the village pond. During the dry season, all the fish, including snakehead fish, rô fish, and catfish, moved from the fields to the ponds and ditches to survive until the rainy season returned when they would go back to the fields to spawn.

The village pond served as the water source for the villagers during the summer, so the banks were built high and surrounded by trâm bầu trees to keep animals away. No one was allowed to drain the pond to catch fish, so it was always teeming with fish of all kinds.

Near the pond's edge stood a large banyan tree, known locally as the "bồ đề" tree. Sometimes, children from the village would come here to carve the bark with knives or pick leaves to collect sap. The sap, a white, sticky substance like glue, would be dried slightly and then used to trap birds by spreading it on tree branches in the rice fields.

Any unfortunate bird that landed there would be stuck and unable to move. The children or adults would then catch the bird, pluck its feathers, and roast it for a tasty snack with homemade rice wine.

Under the bồ đề tree, the villagers had built a small shrine, though Chau didn't know whether it was for a god or goddess. However, on every full moon, someone would bring offerings of bananas or mangoes to the shrine. Many in the village believed in its spiritual powers, but Ut Chau had only heard the stories and didn't know much about their truth.

Carrying a bamboo basket and a long bamboo fishing rod, Chau selected a good spot in the shade of the bồ đề tree to sit down. He carefully baited his hook with a worm and tied a floating cork made from a dried điên điển plant to his fishing line.

The pond was filled with lotus leaves, some dark green and as large as a rice sifting basket. It took some effort for Chau to find a small open space to drop his baited hook. The float bobbed on the surface, and now all he had to do was wait. When the float sank below the surface, he would know that a fish had taken the bait, and then he would pull up his catch.

Fishing requires patience, so while waiting for a fish to bite, he relaxed by resting his rod on the lotus leaves and pulling out a packet of salt mixed with fresh green chilies that he had prepared at home to dip his sour guava fruit.

Sitting under the shade of a tree, eating sour guava dipped in chili salt while waiting for a fish to bite, was one of the simple pleasures of rural life.

Occasionally, the sound of fish splashing in the water made his heart race with anticipation. He imagined that the fish below must be hungry and searching for food. Sooner or later, he was sure to catch a few large rô fish to take home, fry until golden, and serve with nước mắm (fish sauce) and boiled rau nhút (a type of leafy vegetable), or perhaps make a soup with dền, mồng tơi, and okra. Eating it with freshly cooked rice would be so refreshing that he would dream of a carefree adventure, no hat needed!

Suddenly, a loud "splash" sounded, and water splattered onto the nearby lotus leaves, startling him. Chau stood up, looking around. He was sure that the noise wasn't caused by fish splashing or feeding but by someone throwing something to disturb him. Just as he was pondering this, he heard a girl's playful laughter:

"Brother Ut! Why are you so startled? There are no ghosts here during the day!"

Regaining his composure when he recognized Miss Năm Lài, Chau feigned annoyance:

"Heavens! Miss Năm, I thought someone was trying to scare me away!"

She emerged from the trâm bầu bushes growing near the pond and walked toward Chau:

"The other day, you helped me get the bucket out of the pond, and I didn't get a chance to thank you. Today, I came here to thank you, Brother Ut."

"We're neighbors, so there's no need for thanks, Miss Năm."

Lài blushed slightly, speaking softly:

"It wasn't just the bucket... someone fell... remember?"

As she said this, Chau immediately understood her meaning and replied:

"It was my fault for suggesting you tie the rope around your waist, but nothing bad happened, right?"

Lài gave him a long sideways glance:

"You held me, and you say nothing happened?"

Chau smiled, trying to make peace with a playful tone:

"So, did you come here to demand compensation, Miss Năm? How about you hold me, and we'll call it even?"

After Chau's playful suggestion, she quickly walked over and playfully pushed him on the shoulder:

"That's a strange thing to say! Aren't you afraid people will laugh?"

And so, from that day on, Ut Chau and Miss Năm Lài began secretly meeting. Sometimes, they would meet under the banyan tree by the village pond or at the Vó Bridge near his house to enjoy the cool breeze.

Their young love blossomed as vigorously as the rice fields after being fertilized, spreading a vibrant green hue. Miss Năm Lài from the Dừa hamlet brought a new vitality to Chau's life, an intoxicating feeling he had never experienced before. When they sat close, she would shyly pull away, but like a magnet, he would draw closer. It wasn't that Miss Năm didn't love him and didn't want to sit close to him. Sometimes, when he pretended to be upset and moved away, she would slowly inch closer to him.

She loved him very much, but her friends often gossiped about boys who would get girls pregnant and then run away, which made her afraid.

Moreover, she often heard her mother say that "a country girl who gets pregnant out of wedlock is a disgrace in the village, and everyone will know. The rumors will spread, and sometimes you'll have to leave the village to avoid bringing shame to the entire family..."

So every time she thought about it, Miss Năm felt nervous. It seems that girls or women often make decisions more with their hearts than their heads. Miss Năm knew this, but who could stop her from loving?

At eighteen, just blossoming into womanhood, Miss Năm had become even more beautiful, like a flower blooming in early spring, with smooth, tan skin from working outdoors. Her full lips were always moist and inviting, and when she smiled, two dimples appeared, pink like ripe mangoes, making her even more endearing.

Perhaps thanks to Chau's deep affection and care, Miss Năm's figure had become even more shapely since they began their relationship. Her body developed so beautifully that it seemed her chest was about to burst through the buttons of her áo bà ba, revealing the vitality of a young woman. Her waist appeared even smaller and more delicate, making her look even more graceful and lovely.

With her perfect curves and shy demeanor, Miss Năm captivated many young men in the village, each of whom secretly wished to be her lover.

Their love affair remained a closely guarded secret. Although they had been together for five or six months, Miss Năm insisted that Chau not tell anyone, fearing the gossip and rumors that would spread throughout the village. If her parents found out, they might forbid her from leaving the house or going to the market freely, as she had before. To reassure her, Chau swore on his life, with Heaven and Earth as his witnesses, that he would never breathe a word to anyone.

Today was perhaps the saddest day in Ut Chau's life. After dinner, as the clock struck seven, he muttered to himself, "Just half an hour left..." and hurriedly helped clear the dishes from the bamboo mat where the family had eaten their evening meal.

Noticing his unusual behavior and silence, his sister-in-law became curious and asked:

"Why have you been so quiet today, Ut? Is it because of the military draft notice that came yesterday?"

Feigning a grimace, he replied:

"No, it's not that, sister! I just have a slight headache from working in the sun all day without wearing a hat."

"In that case, go take a bath and take some medicine. I'll take care of everything."

Feeling relieved, Ut Chau quickly ran out to the bridge over the canal to fetch water, pouring it over his head to wash away the straw dust that clung to his hair and neck, causing an itchy discomfort. Now he felt refreshed and much more comfortable. He chose a relatively new pair of shorts and a black áo bà ba to wear. Glancing at the clock on the post, he mumbled, "Fifteen minutes left; I should leave now..."

The water level of the canal under the Vó Bridge had risen above the halfway point of the bamboo poles. The Vó Bridge was a small hut made of bamboo and thatched with coconut palm leaves, built by his father nearly a decade ago. It was used for fishing at night.

However, in the past two years, his father had grown old and frequently ill, so he no longer fished here, leaving the task to Ut Chau. Because of this, Chau saw this place as the perfect secret meeting spot, thanks to its location along the canal, hidden by thick clusters of thorny bushes and towering coconut palms that obscured the entrance.

Late this afternoon, the weather was beautiful, with a gentle breeze creating small ripples on the canal's surface. The coconut palms swayed, their leaves rustling together, producing a sound like rain falling on a thatched roof.

After waiting for a while, Chau grew impatient, pacing back and forth on the bamboo platform, which was about four square meters in size. Occasionally, he craned his neck to look through the gaps in the coconut palms as if expecting a very important visitor. He was sure it was already past seven-thirty because the sun had hidden behind the bamboo groves at the end of the village, leaving only a stunning halo of golden-orange light.

Each passing minute felt like an eternity. Feeling weary, Chau walked over to the corner of the hut and pulled on the lác hammock. Just as he lay down, a loud "thud" echoed from the roof of the hut, as if someone had thrown something as a signal or to play a prank. Chau jumped up, relieved and excited, and called out:

"It's okay! Come in!"

This was the secret code that Chau and Miss Năm had agreed upon. He was afraid that someone might catch them meeting in the hut, so to be safe, Miss Năm would throw something as a signal to let him know she was there.

When she heard his voice from inside the hut, Miss Năm cautiously looked around to ensure no neighbors were watching. Then, Chau rushed out, grabbed her hand, and led her inside.

"Heavens! I've been waiting for you so long I almost ran out of breath!"

Miss Năm playfully tugged on his hand:

"Don't you know? My mother kept asking questions. After dinner, I had to pretend to go to Uncle Two's house to borrow some black thread for sewing so I could come here to meet you. So, what's so important that you needed to see me?"

With a slightly sad expression, Chau slowly replied:

"There is something important, which is why I asked you to come here."

He reached out and pulled her onto the lác hammock with him:

"I received a draft notice from the village yesterday. They're calling me for military service."

Miss Năm was startled, staring at her lover:

"You've been drafted?"

"Yes, the village is calling me to report for military service."

"Where will you be stationed?"

"I don't know yet."

"When do you have to leave?"

"Tomorrow, I have to go to the village, and they'll tell me."

Miss Năm's face fell, a deep sadness washing over her. It was as if she was about to lose something she had long cherished. She sat in silence, staring down at the bamboo platform, her hands clasped tightly together. Her long, black hair, shiny with coconut oil, fell over her shoulders, partially obscuring her face. Chau gently slipped his hand through her hair, resting it on her smooth shoulder, and pulled her close:

"Are you sad because I'm going to the army?"

With a slight wriggle, she replied softly, just loud enough for him to hear:

"And then you'll forget about me, won't you?"

To reassure her, he pulled her back onto his lap and kissed her softly on the cool cheek:

"Who said I'd forget about you?"

Caught off guard by his sudden kiss, Miss Năm tried to push him away and sit up, but Chau held her tightly around the waist, so she resigned herself to lying in his arms:

"You'd better be careful! Someone might see us!"

Outside, the night was completely still. The dim light of twilight was gradually being replaced by darkness. A few pairs of white storks hurried back to their nests, and occasionally, the soft "plop" of fish feeding on the surface of the canal could be heard.

The Vó Bridge hut was an inanimate object, but it had witnessed the deepening bond between the two lovers over the past six months. It had seen every moment of passion, every tender caress, and every gentle kiss shared by Chau and Miss Năm.

Their love had blossomed so strongly, like a lush green field of rice, that they had begun to hope for a simple wedding. Perhaps one day, the villagers would whisper among themselves, praising the match: "That Ut boy from Xóm Cầu Nổi is lucky to have married Miss Năm Lài from Xóm Dừa...!"

Today is the graduation day for the new conscripts at the Quang Trung military training camp.

The sun seemed to rise earlier than usual. By seven o'clock, sunlight had already streamed through the gaps in the canvas tents pitched on the concrete yard. Nearby stood the old brick barracks of the officers and instructors.

Everyone was busy checking their uniforms and gear, making sure everything was in order for the flag-raising ceremony and the commander's graduation speech.

Since leaving Xóm Cầu Nổi three months ago, Ut Chau had undergone a complete transformation, almost like shedding his old skin. Even his name had changed; no one in the training camp called him Ut, as his family and Miss Năm Lài used to. They all called him by his given name, Chau. In the early days, he found it strange not to be called by his familiar nickname, but he gradually grew accustomed to it, finding it better than the name Ut, which had been given to him as the youngest child in the family.

Before he left, he remembered how his mother had sold several sacks of rice to buy him a new western-style outfit and a brand-new pair of Japanese sandals. Miss Năm Lài had secretly given him a white handkerchief embroidered with yellow plum blossoms and two birds perched on a branch. She had carefully sprinkled it with jasmine-scented perfume so that he would remember her always, even when he was far away.

When he first arrived at the training camp, he had to go collect his military gear, including a steel helmet, mess kit, uniform, and combat boots. But the hardest part was getting used to wearing the military boots. Having grown up in the countryside, he had only ever gone barefoot. On rare occasions, like weddings, funerals, or the first day of Tết, he might wear sandals to appear more refined, but he had never worn shoes. Moreover, in the rainy season, the mud was so thick that it was impossible to wear shoes or sandals. And in the dry season, the fields were cracked with large fissures and holes big enough to twist an ankle, so going barefoot was the most practical option. As a result, the soles of his feet were thick with calluses, cracked, and stained yellow with acidic soil. Whenever he put on the military boots, he felt uncomfortable, hot, and painful, wanting to take them off immediately. But the drill instructors insisted that everyone had to wear the boots during training unless they were injured or had a doctor's permission.

After about a week, he began to get used to it and even started to appreciate how the yellow stains had almost completely disappeared from his feet, leaving them relatively smooth and much more presentable than before.

After nearly half an hour of standing in formation at the flagpole, listening to the commander's speech about the duties and responsibilities of a soldier, the morning sun had flooded the parade ground. Although the new recruits were eager for the ceremony to end, they stood tall and attentively absorbed the final words of advice from their commanding officer before heading off to join the war.

In just a short while, each new soldier would receive a special leave to visit their family before reporting to their new unit.

Outside the camp gate, a crowd of relatives eagerly awaited their loved ones. Chau knew that none of them were there for him. His parents were too old and weak, and his siblings were too busy with farming to make the journey. They were tied to their land, spending their entire lives in the hamlet. The bustling cities of Saigon and Chợ Lớn seemed distant and daunting to them, filled with deceit and temptation.

More than once in the village, he heard people recount the story of Aunt Hai, the daughter of Mrs. Tám from the upper hamlet, who went to Chợlớn to sell flowers. She was pickpocketed and lost all her money, and had to return home in tears with nothing but empty hands. Not long after that, a few months later, Aunt Hai ran away with a man to Saigon, and to this day, no one knows where she is.

That was all to the story, but the villagers exaggerated it with added details, spreading gossip throughout the entire hamlet. Mrs. Tám, upset and missing her daughter, as well as frustrated by the exaggerated rumors that tarnished her family's reputation, fell ill and passed away within less than a year.

After receiving his special graduation pass, Chau eagerly returned to the camp to pack everything into his military backpack. He slung it over his shoulder and made his way to the gate to catch a three-wheeled taxi to the bus station in Chợlớn, where he would take a bus to the countryside.

The bus on the Chợlớn-Rạch Đào-Cần Đước route was already full. The driver honked the horn loudly to signal that the bus was about to depart.

Chau was still wearing his brand-new green military uniform from the graduation ceremony that morning. The bus conductor invited him to sit in the front row. As we know, on buses heading to the countryside, the front seats near the driver are considered "honor seats" for those who are well-dressed, while the back seats are mostly for vendors or people in shabby clothing.

The bus had just left Phú Lâm, nearing Mũi Tàu, and the houses along the road began to thin out. Fields of rice, about to sprout, swayed gently in the breeze, their vibrant green stretching out to the horizon.

In the relatively soft seat made of white and blue nylon strands, Chau pulled down the brim of his military cap to block the light from outside, trying to catch a short nap.

Water hyacinth is considered a type of aquatic plant like water spinach, water mimosa, and yellow velvet leaf, but it's different from edible plants like water mimosa, which usually need to be planted and anchored in rivers, ponds, or ditches to stay in one place. In contrast, no one has ever heard of people eating water hyacinth; perhaps that's why no one pays attention to this wild plant. However, the slightly strange thing is that when water hyacinth blooms, it looks quite beautiful. The delicate petals, in shades of white, blue, and a hint of purple, rise above the dense green leaves below. Village children often wade into the water to pick these flowers and display them in vases at home.

Where fate would take Ut Chau's life, no one could predict, much like the water hyacinth. It originated from some stream, drifting aimlessly, year after year, through rain and sun, only seeing clusters of thorny bushes or floating patches of duckweed slowly dying over time. Chau was aware of this, but with his basic education and understanding, he couldn't think much beyond accepting his fate, just like all the neighbors around him.

His only dream was that if he had a good harvest, he could save enough money to marry Miss Năm Lài. Then, daily life would follow the footsteps of his ancestors:

"On the dry field and the wet field, husband plows, wife plants, the buffalo harrows."

That would be happiness.

But sometimes life isn't as simple as Chau imagined. As the old saying goes, "Man proposes, God disposes."

People can plan whatever they want, but the outcome is mostly determined by fate and the Universe.

Whether the old saying is entirely true or not, in Chau's life, there was an almost coincidental turn of events. Chau had to temporarily leave Xóm Cầu Nổi to join the army, something he had never thought of before.

Lost in a daydream, Chau imagined meeting Miss Năm Lài again, giving her long kisses on her rosy cheeks, and telling her the hardships of a soldier's life, like in the cải lương (traditional opera) plays he had heard at Mr. Hai Giàu's house, or even more romantically, like the love songs of Hùng Cường and Mai Lệ Huyền he had heard at the district market: "Anh là lính đa tình... Tình anh là tình lính..." (I am a romantic soldier... My love is the love of a soldier...). He thought Miss Năm Lài would admire him and love him even more because being a soldier is heroic, not cowardly or weak like the men dodging the draft at the market.

Just as his dream was about to end, the bus suddenly braked, throwing him forward. The bus conductor, hanging onto the back of the stairs, shouted loudly:

"At Bình Chánh intersection, does anyone want to get off?"

Hearing no response, the driver slowly turned left toward Cầu Tràm. The long road, passing through various villages and connecting districts, was poorly maintained. Each year in the summer, road workers would dig up soil from the fields to fill the sunken spots caused by the rainy season. Because of this, there were many potholes, some larger than a rice winnowing basket, so every time the bus hit one, it felt like the whole bus was about to tip over. Moreover, each time the bus stopped to pick up or drop off passengers, a thick cloud of red dust would rise, forcing everyone to cover their noses and mouths to avoid inhaling the suffocating dust.

However, all the passengers were used to it after traveling on this route many times, and no one seemed to mind or complain.

Over an hour later, after passing Ngả Tư Xoài Đôi, Rạch Kiến, and Cầu Đồn, the bus finally arrived at Rạch Đào bus station. Chau slung his backpack over his shoulder and made his way to the horse-drawn carriage station to hitch a ride back to Xóm Cầu Nổi.

The sun was already low in the sky, around three in the afternoon. He planned to get off near Miss Năm's house first, then walk home afterward. But the difficult part for him was how to arrange a meeting with Miss Năm this evening at the fishing hut, as they used to do. He suddenly came up with an idea: write a note, clearly stating the time and place, then crumple it and toss it to her as he walked by, hoping no one would notice! He smiled to himself, thinking it was the best plan.

As the carriage stopped at the entrance of the alley, he saw the red-tiled roof of Miss Năm's house behind a bamboo grove, a field away. Suddenly, he felt nervous, his heart pounding in his chest, worrying whether Miss Năm was home or not. He tried to stay calm, carrying his backpack straight along the edge of the field, past the bamboo grove in front of her house. A black dog, sleeping on the porch, spotted him and ran out to the entrance, barking loudly.

Fearing the dog might bite, he hurried his pace, but still kept an eye on the house. He didn't see anyone, only hearing Miss Năm's mother scolding the dog for barking at nothing. Returning to the road, he thought, "I can't let a dog get the better of me, can I? Maybe I should try again." He decided to risk it and walked the same path again. The black dog ran out and barked just as before, but this time, he didn't walk quickly. He stopped and stomped his foot, yelling at the dog, hoping Miss Năm would hear him outside. After a moment of tension, Miss Năm's mother yelled at the dog again from inside the house. Hearing her voice, the dog tucked its tail and slunk back inside. Chau didn't see any sign of Miss Năm and felt disappointed as he shouldered his backpack and continued home along the edge of the field.

After crossing a few fields, he noticed someone in the distance, a woman wearing a conical hat and carrying a water jug, walking toward him. Chau stopped to put his bag down, rubbed his eyes, and thought, "Could that be Miss Năm Lài?" Sure enough, it was her, bringing water to the people spreading fertilizer in the fields. She was returning home. What an unexpected and fortunate encounter, he thought. He smiled to himself, remembering how close he had come to being bitten by the dog at the bamboo grove near her house!

Upon seeing Chau, Miss Năm was overjoyed. However, to maintain discretion in the village, she only asked a few simple questions, then took the note from him and hurried home with the water jug.

After a long period of basic military training at the Quang Trung camp, Chau had learned not only the basics of military skills like marching, crawling, assembling and disassembling automatic weapons, and practicing shooting on the range, but he had also been taught military discipline, including military conduct, and the psychology of political warfare.

Thanks to this experience, Chau had become more knowledgeable. Now he understood who the "night men" or "thirty men," as the villagers used to call them, really were. They were actually communist cadres operating in the area, conducting propaganda against the South Vietnamese government at night.

During his leave, Chau impressed Miss Năm, who was amazed and admired him more and more. She was surprised by the knowledge he had gained since leaving Xóm Cầu Nổi. But what she liked most were the times they met, when Chau would wear western-style clothes and polished leather shoes, looking and smelling as clean as the school teachers at the district market, unlike the old days when he wore shorts, a black áo bà ba, and went barefoot, reeking of acidic mud and swamp.

Today was their final meeting. Early tomorrow morning, Chau had to return to camp to receive his orders and report to his new unit. During the family dinner, his brothers and sister-in-law hosted a farewell meal for their youngest sibling. The meal included local delicacies: sour soup with flowers and dried Mỹ Tho fish, crispy fried fish with green chili fish sauce, grilled snakehead fish with basil and boiled water mimosa dipped in fermented anchovy sauce.

But the most special and surprising thing was the rice. Somehow, his siblings had found some "Nang Thơm" fragrant rice for Chau to enjoy. As the family sat around the mat in the front yard, as usual, his eldest brother removed the lid from the earthen pot, releasing steam and a fragrant aroma that filled the air. Even before tasting it, everyone was already praising the rice. Chau, surprised, asked:

"Wow, we've never had this kind of rice before! Where did this come from?" His eldest brother smiled mysteriously:

"You're right, little brother. We can't usually afford this kind of rice. But today, we wanted to give you a memorable meal before you head off to the army, so your sister-in-law went ahead and bought two liters of this rice at the market for you!"

Chau turned to his kind sister-in-law:

"Why did you spend so much on this rice? I would have eaten anything!"

She replied, "You're our youngest brother, and everyone in this family loves you. We wanted to give you something special before you leave for the army. It's not like we can do this every day!"

The family dinner was joyful and satisfying. Chau and his two older brothers had almost finished two quarts of the finest Hóc Môn rice wine, purchased from Mrs. Hai's shop along the dirt road.

Chau had downed over two large "ox-horn" glasses of rice wine. His face was a bit flushed, and his heart was beating faster than usual, but he still remembered the important meeting he had with Miss Năm at seven that evening.

After the family dinner, he pretended he needed to visit Mrs. Hai's shop to buy a few things for his early departure the next day.

The vast fields in front of him, with their green rice stalks just beginning to bud, made him feel calm. The evening was still early, and he leisurely strolled through the fields toward the riverbank. As he walked, his thoughts drifted. He realized that his life was about to change, leaving behind the plows, the vegetable patches, the okra rows, and the water buffaloes he had known so well. And then there was Miss Năm, his first love from Xóm Dừa! What force was pulling him away from all these things that had bound his family and ancestors to this remote village for generations?

To push aside these vague, unsettling thoughts, he bent down and carefully plucked a large rice stalk, its swollen belly as thick as a finger. He gently peeled away the husk, revealing a cluster of tender, green rice grains full of sweet juice. He popped them into his mouth, savoring the sweetness and delicate aroma of the young rice, still in its early stages of development. The taste brought him a sense of joy and comfort. Just as he was about to pick another stalk, he heard a soft voice behind a nearby cluster of thorny pineapple plants teasing him:

"Picking young rice? I'm going to tell the landowner!"

Like a child caught stealing candy, Chau quickly stood up straight and turned around. When he heard the voice, he knew it was Miss Năm, playing a trick on him, so he pretended not to know:

"Who's hiding back there? If you don't come out, I'll throw mud at you!"

Miss Năm had been hiding to scare him, but when she heard the threat of mud-throwing, she stepped out from behind the pineapple plants and playfully scolded him:

"Stealing my rice and you're not even ashamed, now you want to throw mud too?"

Chau pretended to deny it:

"I thought it was someone else! Who would have thought it was you hiding in there?"

After a long, playful glance, she retorted:

"Hmph! You knew all along..."

The fishing hut once again served as the silent witness to the secret love between the village girl and the neighboring boy.

Chau gently guided her to lie down on the rattan hammock where they had first met during their initial secret rendezvous. As soon as she lay down, she pulled him down to join her on the hammock. Not paying attention, he lost his balance and fell on top of her. The hammock tilted to one side, causing the buttons on her blouse to accidentally pop open, as Chau instinctively grabbed at them to keep from falling to the floor.

Miss Năm blushed with embarrassment. She quickly tried to cover herself, but it was too late. Her ample, smooth breasts, untouched by anyone, were now fully exposed as if challenging Chau.

What had just happened was so sudden and unexpected that Chau was a bit flustered. However, he couldn't deny the wonderful sensation of having his face nestled against those warm, soft mounds.

He tried to make amends:

"It's your fault for pulling me down without warning!" Miss Năm remained calm, lying there with her cheeks turning a lovely shade of pink:

"Are you sure it wasn't your fault?"

Chau smiled and whispered softly into her ear:

"Let me help you button up!"

Miss Năm didn't say a word. She simply reached up, wrapped her arms around his shoulders, and pulled him closer.

Although Chau was two years older than Miss Năm, he was still shy and reserved when it came to girls. To be honest, he was afraid of getting a girl pregnant. What would he do if that happened? The neighbors would call him a scoundrel and a fool, so even though he desired it, he held back in the face of temptation.

Miss Năm's feelings weren't much different. Her parents constantly reminded her, "A girl must wait for a proper proposal, with betel leaves and areca nuts brought by a matchmaker. You can't meet boys at the crossroads or in the alleys, or if you get pregnant out of wedlock, the whole village will gossip about our family, and we'll be disgraced!"

There was even a time when a rooster crowed at noon in the village. That alone sparked enough gossip for the entire village. People said that a rooster crowing at noon was a bad omen, a sign that some girl in the village would soon get pregnant out of wedlock! Then they would whisper and speculate, saying things like, "I bet it's that girl Tư, she's always giving the teacher those flirtatious looks," or "That girl Sáu has been fooling around with that guy Hai."

Remembering these stories made Miss Năm nervous. She didn't want to go any further with Chau beyond the tender kisses, caresses, and embraces they shared when they were together.

The hammock swayed gently back and forth, creaking rhythmically as it rocked. Chau remained silent, his eyes closed, with his face resting on the warm, bare chest of his beloved. Feeling the quickening heartbeat and breathing, Miss Năm softly reminded him:

"When you're far away, don't forget about me, okay?"

Lost in the sensation of holding this rare, precious land in his arms, Chau could only nod gently in response. This didn't satisfy Miss Năm, as she wanted him to look her in the eyes and give her a more reassuring answer. She quickly pushed him up, catching him off guard:

"What are you doing?"

Playfully, she scolded him:

"Did you even hear what I said?"

"Yes, of course! I promise that no matter where I am, I will always remember you. I swear by the heavens, if I don't keep my word, may I be struck down by lightning!" he replied, trying to reassure her.

Miss Năm, now satisfied, smiled:

"Really?"

"Really!"

After rebuttoning her blouse, she said:

"It's getting late. I should go. When will you get leave again?"

"I don't know, but I'll let you know when I do."

As she got off the hammock, she tucked a strand of hair behind her ear to keep it from getting tangled. Chau, standing behind her, wrapped his arms around her slender waist, holding her tightly:

"When I finish my service, I'll go home and tell my parents to sell a calf so I can marry you."

Miss Năm turned around and gently patted his back, showing her gratitude.

After reporting for duty in Saigon, Chau was given an assignment, along with a group of ten other new recruits, to go to Ban Mê Thuột and report to the personnel office of the 23rd Infantry Division stationed there. Chau was originally supposed to be assigned to a unit in the third or fourth military region near his hometown, but during his training at the Quang Trung camp, he became close friends with Võ Văn Tấn, another recruit like him. Tấn had been a coffee farmer in Ban Mê Thuột before being drafted, and they met at the Quang Trung Training Center.

Because they shared similar backgrounds in farming, they quickly became friends. Tấn was kind and generous, often sharing dried fruits and snacks sent by his parents from Ban Mê Thuột. Sometimes, Tấn would tell Chau about mischievous but exciting stories, like sneaking around to watch Montagnard girls bathe in the streams. Chau's favorite story was about the half-French, half-Thai girls, with their creamy white skin, shaped like the most beautiful flowers in the forest. Whenever Tấn talked about them, Chau couldn't help but think of Miss Năm Lài, his girlfriend back home.

But all of this was just a small part of the reason he decided to choose this unit. The most important reason was that he wanted to see a place different from the vast fields where there were only two seasons each year: the rainy season, when the fields flooded and the dry season when the land cracked. It was the same cycle every year, with nothing new or exciting. He wanted to see mountains, forests, tigers, leopards, deer, and bears rather than the old buffaloes, black dogs, and tabby cats back home.

At twenty, the desire for adventure and discovery was strong in him, like an eagle wanting to leave its nest and spread its wings to explore the vast sky.

Ut Chau was a reflection of that awakening.

His family had lived in the same place for generations, from his great-grandfather to his grandfather to his father, none of whom ever thought of leaving the bamboo groves and the old shrine by the village pond to venture into what people called "a place where the wild birds and animals live... a place of jungle and wilderness..." Perhaps it was due to an inherited timidity, a mindset that valued "safety and comfort," typical of the rural farming folk. They feared "throwing away the bird in hand to catch the one in the bush," so they preferred to stick to their muddy fields where they had a guaranteed livelihood rather than wander off to an unknown land.

Thanks to the military, Ut Chau now had the opportunity to break away. To be honest, he wasn't completely unafraid of leaving his homeland and village, especially leaving his girlfriend, Miss Năm Lài. But he believed that after his military service, he would be discharged, return home, marry, and live his life there.

Still lost in thought about the journey ahead, Tấn suddenly came up from behind and patted him on the shoulder:

"Hey Chau! Put that assignment order away and come on, the military truck is waiting outside!"

Curious, Chau asked:

"We're going by truck? Is Ban Mê Thuột far?"

"Ban Mê Thuột is really far! I don't know if we're going by truck or by plane. Let's get on the truck, and the squad leader will let us know."

With his backpack on his shoulder, Chau followed his friend. A convoy of ten-wheeled military trucks was lined up outside the camp gate. Chau nudged Tấn:

"Hey Tấn! How do we know which truck is ours?"

"I don't know! We'll ask the driver when we get there!"

Just then, they noticed a senior soldier, a sergeant, looking very dignified, stepping down from the front seat of a truck. He held a stack of papers in his hand and looked at the two new recruits approaching:

"Are you Võ Văn Tấn and Nguyễn Văn Châu?"

"Yes, sir!"

"Get on the truck. We've been waiting for you two to complete the group!"

The military truck left the convoy and drove through the city streets. The group of over ten new recruits sat silently, each lost in their own thoughts, looking around the city as if they were soulless bodies.

After about fifteen minutes, the truck stopped in front of a camp gate. The military policeman on guard leaned out from the sentry post as the driver handed him a piece of paper, probably the military transportation order. After a quick glance and a look at the new recruits in the back, he returned the paper and waved them through.

Chau, looking confused, turned to Tấn:

"Are they taking us to another camp?"

"I think we're going to Ban Mê Thuột. We're being transported there!"

The truck continued past rows of old brick buildings, probably built during the French colonial period. On both sides of the road were lush trees with branches spreading out over the pavement.

Suddenly, Chau noticed three soldiers on the left side of the road, wearing army-green jumpsuits, with small guns and ammunition belts strapped around their waists.

Chau lightly nudged Tấn's thigh:

"Hey Tấn! Look at those soldiers! Their uniforms are weird; the shirt and pants are connected!"

After a moment of observation, Tấn shook his head:

"I don't know! Maybe they're special forces! That's why they wear jumpsuits for easy movement!"

The military truck slowed down and eventually stopped on an empty concrete lot. The sergeant in front opened the door and ordered:

"We're here, get off the truck!"

Over a dozen new recruits, moving like machines, shouldered their backpacks and jumped off the truck, following the sergeant's lead.

An unusual sight caught Chau's eye as he stared at the C-47 aircraft parked nearby. Tấn, carrying his backpack, slowed down to let Chau catch up.

"Hey Chau! We're flying, not taking a truck!"

Chau hurried forward, excited:

"We're flying? It must feel strange when it takes off and lands, like a roller coaster, right?"

Tấn, who seemed more knowledgeable because he had worked on fields near the Phụng Dực airfield, where he occasionally saw Vietnam Airlines' civilian planes taking off and landing, dreamed of one day saving enough money to take a flight to Nha Trang just for the experience.

"Yeah! Flying is like riding a bus, going up and down hills; it makes your stomach drop!"

"I've never seen a plane this big before! I've only seen them flying high in the sky. Sometimes, in my village, we see old DC-3s flying low, dropping leaflets, but nothing this big!"

Still puzzled, Chau scratched his head:

"Hey Tấn! It looks like it's made of metal, probably aluminum or steel. How can something so heavy fly in the sky? I always thought planes were made of paper or fabric to keep them light! The Americans and the French are amazing, aren't they?"

"Yeah! It looks like we're flying in this one!"

As the two young soldiers continued their conversation, the sergeant ordered the group to halt while he ran to the plane to check if they were on the correct flight.

A moment later, he returned and lined up the group in pairs, leading them to the plane that Tấn and Chau had just been discussing.

Once the flight attendant seated everyone on the plane, Chau noticed the three soldiers in jumpsuits, whom he had seen earlier, also boarding the plane. One of them asked the flight attendant, who was busy inside:

"Are all the passengers on board, Mr. Hùng?"

"Yes, sir, everyone's on board."

"Signal the runway that we're ready to start the engines!"

After that, the three soldiers entered the cockpit. Chau, surprised, whispered to Tấn:

"Are those guys the pilots?"

"Yeah, probably. But they're officers, so why don't they wear any rank insignia? The officers we've seen wear their ranks and medals proudly. These guys are so casual, wearing jumpsuits!"

Trying to show some understanding, Chau added:

"The officers we've seen were combat soldiers, while these guys probably just fly planes and don't engage in combat!"

Chau had barely finished his sentence when Tấn objected:

"They do fight! Haven't you heard of fighter planes dropping bombs during the war with the Japanese and the French?"

Their conversation was interrupted by the roar of the engines starting. The entire plane vibrated slightly with the high RPMs of the propellers outside.

Chau looked out through the small window as the plane began to taxi to the runway.

A few minutes later, the plane braked suddenly, pushing the passengers forward. As everyone regained their balance in their seats, the engines roared even louder. The plane, now like a powerful beast, charged down the runway. The passengers were pushed back by the centrifugal force.

Chau closed his eyes and covered his ears with his fingers, bending down to the floor of the plane. The "whooshing" sound of the wheels against the runway stopped. The entire plane lifted gently, like a paper kite. The plane calmly ascended into the sky like an arrow shot from the ground.

A moment later, the plane leveled off, and Tấn, sitting next to Chau, leaned close to his ear and whispered:

"Hey! Look out the window!"

Chau sat up and leaned to look down below:

"Wow! The houses and fields are so tiny! I can't see anyone at all!"

"We're too high up. You can't see anyone from here."

This was the first time in Chau's life that he had flown on a plane, so everything seemed strange and impressive to him.

"I always thought only the Americans and the French flew planes, but I never knew that Vietnamese people could fly too! What kind of soldiers are these pilots?"

Tấn, seizing the opportunity to show Chau that he knew a bit more about the military, replied:

"I heard they're from the Air Force at Tân Sơn Nhất!"

"Oh, Air Force, Navy, and Army, just like we learned in training at Quang Trung. I've forgotten most of it already! Air Force flies planes, Navy sails ships, and Army fights on land, like us."

After explaining this to himself, Chau was still curious and asked:

"Is it hard to become a pilot?"

"Who knows! I've always heard about pilots flying planes, but I've never seen one before! This is my first time seeing them too!"

"I noticed they dress differently and seem to be bigger than us!"

The two young conscripted soldiers, fresh from the fields, were experiencing life outside their village for the first time, like fawns exploring a new world. Everything was new and fascinating to them, contrary to all the fears that people in their village had talked about. They had heard that being a soldier was dangerous, with the risk of being crippled or killed.

Many families, grandparents, and parents would refuse to marry their daughters to soldiers, fearing they would become widows too soon. They preferred to find sons-in-law who were teachers, exempt from military service, or at least the Chinese shopkeepers at the market, or even better, farmers.

But sometimes, the reality was quite different. Many country girls didn't agree with their parents' choices. Some were actually attracted to soldiers. With their short haircuts, neat uniforms, and pistols at their hips, they looked heroic. Especially the young officers with their shiny gold epaulets on their collars, who were charming and often had many girls pursuing them like flies to honey.

The noise from the engines outside seemed to have diminished as the plane banked to the left, avoiding low clouds near the mountain peaks. The road from Khánh Dương to Ban Mê Thuột was clearly visible below. A few minutes later, the C-47 landed safely, taxiing to the military aircraft parking area at Phụng Dực airfield.

The sergeant, who was in charge of the group, unbuckled his seatbelt and jumped down first. He ordered the new recruits to stand in one place while he arranged for transportation to take them to the 23rd Infantry Division.

Chau slung his backpack over his shoulder and followed Tấn into the shade of a corrugated iron building:

"It's really hot here!"

"Yeah, we're in the mountains and forests! No flooded fields like back home!"

Tấn pointed toward Khánh Dương:

"See all those mountains? Beyond those ridges is Ninh Hòa, and then Nha Trang. The climate is more pleasant there because it's near the coast."

This was the first time Chau had ever seen a landscape of mountains and forests stretching endlessly before him. Up until now, he had only imagined such a scene from colorful landscape paintings sold at the market.

"Hey Tấn! There must be a lot of elephants and tigers here, right?"

"I've never seen tigers, just heard about them. But I've seen elephants occasionally, used for logging. I heard that King Bảo Đại used to have a vacation home near Lạc Thiện Lake, and he would come here to hunt and relax."

As the two friends talked to pass the time, the sergeant returned and informed them that in a little while, a military truck would come to take them to the 23rd Infantry Division's headquarters.

After reporting to the division's personnel office, Tấn and Châu were assigned to the Ban Mê Thuột subdistrict to reinforce the militia unit stationed at an outpost protecting the perimeter of the L-19 airfield, right on the edge of the city.

Second Lieutenant Trần Văn Tiến was their platoon leader. Tiến was a young reserve officer, newly graduated from the Thủ Đức Military Academy, who had been transferred here just over two months ago. Tiến was quite surprised when Châu reported in, noting that he was from Rạch Đào village, Cần Đước district, Long An province. Tiến pointed to a stool in front of the desk:

"Châu, you can sit down there. You said you're from Rạch Đào village?"

"Yes, Lieutenant! I'm from Rạch Đào village, but I live in the 'Cầu Nổi' hamlet and work in the fields, not in Rạch Đào market."

"So why didn't you request a transfer to the south? Why did you come here?"

Hesitating for a moment, with his hands intertwined, he replied slowly:

"I wanted to see new places and gain new experiences!"

Tiến laughed lightly:

"You have an adventurous spirit! My hometown is also Rạch Đào village, like yours. My parents own a fabric and clothing shop at the market. When I was young, I attended elementary school there, and later, my father sent me to Saigon to study at Cao Thắng Technical School. After graduating from high school, I was drafted and went to Thủ Đức, and after graduation, I was posted here."

With Tiến's open and friendly manner, Châu felt very comfortable and wanted to learn more:

"So, Lieutenant, when did you attend Rạch Đào school?"

"I think it was around 1947 to 1952."

As if he had just realized a surprising coincidence, Châu happily exclaimed:

"I also studied during that time!"

Then Châu suddenly paused. It seemed there was a gap between him and his platoon leader, Second Lieutenant Tiến. Châu's voice softened and became more subdued:

"But after finishing elementary school, my parents were poor, so I had to quit and return home to work in the fields."

Sensing Châu's feelings of inferiority due to his poor circumstances, Tiến quickly changed the subject to avoid deepening the embarrassment of his less fortunate fellow villager.

"Have you married yet, Châu?"

"No, Lieutenant, I haven't!"

Lieutenant Tiến laughed, half-joking:

"In that case, one of these days when I'm free, I'll introduce you to some girls from the rear to get to know. The school girls here are very fair-skinned and lovely, but be careful, or you might get stuck with one for good!"

Châu found Tiến's joke amusing:

"Lieutenant, you're kidding! I don't dare to approach them; they'd likely scorn a low-ranking soldier like me and embarrass you!"

Remembering some unfinished paperwork on his desk, Tiến stood up and stepped forward:

"Oh, have you and Tấn received your bedding and everything else you need?"

"Yes, we have!"

"Good, now go report to Sergeant Minh, my deputy platoon leader, who will assign you both your daily duties. If you have any questions, feel free to come to see me anytime."

Châu respectfully stood up straight and saluted his commanding officer before heading out.

###

The weather in Ban Mê Thuột was beautiful today, as if to make up for the drizzling rain that had soaked the ground for days on end. During times like this, Châu finally understood why Lieutenant Tiến called Ban Mê Thuột the "land of eternal sadness" or the city of "dust in the sun and mud in the rain."

There were days when Châu sat alone with his rifle, keeping watch in a guard tower. Below him, rows of rubber trees stretched out in a lush green expanse, blending with the misty mountain ranges that appeared and disappeared through the fine mist, giving the landscape a ghostly, cool feel characteristic of the Central Highlands. These moments often reminded him of home, his family, friends, and especially Miss Năm from his hamlet, "Xóm Dừa."

It had been more than seven months since he left his love at Cầu Vó. When he inquired with his comrades, he learned that it might be possible to visit home once a year, if at all. Sometimes, they were put on alert and could not leave the unit. Châu thought that with time, he would get used to it. He mused that maybe fate would have him grow roots here, like some of his friends who were from Cà Mau but now had wives and children here.

But Châu believed he wouldn't follow in their footsteps because he had promised Miss Năm that he would marry her after his military service ended. He was sure she would remember and wait for his return.

Châu had even asked Lieutenant Tiến about the duration of military service. Tiến had only shaken his head, unable to give a precise answer. He had promised to inform Châu as soon as any discharge papers arrived, and that was all.

Seeing the warm sunshine, Châu took his army blanket out to dry on the fence. At that moment, Lieutenant Tiến drove up in a Jeep and stopped next to him:

"Hey, Châu! Do you have anything to do this morning?"

"No, Lieutenant, I'm off duty today."

"Want to come with me to the L-19 airfield?"

"If you want me to, I'll go!"

"Go put your boots on; I'll wait."

About ten minutes later, the two were at the helicopter and L-19 airfield in the city. Tiến stopped the Jeep near the guard station and told Châu to stay with the vehicle while he went into the communication center for some business.

Nearby, Châu noticed a few soldiers who seemed to be preparing an observation aircraft parked there. Curious, he wandered over to take a closer look.

After a quick inspection of the plane, he struck up a conversation:

"Is this what they call the 'Old Lady' plane?"

One of the aircraft mechanics turned and smiled:

"Yeah! The locals used to call this the 'Old Lady,' but we in the Air Force call it the L-19 observation plane."

"So, you guys are Air Force mechanics?"

"Yep! We're non-commissioned officers and aircraft mechanics."

Châu looked surprised:

"All this time, I thought only the French or Americans did the repairs, not our Vietnamese."

"In the old days, the French did the repairs, but now it's our Air Force."

"I've heard people go to Saigon to learn car mechanics, but where do you learn to fix airplanes?"

Seeing that the young soldier was genuinely interested in their profession, the Air Force mechanic wiped his oil-stained hands with a rag and introduced each of his colleagues:

"That guy over there is Senior Sergeant Thành, a radio specialist. The one over there working on the engine is Sergeant Tư, an engine specialist, and I'm Master Sergeant Hùng, the team leader on the airfield."

When Hùng finished the introductions, Châu felt a bit embarrassed about addressing them casually as "brother":

"Sorry, I didn't know your rank, Master Sergeant, so I called you 'brother.'"

Hùng patted Châu on the shoulder, smiling warmly:

"In the Air Force, we care more about mutual respect than about strict formalities like other branches. You can call me 'brother' or 'Master Sergeant'; it's all good."

"Thank you, Master Sergeant."

Hùng continued answering Châu's earlier question:

"Most of us in aircraft maintenance are sent to the United States for specialized training for about a year or longer, depending on the field. Then we return to Vietnam to serve in the Air Force."

Hùng pointed to each of his colleagues and added:

"Like Senior Sergeant Thành over there—he just returned from the U.S. a little over a month ago. Sergeant Tư and I are scheduled to go next year."

"So, the Air Force doesn't fight like we do, right, Master Sergeant?"

"Of course they do! The pilots of helicopters, fighter jets, and L-19s fly into battle all the time. The difference is that you fight on the ground, and they fight in the air. As for us, our job is to maintain the planes in top condition so they can fly out and fight or provide air support when under attack."

Feeling like he was talking too much, Hùng patted Châu on the shoulder again and asked:

"And you're serving in the militia here, right?"

"Yes, I'm doing my military service and was transferred here. I'm stationed at an outpost along the airfield perimeter. The Lieutenant you saw earlier is my platoon leader."

"Yes, I know Lieutenant Tiến. He often drives in here to inquire about the security situation around the airfield."

On the way back to the outpost, they stopped at a café for breakfast. While waiting for their food, Châu shared his earlier conversation with Tiến.

"When you went into the communication center at the airfield, I made friends with some Air Force mechanics. They're really skilled; they even went to the U.S. for training. I remember that back home, only the rich and powerful, like the village chiefs or large landowners, could afford to send their children to study in France."

Lieutenant Tiến nodded as he sipped his hot coffee:

"Yeah, you're right. Most Air Force personnel are non-commissioned officers or skilled technicians. They have specialized skills, not like us infantry soldiers."

"So why didn't you join the Air Force, Lieutenant?"

"After graduating from high school, I applied for flight training in Nha Trang, but my mother raised a fuss, saying it was too dangerous—planes get shot down easily and all that. She insisted that I continue my studies to get a draft deferment. In the end, I had to listen to her, giving up on the Air Force and ending up at Thủ Đức."

After answering and seeing that Châu was interested in the Air Force, Tiến suddenly asked:

"And what about you? Would you like to join the Air Force?"

Châu hesitated, scratching his head:

"How could I, Lieutenant? I don't have the education they require."

Tiến put his coffee cup down gently:

"I heard that most Air Force personnel are career servicemen who can make a lifelong career in the military. If you're interested, I think you could apply to transfer to the Air Force's security force. I'm pretty sure you could get it."

"But I don't know much about paperwork or applications, Lieutenant."

"I can help you with the paperwork and submit it through the chain of command to the General Staff. I've found that the officers here are pretty good; I'm sure they'll approve your transfer."

Châu's face lit up with joy:

"If that happens, I'll be deeply grateful to you, Lieutenant."

After saying that, Châu bowed his head slightly and changed his tone:

"But if I leave, who will replace me here?"

Hearing Châu's naive question, Tiến couldn't help but laugh out loud:

"When you leave, they'll assign someone else to replace you. No need to worry about that!"

That evening, Tấn was scheduled for night watch duty at the guard post on the edge of the rubber forest. He was bent over, stuffing a blanket into his backpack to take with him to keep warm during the night watch, when Châu approached from behind:

"Hey Tấn, when are you heading out there? I'm brewing some coffee to bring along so we can sit and chat."

Tấn straightened up and turned around:

"In about fifteen minutes. You're off duty tonight!"

"Yeah, but the weather's nice, so I thought I'd join you out there for some company."

By seven-thirty in the evening, it was getting dark. The clouds in the west had turned gray. A few lights outside the perimeter fence had been turned on, casting a faint, weak yellow glow like candles burning in the vast darkness.

The surrounding forest was completely silent, a silence like that of someone slipping quietly toward death.

Châu reached for the thermos and poured more coffee into a cup, then handed it to Tấn:

"Doesn't it seem particularly gloomy tonight?"

"Out here in the forest, how could it not be gloomy? But I've lived here all my life, so I'm used to it! You've been here for over seven months now; how does it feel?"

"I miss home, but I can't get leave to go back."

Tấn laughed teasingly:

"Do you miss home or Miss Năm? Just say it straight!"

"Yeah, I miss Năm a lot! But I don't know if she misses me."

Tấn, trying to sound worldly, advised:

"Forget it! Girls these days are hard to trust. They're very picky; they'll go for someone rich and successful, not a poor soldier like us! You know, I once dated a girl named Thoa near my neighborhood. At first, she claimed to love me a lot, but a few months later, she started seeing this guy Tư because he bought a brand-new Gobel motorcycle to take her out."

"I don't believe that. I think Năm really loves me. She embroidered a handkerchief for me as a keepsake when I went into the army."

"Oh, come on! Ten handkerchiefs like yours are worth nothing. Have you ever done anything with her?"

"I've only kissed and touched her a bit; I didn't dare do anything improper!"

"Well, maybe we guys are shyer than girls about that sort of thing."

"Okay, let's drop that and move on. I want to tell you about what happened this morning."

Tấn quickly teased:

"What now? Did you meet another girl?"

"Man, your mind is always in the gutter! It's about me and Lieutenant Tiến this morning."

Tấn stopped laughing and looked at Châu with surprise:

"Something between you and Lieutenant Tiến?"

"Yeah! He promised to help me apply for a transfer to the Air Force security unit."

"You're serious, Châu? Not kidding?"

"This is serious business, man. I'm not joking."

Tấn, still skeptical, scratched his head:

"Didn't you say you only went to elementary school? How can you join the Air Force?"

"Yeah, I told him that, but he said I'd be in the Air Force security unit, not as a technician or anything requiring high qualifications."

"Air Force security means guarding the airfield, kind of like what we're doing now. I have an old school friend who's in the Air Force security unit in Nha Trang. Sometimes, he gets leave and catches a ride on a helicopter to visit home. He looks a lot cleaner than we do!"

Suddenly remembering something important, Châu slapped Tấn on the thigh:

"Hey! Why don't you ask Lieutenant Tiến to help you apply for a transfer too, so we can go together?"

Tấn shook his head:

"My parents, siblings, and home are all here. When my service ends, I'll go back to help with the farm. I don't want to go far. If you like it, then you go!"

More than two months passed, and Châu seemed to have forgotten about the transfer application he had sent to the Personnel Office of the General Staff of the Army of the Republic of Vietnam.

This morning, as usual, Châu invited Tấn to ride their bikes to Mrs. Tư's café for breakfast and coffee. The café was actually just four tin sheets forming a roof against the rain and sun, with two low tables and five or six chairs made from wooden planks from ammunition boxes brought by the artillery soldiers.

It was said that Mrs. Tư's husband had been a National Army soldier who died when he stepped on a Viet Cong mine during a patrol. For several years, she had diligently run her small business to support her children. Every morning, she cooked sticky rice, corn, and coffee for the local laborers and soldiers, offering good food at reasonable prices. Tấn and Châu were her most loyal customers.

Besides running her business, Mrs. Tư had a daughter named Hằng, about eighteen or nineteen years old, with a fair complexion and very pleasant appearance. Every morning, Hằng would help her mother set up the shop and assist with serving customers when it got busy.

Hằng wasn't a beauty queen, but she had won the hearts of many young men with her sweet smile, featuring two deep dimples when she greeted customers. Tấn was one of those guys who had a crush on her and had practically rooted himself in front of the café, so when he and Châu arrived and parked their bikes, Mrs. Tư greeted them cheerfully:

"Come on in, you two. Today, Hằng made some sticky rice with mung bean filling and grilled sticky rice cakes. You should try them!"

"Yes, please, and also two cups of black coffee."

Tấn looked around, searching for the young lady. Noticing this, Mrs. Tư said as she poured the coffee:

"I just sent Hằng inside to get more rice cakes; she'll be out soon."

To tease his friend, Châu chimed in:

"Next time, Mrs. Tư, if you need someone to run errands or pick up supplies from the market, just ask Tấn. He has a bike and can get there faster than Hằng."

Tấn knew Châu was teasing him, so he sat there stiffly, blowing on his hot coffee to hide his embarrassment.

With her usual cheerful disposition, Mrs. Tư playfully responded:

"Really, Tấn?"

Unsure how to reply, Tấn hesitated, but Châu quickly interjected:

"You don't even need to ask—he'd love to do it!"

This hit the mark, making everyone burst into laughter.

Just then, Hằng entered with a stack of rice cakes. Seeing everyone laughing, she asked:

"What's so funny, Mom?"

"Oh, Châu here just said you won't need to run errands anymore; Tấn has volunteered to handle it."

Hearing this, Hằng blushed:

"Mom, you're making this sound weird. It's our own business, not theirs. People will laugh at us!"

The casual banter between the young soldier and the charming café girl could sometimes last for hours, covering everything from mundane matters to wild fantasies. But today, it was cut short by the unexpected arrival of Lieutenant Tiến.

That morning, as usual, the mail clerk delivered orders and personal letters from the district to the platoon. Tiến found a dispatch from the General Staff to the Ban Mê Thuột sub-district, requesting Private First Class Nguyễn Văn Châu to report to the General Staff Personnel Office to complete paperwork for a transfer to the Air Force Personnel Office in Saigon.

Tiến immediately went down to the barracks to inform Châu, but the soldiers there told him that Châu and Tấn had gone out for coffee earlier. Folding the dispatch and putting it in his pocket, Tiến returned to his office to get his Jeep and head to the café. He knew the two of them would likely be hanging out at Mrs. Tư's, flirting with Hằng.

Châu and Tấn were chatting with Mrs. Tư and Hằng at the counter when they heard a Jeep pull up outside. Tiến entered the café just as Châu stood up and greeted him:

"Lieutenant, would you like to join us for some coffee?"

"Thanks, but I just had breakfast. Oh, by the way, Châu! I received a dispatch from the General Staff calling you to report for your transfer to the Air Force."

Surprised and excited, Châu asked:

"Really, Lieutenant? The Air Force accepted me?"

"Well, they wouldn't have called you back otherwise. I'll have the clerk prepare the travel orders for you to take a military flight to Saigon. So after breakfast, go back to the barracks and start preparing for your departure."

"Thank you, Lieutenant. I'll head back right after this."

After handing Châu the dispatch, Tiến left the café:

"I've got some work to do at the office. You two can head back whenever you're ready."

As Tiến drove away, Tấn turned to Mrs. Tư and said:

"See what I mean? They say 'fortune favors the slow and steady.' Châu just got transferred here, and now he's already being reassigned somewhere else."

"Yes, Tấn, you're right! People who are calm and steady often get blessed by fortune!"

Now, Châu began to feel a bit anxious. He wasn't sure if this was good or bad, lucky or unlucky. What would he be doing in the Air Force security unit? He only knew what Lieutenant Tiến had told him—that Air Force personnel were stationed in big cities, with airports and planes. But after hearing what Tấn and Mrs. Tư said, he started to feel more confident.

"Everything is thanks to Lieutenant Tiến's help; I wouldn't have known what to do otherwise. By the way, Tấn, how will I find the General Staff headquarters when I get to Saigon?"

Châu was in the middle of asking when Mrs. Tư, who seemed knowledgeable about such things, chimed in:

"Don't worry, Châu. When my husband passed away, I had to go there once to process the paperwork for his benefits. I didn't know where it was either, but I just asked one of the cyclo drivers, and they took me straight there. So when you get to Saigon, just ask a cyclo driver or a motorized cyclo driver, and they'll take you straight there. No need to worry about getting lost—they know Saigon and Chợ Lớn like the back of their hand!"

The plane had just taken off from Phụng Dực Airport in Ban Mê Thuột, heading toward Saigon. Châu looked out the small window. The forests below lay still, their green expanse flat like a painting on a canvas. The mountain ranges ahead seemed to stand still, just as they had when he first arrived.

As the days and months passed, Châu felt as if everything around him remained unchanged. If we measure a person's life by the shadows cast through a window, the time of nature is measured by the changing seasons of the earth and the heavens.

Having spent nearly nine months here, Châu had grown a bit older, gained some experience, and become more worldly through his interactions with people outside his small village. He was now more mature and confident, no longer the 15 or 16-year-old boy herding buffaloes in Xóm Cầu Nổi. Gone were the days of being scolded or chased by landowners for letting the buffaloes eat their rice crops, or wandering by the road just to breathe in the smell of gasoline, imagining himself as a modern man interacting with technology and science.

Time flowed like water, carrying the lone water hyacinth in a small stream toward a new and unfamiliar horizon. Perhaps one day, like that water hyacinth, he too would blossom into something new, just as his life was changing now.

After nearly an hour, the C-47 transport plane landed safely on the runway at Tân Sơn Nhất Airport. The passengers were transferred onto military trucks from the airfield liaison station to the gate.

It was still early in Saigon. The morning sun had not yet reached its peak, and Châu estimated it was just after 10 a.m. He decided to make use of the time by heading straight to the General Staff Personnel Office, leaving the rest of the day for other matters.

As the military truck from the liaison station dropped the passengers off in front of Phi Long camp, near Lăng Cha Cả, Châu noticed several cyclo and motorized cyclo drivers approaching to offer rides.

Struggling with his backpack as he got off the truck, a nearby cyclo driver quickly offered to help:

"Young man, get on my cyclo. I'll take you home cheaply!"

"I'm not going home right now. I need to go to the General Staff Personnel Office."

"I can take you there too. I know you're a soldier, so I won't charge much!"

"Okay, take me there. You know where it is, right?"

"The Personnel Office is inside the General Staff headquarters. I can't take you all the way in, but I can drop you off at the gate, and you can walk in. Just ask around, and they'll direct you—it's not difficult!"

After presenting his travel orders to the military police at the gate and following their directions, Châu found the Personnel Office easily.

An old master sergeant was sitting in the clerical office when he saw Châu enter and inquire about the dispatch from the General Staff. After glancing at the document, he nodded knowingly:

"Ah! Châu, I just saw your file this morning. Wait a moment; I'll check to see if anything is missing, and then I'll prepare the paperwork for you to report to the Air Force Personnel Office. They'll decide where to assign you."

Seeing that the master sergeant seemed approachable, Châu asked:

"Sergeant, once I'm done here, can I visit my family?"

"Where do you plan to go?"

"To Cầu Nối, Rạch Đào village!"

Upon hearing this, the sergeant looked surprised, quickly setting down the file and looking directly at Châu:

"I just read in the newspaper yesterday that there were some skirmishes in that area. I think you shouldn't go there. It's better to find a place to stay here and have your family come to visit you—it's too dangerous for you to go back!"

133

Châu seemed puzzled:

"Really? About nine months ago, I visited home and stayed there for a few days, and I didn't see any trouble."

"A few months ago, it was calm there, but it's only been in the last few weeks that things have heated up. The Viet Cong in the area are trying to disrupt President Ngô Đình Diệm's hamlet development program."

"Is the fighting intense, Sergeant?"

"It's mostly small-scale guerrilla attacks. At night, they sneak in and round up young men of conscription age to take them into the jungle. Anyway, wait outside for a bit while I take your file to get signed."

Châu stepped out to the veranda, sat down on the brick steps under the shade of a frangipani tree, and reflected on how his plans had completely unraveled. After being away from home for months, he had saved up a little money and planned to buy a bottle of imported perfume from the market, instead of jasmine oil, to give to Miss Năm. He was also going to ask his parents to arrange for a proposal to ask for her hand in marriage. But now, with the situation as it was, he didn't know how to get word to his family. Feeling disheartened by the unexpected turn of events, he leaned back against the brick column and closed his eyes, thinking, "Whatever happens, happens."

The midday Saigon breeze was gentle, just enough to carry the faint fragrance of the frangipani blossoms opening wide on the branches.

"Châu! Châu!..."

Hearing someone call his name, he quickly stood up:

"Yes, Sergeant, are you calling me?"

"Yes! Your paperwork is done! We'll send your file over later. Now, take this introduction letter to the Air Force Personnel Office so they can process your paperwork and pay."

The sergeant checked his watch and continued:

"It's almost 3 p.m. now, so you can report there tomorrow morning if you want. Do you need travel orders?"

"Yes, I do!"

"Good! Find a place to stay tonight, then!"

Châu hesitated slightly, adjusting the military cap on his head:

"Sergeant, I don't know where the Air Force Headquarters is."

The old sergeant, being easygoing, walked over and patted Châu on the shoulder:

"You flew in this morning, right?"

"Yes, I did!"

"Well, the place where you got off is within the Air Force Headquarters area, but to be sure, just take a cyclo—they'll get you there without any trouble."

Hearing the sergeant's advice reminded Châu of what Mrs. Tư, the café owner in Ban Mê Thuột, had said—that when he got to Saigon, if he needed to go anywhere, just ask a cyclo driver, and they'd take him straight there. This made Châu feel a bit embarrassed, as he recalled a fable from his third-grade reading book about a man who couldn't read street signs in the city and had to follow a cow to find his way.

After saying goodbye to the sergeant in the Personnel Office, Châu slung his backpack over his shoulder and walked out the gate. He stopped at a sugarcane juice stand under the shade of a tree, planning to ask about a place to stay for the night.

After a brief conversation, the juice vendor, realizing he was a soldier from far away, kindly suggested:

"There are plenty of guesthouses and eateries around here, but they're expensive. I've seen people eat at the bus station and then rent a cot to sleep on overnight—it's much cheaper."

Châu smiled at this information, remembering overhearing similar conversations among flower sellers at the bus station when he was heading home earlier.

He stood up, paid for his drink, and thanked the vendor for her advice.

It was nearly 6 p.m., and at the bus station for the provincial routes, the sun was already casting long shadows due to the tall buildings on either side of the street blocking the evening sunlight. The station was nearly empty, with only a few buses remaining for early departures the next morning. A few street sweepers were busy clearing the sidewalks.

After finding a place to rent for the night, Châu felt a bit more relaxed. He leisurely walked toward the end of the street, where there was a makeshift food stall under a tree, catering to the vendors and bus drivers who spent the night at the station.

The food stall owner, seeing him approach, quickly invited him in:

"Young man, come in for dinner or a beer."

Châu nodded and sat down:

"Please give me a plate of stir-fried rice and a glass of iced tea."

Noticing that Châu was wearing a military uniform, the stall owner cheerfully struck up a conversation:

"You just got back from leave and missed your bus, didn't you?"

"No, I came back to report for a transfer to another unit!"

Understanding his situation, she replied:

"Oh! So you don't have any acquaintances here and need a place to stay! Where are you from?"

"I'm from Cầu Nổi, Rạch Đào village."

"I don't know Cầu Nổi, but I know Rạch Đào village. It's close enough that you could visit home in a day."

"I wanted to visit my family, but I heard there's some trouble there, so I'm afraid to go."

The stall owner's face lost its initial cheerfulness, and she sighed:

"Yeah, I've heard the same from some bus drivers. The situation there has gotten worse lately. They're stopping buses to collect taxes and digging up the roads constantly. It's dangerous for a soldier like you to go back!"

After dinner, Châu felt a bit melancholic. He returned to his cot, set up the mosquito net, and went to bed early. The ceiling fan above swayed back and forth, producing a rhythmic "click-clack" sound. More than an hour had passed, but Châu still lay there with his hand resting on his forehead, unable to sleep. His thoughts were more occupied with Cô Năm Lài than with his parents and siblings back home.

Cô Năm had stirred deep emotions and sensations within him. Their rendezvous at Cầu Vó were moments when their bodies touched, and their breaths nearly stopped as they pressed together like magnets, almost becoming one. In those moments of desire, he held and caressed her as if she were a ripe star apple, waiting to be gently bitten to release its sweet nectar. As he lay alone, these old memories came flooding back.

A gentle night breeze blew through the mosquito net, making it sway slightly. Feeling a bit cold, he pulled the blanket up to his neck.

The alarm clock belonging to the flower vendor behind the room suddenly went off loudly, its sharp ringing cutting through the night like a thousand knives. The room's light flickered on, signaling the dawn of a new day.

Many people were already sitting up, folding their nets. Outside, the engines of a few motorized cyclos could be heard as they started up to take vegetables to the early market. The mother and daughter selling coffee by the wall outside were already bustling around. Châu pulled up the blanket and glanced at his watch—it was only 4 AM. Still early, he thought, and he snuggled back under the blanket, listening to the city slowly awaken after a quiet night.

Châu suddenly remembered that back in his village, during the planting season, around 4 AM, he would hear the "tù và" (buffalo horn) being blown by Bác Trùm Vạn, echoing throughout the neighborhood to wake the workers for the day's rice planting or seedling pulling.

What Châu enjoyed most back then was going to eat with the planting crew. It wasn't that the food provided by the landowner was particularly delicious, but the joy came from eating together with a large group, sitting on mats spread out in the courtyard or on a high mound in the fields. Most of the workers were women, ranging in age from middle-aged to young girls of sixteen or seventeen.

During lunch, they would chat and tease each other, joking about couples or potential matches, as a way to forget the fatigue from hours of bending over to plant rice or pull seedlings. They worked hard, and their appetites were hearty. Those who weren't quick enough at serving rice from the pot would have to keep passing bowls for others until they missed out on their share. Knowing this, landowners usually placed a family member near the rice pot to serve everyone.

Now, the sounds of roosters crowing at dawn or the "tù và" of Bác Trùm Vạn seemed to belong to a distant past. Around him now were strange noises—a mix of car engines, honking horns, the shouting of bus conductors, and the street vendors hawking their goods. All these chaotic sounds were part of what people called the bustling life of the city.

As daylight began to break outside, Châu tried to stay in bed for a few more minutes, but his plan was foiled by the boarding house owner, who approached him and yanked the mosquito net:

"Get up, young man! It's past 6 AM already! I need to fold up the cots and put them away!"

The woman's loud, sharp voice, typical of a bus station regular, startled Châu fully awake. He quickly sat up, rolled up his blanket, and packed it into his backpack, ready to start a new day.

The motorized cyclo stopped at a corner near Lăng Cha Cả. The driver pointed to Châu:

"Do you see the gate of Phi Long camp? Just go in there, and they'll direct you."

After paying the driver, Châu checked his wristwatch—it was only around 8 AM. Realizing it was still early, he remembered a piece of advice from some of his friends who had been in the military for a long time. They had told him: "If you want to get anything done in an office, go after 9 AM. If you go too early, those officers might still be grumpy from dealing with their wives at home."

Châu thought this made sense, so he decided to pass the time at a nearby café before reporting in.

The café was almost full. Every table was crowded, mostly with soldiers in uniform like himself. The petite, charming café owner was busy serving food. Seeing Châu looking for a seat, she quickly called out:

"Hey there! There's a spot at this table. Have a seat!"

Châu quietly made his way to the empty seat.

"What would you like to eat or drink?"

"I just had breakfast. Could I get an iced coffee?"

The two soldiers sharing the table turned to Châu and struck up a conversation:

"Did you just come back from leave?"

"No, I've been reassigned from Ban Mê Thuột and am waiting to report to the Air Force."

"Were you an Air Force soldier assigned to Ban Mê Thuột?"

"No, I was a militia soldier, but I applied to transfer to the Air Force security unit."

Upon hearing this, the two soldiers seemed pleased to have found a new friend. One of them, smiling, pointed to his companion:

"We're both Air Force security soldiers at Tân Sơn Nhất. So where are you stationed?"

"I'm not sure yet. I'll find out when I report in."

"When do they expect you? Do you have a ride there?"

"They said I could report anytime during office hours. I don't have any transportation."

"In that case, after we finish our coffee, I can give you a lift there!"

Châu was grateful:

"That would be great! Thank you so much."

After reviewing the dispatch and introduction letter from the General Staff Personnel Office, the master sergeant at the Air Force Personnel Office looked at Châu and asked:

"Are you Private First Class Nguyễn Văn Châu?"

"Yes, sir."

"Please wait a moment while I present your documents to the head officer. Then you can go in to report."

Châu followed the sergeant's instructions and waited outside on the veranda. About five minutes later, he was called in to meet the captain in charge.

After a brief conversation, the captain informed him that he would be sent to Nha Trang to bolster the security forces at the airfield there. He would be issued new travel orders while awaiting a flight to Nha Trang.

After saluting and bidding farewell to the captain, Châu stepped out and met the sergeant, who had just finished processing his paperwork.

"Here are your new travel orders for your flight. Your file will be sent to Nha Trang later. Do you have any other questions?"

"Sergeant, where do I take these orders to get a flight?"

"Take these orders to the military liaison station at the Phi Long camp gate. They'll schedule your flight."

Châu was about to ask more but then remembered the advice from Mrs. Tư, the coffee shop owner in Ban Mê Thuột: "The way is in everyone's mouth; if you want to know, just ask people."

He thanked the sergeant, tucked the orders into his pocket, and headed out to catch a three-wheeled taxi to the main gate.

Over a year had passed since Cô Năm last saw Út Châu during a rendezvous at Cầu Vó, when he was on leave after completing basic training at the Quang Trung Training Center. After that, she had heard nothing from him and had no idea where he was. Several families in the village had sent matchmakers to propose marriage, but she had found excuses to refuse, holding on to the hope that Út Châu would return after his military service.

A young woman who has felt the presence of a man is like a plant that receives rain after a drought. Perhaps this is a natural affinity between humans and the environment around them.

Indeed, it seemed that the memory of Út Châu had somehow made her even more beautiful. Her curves were accentuated by the snug áo bà ba and cool, black silk pants she wore to the market. Cô Năm was blossoming like a piece of fatty pork hanging in a kitchen, with the village boys looking at her like cats licking their chops.

Among the young men who often glanced at her or passed by her bamboo fence just to catch a glimpse of her, one stood out to Mrs. Sáu, her mother: a young schoolteacher who taught at the local elementary school in Rạch Đào. Mrs. Sáu considered him the best among the suitors.

The teacher was clean-cut, wearing neatly pressed slacks and a white shirt. He had a respectable profession, unlike the other boys who spent their time drinking rice wine at Mrs. Hai's shop, with no real jobs, always teasing girls and speaking rudely. The teacher, in contrast, was polite and dignified, earning much favor with Mrs. Sáu, who often mentioned him when discussing marriage with Cô Năm Lài.

Mrs. Sáu's views were typical of the older generation in rural areas—they held scholars in the highest regard: "First, the scholar; second, the farmer..." Any family with a son-in-law who was a teacher or a clerk was considered prestigious in the community.

Though from the countryside, Cô Năm didn't share such narrow views. Whenever her parents talked about marriage, even though she couldn't find a fault with the teacher, she would say she wanted to stay home and help her parents a little longer before getting married. Mrs. Sáu, thinking her daughter was filial, decided to wait, not wanting to force her.

Sometimes, she would remind her daughter that girls who didn't marry by twenty were considered old maids, and people would gossip about them. In previous generations, girls over twenty who hadn't received any marriage proposals were considered past their prime, bringing shame to their parents.

In those days, a girl's status was likened to a market commodity or a ticking time bomb—if someone offered the right price, the deal was done, lest she be left on the shelf or, worse, become pregnant out of wedlock, bringing disgrace to the entire family.

In contrast, a young man was seen as a buyer who could take his time, always able to choose the best option. However, if he was too picky, he might end up alone, a bachelor for life.

Despite the passage of time and the lack of news from Út Châu, Cô Năm remained faithful to his promise to marry her after his military service. She often wanted to tell her mother about their relationship, but each time she tried, she feared being scolded for secretly meeting a man, bringing shame to the family. So, she kept quiet.

After submitting his travel orders at the military air liaison station, two days later, Châu was called to board a flight to Nha Trang. It was a Friday, and while processing his paperwork at his new unit, he met a young soldier named Lê Văn Phước, who was also part of the security team at the Nha Trang airfield. Hearing that this was Châu's first time in the province and that he didn't know anyone, Phước offered to help:

"I've rented a small room in town. If you don't have a place to stay, you can stay with me for a few days until you find something."

Châu, relieved, thanked Phước profusely:

"Thank you so much! During the wait for my flight, I had to rent a cot at the Chợ Lớn bus station."

"Didn't you go home?"

"My home is in the countryside, where there's fighting, so I didn't dare to go back."

"I've heard that the Chợ Lớn bus station is full of thieves and thugs. Weren't you afraid?"

"I don't have anything valuable other than this backpack, so I wasn't too worried."

"Did you finish all your paperwork? When do they expect you to report in?"

"The master sergeant at the clerical office told me to report to the captain on Monday."

"Alright, grab your backpack, and I'll take you home!"

Riding on the back of Phước's motorcycle, Châu held onto his shoulders to keep from falling. The wind from the ocean blew in, cooling him down. Châu felt light-hearted, remembering what Tấn had told him back in Ban Mê Thuột about the beauty of Nha Thành, with its golden sandy beaches under tall coconut palms and rows of pines.

The blue sea rolled in gentle waves against the sandy shore, creating sounds that invited and welcomed visitors. Tấn had even joked, "The girls in Nha Trang are beautiful and sweet. You better watch out, or they'll kidnap a country boy like you."

From what Châu had seen so far, this city was indeed more beautiful and cleaner than Ban Mê Thuột. The two places were nearly opposite in character. Ban Mê Thuột, the capital of the former highland kingdom, was historically a land of ethnic minority villages nestled among the forests near the Cambodian border.

The sounds of gongs and drums, calling to the wind and clouds, still seemed to echo in the air, remnants of the native people's rituals. To soldiers far from home, Ban Mê Thuột was often called the "land of eternal sorrow" because of its mountain winds and persistent drizzle that could last for weeks without pause.

In contrast, Nha Trang was a famous tourist city, blessed by nature, lying in a vast valley along the coast from south to north. Across from it, mountain ranges formed a natural fortress, protecting it from the misfortunes of storms rolling down from the highlands. Nha Trang had warm sunshine, Hon Chong, Hon Vo, Nui Ba, Hon Rua, Xom Bung, and Suoi Tien.

Visitors to Nha Trang could sit under the trees along the beach, sipping fresh coconut water and gazing out at the sea. Even the most disillusioned or pessimistic would find life still had meaning.

Phước turned left onto the boulevard running along the coast, just as Châu emerged from his daydream, having only known the city through others' vivid descriptions, like a delicate painting on fine silk.

The booming, continuous roar, like thunder announcing an impending storm, snapped him back to reality. Châu looked up at the clear blue sky, puzzled by the noise. He tapped Phước on the shoulder:

"What's that thunderous sound?"

"Oh! That's the waves crashing against the shore. The wind must be strong today, blowing in from the open sea!"

Châu felt a bit embarrassed by his naive question. He had never seen the ocean before. He had only known mountains and the highland people carrying their children in Ban Mê Thuột, and before that, his life had been confined to fields, small rivers, and canals in his hometown. He had never seen such vast mountains or open seas!

Sensing that his new friend had never been to Nha Trang before, Phước decided that after they settled in, he would take Châu on a tour of the city.

After a few days of staying together, Phước found that he got along well with Châu and suggested they share the rent and live together permanently. Châu happily accepted.

Châu thought back to the fortune-teller who had read his palm at the Chợ Lớn bus station while he was waiting for his flight to Nha Trang. The fortune-teller had predicted that he would meet good fortune and have benefactors helping him.

The idea of becoming rich seemed far-fetched, given the modest salary of a soldier. At best, it would be enough to live on, but certainly not enough to become wealthy. However, the part about meeting benefactors had proven true. In Ban Mê Thuột, he had Tấn and Chuẩn Úy Tiến to guide him. Now in Nha Trang, he had Phước to rely on, so the fortune-teller's prediction wasn't entirely off.

With the daily responsibilities of a security soldier, Phước and Châu would occasionally take a day off to visit the outskirts of Diên Khánh district, enjoying the gardens and fruit trees.

But this Sunday morning, Phước woke up earlier than usual. It was only around 6 AM when he called out:

"Hey, Châu! Get up and get ready—we're going to a death anniversary!"

Still half-asleep, Châu grumbled:

"Whose death anniversary? Why are we going?"

"My girlfriend invited us yesterday when you were on duty!"

"She invited you, so you should go. Why do I have to go?"

"She said I should bring you along for fun!"

The two friends rode together on Phước's motorcycle, heading toward Ninh Hòa district. The morning breeze was cool as it blew in from the sea. The road ahead stretched long, undulating as it rose and fell, sometimes winding through sparse forests or along the mountain slopes by the sea.

Time passed indifferently; as one month ended, another year arrived.

The sun rose and set before the front porch like drops of water steadily falling on a stone, and before long, those gentle, seemingly harmless drops had eroded and altered the once-solid stone.

No matter how determined she was to keep her promise to Út Châu, Cô Năm Lài's resolve began to waver under her mother's persistent reminders.

"Lài, I've been hearing rumors about you from the neighbors!"

Hearing her mother's voice, Cô Năm knew what was coming.

"What did I do for them to gossip about?"

"They say you're getting old and still haven't married!"

"I'm only twenty! That's not that old!"

"Sure, but before you know it, you'll be past your prime! Your father and I can't keep worrying about you forever! I think that teacher Hai in town is a good match. He's exempt from military service, so what are you waiting for?"

It wasn't that Cô Năm disliked teacher Hai; on the contrary, she had more affection for him than for the other village boys who spent their time drinking and gambling. Moreover, having a husband who was a teacher would bring pride to her family, as she would be addressed as "Cô giáo Hai" (Mrs. Teacher Hai). But she still couldn't forget Út Châu, who had captured her heart long ago.

People often say, "Out of sight, out of mind," and she didn't know if Út Châu still remembered her or if he had found someone else. She had heard people say, "You can't trust soldiers; wherever they go, they have a girl or a wife there..."

Thinking of this made her sad, and so when her mother brought up teacher Hai, she no longer objected as she used to.

Seeing her daughter's silence, Mrs. Sáu continued:

"Yesterday, I went to the market and met teacher Hai. I asked him to help us with a request for a tax reduction after the recent flood ruined our crops. He promised to come by this afternoon to get your father's signature before taking it to the village office. So, you should dress neatly—don't let people talk behind your back."

Hearing this, Lài quietly went to the kitchen. She didn't know whether to be happy or sad at that moment.

"Your friends like Hoa, Nở, and Nhạn are all married with kids by now. You're the only one left. A girl is like a flower—if you wait too long, it'll wither, and no one will want it."

This was the mantra Mrs. Sáu repeated whenever she discussed her daughter's future.

Around 4 PM, the dog began barking loudly at the bamboo fence. Mrs. Sáu knew it was teacher Hai arriving and called out:

"Lài, go quiet the dog so the teacher can come in!"

Lài, who was boiling water in the kitchen, quickly slipped on her sandals and dusted off the back of her pants:

"Yes, I'm going now, Mom!"

"Go invite the teacher inside!"

Seeing Lài come out, the black dog stopped barking and ran back into the house.

Teacher Hai, ever the gentleman, was dressed impeccably: a long-sleeved white shirt, blue trousers, and well-polished leather shoes, with his hair neatly combed in the "mui xe ngựa" (horse-drawn carriage style).

Lài clasped her hands together in front of her waist and bowed slightly:

"Please come in, teacher Hai!"

"Thank you, Cô Năm! Is Bác Sáu home?"

"Yes, my parents are out back feeding the pigs."

Teacher Hai was the eldest son of Bác Trùm Vạn, who lived near the pig slaughterhouse in the upper part of the village. After completing elementary school at the village school, he went to study in Saigon-Cholon for three or four years before returning to the village to become a teacher.

Still single, teacher Hai held a relatively high position in the village, so many families wanted to marry their daughters to him. But teacher Hai was still picky, determined to find a truly beautiful wife, so he kept stalling.

But this year, after seeing Cô Năm Lài and her mother at the market, teacher Hai became smitten. He couldn't stop thinking about her captivating eyes and graceful walk, which haunted him day and night. But being a teacher and a role model in the village, he tried to maintain his dignity, not daring to flirt as the other boys did.

However, his eager gestures and stolen glances didn't escape the experienced eyes of Mrs. Sáu. To her, having teacher Hai as a son-in-law would be a source of pride in the village. The only problem was that Cô Năm Lài hadn't yet warmed to the idea, as she would usually find a way to slip out the back whenever she saw teacher Hai coming to visit.

But today seemed to be a lucky day. Teacher Hai felt his spirits lift as if a sail had caught a favorable wind when he saw Cô Năm Lài warmly greeting him at the door.

The path from the bamboo gate to the house wasn't far. Lài walked silently in front, while teacher Hai followed closely behind, making small talk:

"Didn't you go to the market today, Cô Năm?"

"I went this morning!"

"No wonder I didn't see you and Bác Sáu pass by the school!"

At that moment, Bác Sáu emerged from the back:

"Please come in, teacher Hai. I'm so grateful for your help with the paperwork, and now you've taken the trouble to bring it to our house!"

"It's nothing, Bác. I was on my way home from school and thought I'd stop by."

As teacher Hai finished speaking, Bác Sáu called out to his wife in the kitchen:

"Lài, bring up the fresh pot of tea I brewed earlier!"

Lài brought the teapot, encased in a coconut shell holder, to the table. Bác Sáu's wife, having finished feeding the pigs, joined them:

"Ah, teacher, you're here."

She glanced at Lài, who was pouring tea into cups for the guest.

"Lài, go get the jar of dried bananas from the grain bin and bring it up to offer the teacher with his tea."

Teacher Hai stood up, trying to politely decline:

"Bác Sáu, please don't trouble yourself. I just came by for a quick visit."

"Nonsense! It's not often you visit. We have more bananas than we can eat, so Lài dried some in the sun. They're delicious with tea."

The afternoon breeze blew in from the golden rice fields in front of the house, cool and refreshing. The sow, greedy for food, led a litter of piglets as big as calves across the yard, their snouts snuffling as they demanded more food.

Bác Sáu's wife grabbed a broom and chased them away, scolding:

"I just fed you at 4 o'clock, and now you want more? You gluttonous creatures!"

Teacher Hai glanced at his watch. Seeing that it was already past 5 PM, he quickly stood up to leave:

"Thank you, Bác. I've stayed long enough; I should head back now."

Teacher Hai turned to Bác Sáu, who was sipping tea:

"If there's any issue with the paperwork, please let me know."

"Thank you, teacher. I'll take it to the village office tomorrow for approval, then send it to the province. Whenever you have time, feel free to visit and chat with us."

"Thank you, Bác! I'll stop by whenever I can."

After chasing the pigs to the back of the house, Bác Sáu's wife called out to her daughter:

"Lài, walk the teacher out to the gate, or the dog might chase and bark at him."

So, teacher Hai was once again escorted to the bamboo gate by Cô Năm Lài. He planned to use this opportunity to make a deeper connection with her, beyond the usual greetings on the street. But after much internal debate, he couldn't find the right words and ended up complimenting something trivial:

"Cô Năm's dried bananas are delicious!"

"Thank you, teacher. They're made from the bananas we grow in our garden. They go well with tea."

"Did my visit bother you?"

"No, you were just chatting with my parents; why would it bother me?"

Their conversation ended as they reached the bamboo gate. Lài stopped, planning to turn back. Teacher Hai also paused, looking down at the ground as if he wanted to say something more, but in the end, he just nodded and walked away.

Lài thought to herself: "What a timid man! He wants to say something but doesn't even dare!"

Suddenly, she remembered Út Châu, recalling the first time they met by the village pond when her rope snapped, and she asked Út to retrieve it for her. Despite the mishap that left them both covered in mud, Út didn't care about himself; he was more concerned with cleaning the mud off her face and clothes.

It was this kind and thoughtful gesture that made Cô Năm fall in love with Út. Compared to teacher Hai, Út was more of a man—bold and decisive, not timid and hesitant like a girl.

After President Ngô Đình Diệm was overthrown, the war intensified. The strategic hamlets in rural areas became effective operating bases for communist guerrillas. Transportation along main roads was disrupted. Bridges and roads were mined. Airports and cities were frequently shelled by the communists. The national army was often placed on high alert.

Nha Trang Airport, being close to the Đồng Bò base area, required even more stringent security measures. As an air force security soldier, Châu shared the same fate as other combat soldiers, being confined to the base and constantly on guard duty to protect the airport perimeter. About six years ago, he was granted annual leave from his unit.

He took a helicopter to Saigon, intending to catch a bus home. When he arrived at the Chợ Lớn bus station, he was stunned to find that the buses running the Rạch Đào-Cần Đước route were no longer operating. Upon asking around, he learned that the bridges and roads had been nearly destroyed by the Viet Cong and that the only way to travel was by bicycle or on foot. Since then, whenever he got leave, instead of visiting his old village and his former lover, he would fly to Ban Mê Thuột to visit his old friends: Tấn and Tiến.

Chuẩn úy Tiến had now been promoted to captain, and Tấn had been discharged after his military service. He returned to his village and farmed near the Phụng Dực Airport in Ban Mê Thuột. Today was the third time he returned to attend Tấn's wedding, where the honorary host was Captain Tiến.

Two HU-1 helicopters from Squadron 215 of the Thần Tượng (Idol) Battalion had just landed at the L-19 airport in the city. Through the side window, Châu noticed Captain Tiến's Jeep parked near the guard post. He quickly grabbed his suitcase and jumped down first. As he was about to head toward the gate, a voice called out from behind:

"Hey, Châu! Is there anything else I can help carry?"

Châu turned around to see Tấn approaching:

"My gosh, Tấn! I didn't recognize you in civilian clothes!"

"I've been discharged, remember? Captain Tiến is waiting for you in the car!"

The two friends embraced warmly as they had in the old days when they were still in the military. Tấn continued:

"We've been waiting for you for almost half an hour. Poor Captain Tiến is so kind. He often visits my house for meals and always mentions you. He said you're from the same hometown, right?"

"Yes, I'm from Cầu Nổi, and he's from the Rạch Đào market area. We're from the same village."

At that moment, Captain Tiến got out of the car and walked over:

"Hey, Sergeant Châu! You're looking younger and more handsome this time!"

Châu stopped and saluted in military fashion:

"Hello, Captain!"

Tiến walked over and patted Châu on the shoulder:

"Tấn and I have been waiting for you all morning. Let's go into town for breakfast; you're paying for the pho and coffee, okay?"

"No problem! A single air force sergeant—no worries, Captain!"

Châu's joking remark made Tấn laugh heartily.

"Did you hear that, Tấn? Châu's gotten much quicker and sharper with his comebacks than before!"

"That's right, Captain. Now he's an air force non-commissioned officer, all sharp and polished, not the clueless draftee he used to be."

Tấn got into the car first, sliding into the back seat. As Captain Tiến was about to start the engine, he seemed to remember something and turned to Châu:

"Oh, I received a letter from you asking me to check on your family back home. I couldn't go myself, but I had my uncle in Saigon visit your family. He stopped by and reported that your parents and siblings are all doing well. Your parents sent their regards and advised you not to come home now because it's too dangerous."

"Thank you, Captain. I've heard that things have been very chaotic down there these past few years."

From the back seat, Tấn leaned forward and interjected:

"Captain, did you ask about his girlfriend down there?"

"I didn't hear him mention any girlfriend!"

"He had a girlfriend before he joined the army. He said they had a very passionate relationship."

Tấn playfully slapped Châu on the shoulder:

"Hey, what's your girlfriend's name again, Châu? You told me before, but I forgot."

"Her name is Lài!"

"Oh, I remember now—Cô Năm Lài."

Captain Tiến chimed in:

"Lài is a lovely name, isn't it? But did you have an engagement or anything?"

"No, Captain. We were just dating."

"Well, in that case, nothing is certain. It's been three or four years since you've been back, so she's probably married by now. No girl stays single for that long, waiting for someone to come back and marry her. How about we find someone here for you—Tấn can introduce you to one of his wife's cousins."

Captain Tiến had barely finished speaking when Châu turned to Tấn:

"Hey, Tấn! Who are you marrying? I came here for your wedding, but I don't even know the bride's name!"

Captain Tiến jumped in again:

"Who else could it be? The cute little café owner you two used to visit for breakfast!"

Châu clapped his hands in surprise and delight:

"Oh! Cô Hằng, Auntie Tư's daughter who ran the café. Tấn, you've done well! You're marrying Hằng, who's beautiful and good at business too. So, after the wedding, are you planning to take her back to the farm?"

"No, my parents are too old to work the farm anymore. We plan to sell the land so they can retire. After I get married, I'll open a small coffee shop for the neighbors and the farm workers. What about you? Do you have a girlfriend in Nha Trang?"

"I'm still unattached. I go out and meet people occasionally, but there's nothing serious. If I ever do find someone, I'll need Captain Tiến to officiate the wedding."

Châu's remark made Tiến laugh out loud:

"I'd be happy to help, but remember, when things are good, you won't say anything. But if things go bad, don't blame me!"

"You don't have to worry, Captain! Tấn and I aren't like that. Who would complain about someone who did them a favor? By the way, how are things going with you and Ms. Thảo?"

"It's been years, and you still remember that?"

"How could I forget? It was my first leave, and you took me to visit her house and have dinner, right?"

"You have a good memory!"

A slight sadness crossed Tiến's face. After a deep sigh, he continued:

"Maybe soldiers like us don't get as lucky as others. Women love soldiers as boyfriends or as accessories to show off in public or at parties, but when it comes to marriage, they have second thoughts."

Tấn, a bit puzzled, scratched his head:

"But Captain, you're an officer with a respectable position in society. I thought women would be all over that!"

"You're right. They love being taken out for ice cream, a stroll, music, or dancing, but marrying is a different story."

"Why's that?"

"Because they're afraid of becoming widows too soon."

Tấn and Châu both understood the implication. Châu then continued his unfinished question:

"So, you and Ms. Thảo aren't in touch anymore?"

"The Ministry of Education reassigned her to Saigon over two years ago. I've been too busy with operations to keep in touch. I heard she got married to some professor in Saigon."

After Tấn's wedding, Châu took a helicopter back to Nha Trang, where Phước was waiting for him at the airstrip to take him home.

"You picked the perfect time to go on leave. Last night, they shelled us with five 122mm rockets. We were up all night, and now I'm exhausted. Let's go home so I can get some sleep."

Châu grabbed his suitcase and sat in the back. He suddenly recalled Captain Tiến's words, which were so true. As a soldier, you could die or become crippled at any moment. Life offered no guarantees.

People praised or honored you to encourage bravery and heroism, but often, those praising you were the ones who wanted to command and control you, not necessarily those who would make the same sacrifices as you.

For example, just a few days ago, during the flag-raising ceremony, after the national anthem was the song "Suy Tôn Ngô Tổng Thống," praising Ngô Đình Diệm as a great leader. Then, the next day, there was a coup, and he was treated as a mortal enemy.

With his limited understanding, Châu often found himself doubting the statements of contemporary political leaders. He observed how Nha Trang had changed in recent years, with the presence of too many foreign troops. The city had become dirty and chaotic, with bars and nightclubs springing up everywhere. Both in the city and on the beaches, one could see women with heavy makeup and teased hair, clinging to American men, leisurely strolling around.

The rapid changes irritated both him and Phước, as landlords kept raising rent prices. Phước frequently grumbled and cursed:

"Damn it! Our soldier's pay doesn't go up, but rent goes up twice a year. The landlords want to kick us out so they can rent to American prostitutes. What kind of society is this? Soldiers can't even buy groceries, but when those American women go shopping, they get everything instantly."

Châu could only chuckle in response:

"Forget it, man! There's no point in comparing. They have money and don't haggle when they shop; sometimes, they even give extra. That's why the sellers save their goods for them."

Châu still remembered when he first transferred here, he would often ride his bike along the beach to work or, in the evenings, take leisurely rides down Độc Lập Street. But now, whenever he went out, he had to be extra careful, as the American military vehicles would speed recklessly, sometimes even hitting and killing people without stopping, as if they were running over a dog. If someone tried to file a complaint, it would either be covered up or dismissed.

Just as he was thinking about this, Phước suddenly made a sharp turn into the alley before the house, startling Châu, who grabbed onto his friend's shoulder tightly.

"Take it easy, man!"

"Were you dozing off back there?"

"Not really! My ears are still ringing from the helicopter, so I just closed my eyes for a bit. I didn't sleep at all! I bought some dried bananas for you!"

"Oh, my girlfriend in Ninh Hoà loves the dried bananas from Ban Mê Thuột. The last time we went to her place for a memorial, she asked if I wanted to be introduced to her friend. I told her you already had a girlfriend in the South, so she dropped it."

Châu remained silent as he opened the door and waited for Phước to push the motorbike inside.

Phước's casual mention of his old girlfriend made Châu suddenly remember her. He had promised to return and marry her after completing his military service. Now, he was no longer a draftee but a regular volunteer in the air force.

Sometimes, Châu confided this personal dilemma to Phước. He was worried about Cô Năm waiting for him day by day, month by month, back in their village. Phước would just smile and say:

"Châu, you're a soldier but not realistic at all! You're too romantic and dreamy! No girl is going to wait for you until she's old! Once they hit their twenties, they're terrified of being left on the shelf! If they find a decent guy, they'll latch onto him immediately, not sit around waiting like you imagine!"

Châu felt annoyed by Phước's sweeping generalization and retorted:

"You always generalize, which is why you're almost thirty and still single."

"Me? I can get married any time I want, but who would be stupid enough to put their head in the noose early just to be dragged around?"

"So, are you saying all the guys who got married are crazy?"

"I wouldn't go that far, but if you ask them after ten years of marriage, they'll tell you what they really think."

"You talk like you've got a dozen kids already, full of experience."

"I may not be married, but I've seen what happened to my oldest brother. For the first five years, he said his wife was as sweet as a nun, always agreeing with him. But after having two or three kids, she became more jealous and fierce than a shrew. Now, they don't talk anymore, just argue. He's so fed up he wants to leave, but he stays because of the kids.

So he puts up with everything to keep the peace, while she gets more aggressive and difficult every day."

"Your brother's story is giving me goosebumps, but maybe it's because they live in the city, where people are more likely to change. Things aren't like that in the countryside."

"Oh, please! Whether in the city or the countryside, it's all the same. The only difference is that some women are more refined and tactful, while others are rough and crude. If you marry a clever woman, she'll sweet-talk you into working yourself to the bone, and you'll be happy to do it. But if you marry a rough one who always speaks harshly, where's the joy in that?"

"So, according to you, all men should just stay single?"

"I don't think so. My point is that you have to choose carefully, not rush into it and end up regretting it for life."

"So, what's your ideal wife like?"

"Me? I want someone reasonably attractive, not too plain but not too flashy either. She doesn't need to be highly educated, but she should be smart, simple, compassionate, cheerful, and a bit humorous."

Their conversation about marriage and family seemed to have run its course. Châu stood up, stretched, and headed out.

"Well, I'll wait for you to pick a wife, then I'll marry her sister or cousin to be safe!"

Châu's offhand remark made both of them laugh it off.

Through these discussions with Phước, Châu gradually felt less guilty about his old promise. Sometimes he secretly wished for Cô Năm Lài to find someone else and settle down rather than wait for him.

Since teacher Hai had come to the house to help with the tax exemption paperwork, Cô Năm Lài no longer resisted or avoided him as she had before, which pleased Bác Sáu. Teacher Hai also felt more encouraged as Cô Năm now welcomed him more warmly and treated him kindly.

Whenever she went to the market, Bác Sáu would stop by the school to give him a bunch of bananas or half a dozen mangoes. In return, teacher Hai often visited the house, bringing Bác Sáu tobacco, a few packs of Thai Duc tea, and for Cô Năm, a few hair clips adorned with yellow chrysanthemums to wear in her hair.

The memory of Út Châu had faded with time. Several harvest seasons had passed. The mango tree in front of the house had bloomed multiple times, yet there was no sign of Út returning. Thus, Cô Năm could no longer wait and risk becoming an old maid, which would bring shame and ridicule upon her family. She reluctantly agreed to marry teacher Hai.

The neighbors no longer called her "Năm Lài," but instead "Mrs. Teacher," the wife of teacher Hai, who taught in the village.

Six years had passed since Cô Năm had married teacher Hai. She had given him three children. With a modest but sufficient income and the status of a teacher's wife, she stayed home, taking care of the household and raising the children, no longer working as a hired laborer like in her youth.

Today was the annual death anniversary of her father, Bác Sáu. She and her husband, along with their three children, arrived early to help her mother with the cooking and preparations for the offering.

Seeing her three grandchildren running around the yard near the moat, Bác Sáu called out to her daughter:

"Lài, we have enough help here. Why don't you take the kids to the main road and buy some garlic and pepper from Mrs. Hai's shop to marinate the meat?"

Teacher Hai, who was helping his father-in-law polish the brass candle holders on the altar, also chimed in:

"While you're there, pick up a pack of Ruby or Basto cigarettes for me to offer to the guests later."

Cô Năm walked out from the back, filled a scoop with water from the rain jar, and washed her hands. The children, seeing their mother, quickly gathered around, asking to play with the water.

Hoa, the eldest at about five years old, scolded her younger siblings:

"If you splash water, Mom won't take us to the shop to buy candy!"

Hearing their sister's warning, the younger ones stepped back. Tý, the youngest, looked up at his mother and said in a mix of words and sounds:

"Mom... buy candy... eat..."

Seeing Tý's dirty face and muddy hands, Cô Năm beckoned him over:

"Tý, come here, and I'll wash your face before we go buy candy! Hoa, wipe Tèo's face with a towel, okay?"

###

This year's dry season had come earlier than usual. The fields in front of the house had already been harvested. The stubble still smelled fresh with the scent of new hay, standing upright in neat rows like a ruler's edge.

The sun was still low in the eastern sky. A few black buffaloes grazed slowly on the early morning grass near the field's edge. Hoa led the way, bending down to pick up dirt clods to toss at the buffaloes to move them out of the path.

Carrying Tý, Cô Năm followed behind and suggested:

"Why don't we cut across the field to get there faster?"

"No, Mom! The dry fields are hard on our feet. Let me chase the buffaloes into the field so we can walk on the soft grass instead."

Seeing the logic in her daughter's suggestion, Cô Năm stood quietly and waited. Mrs. Hai's shop, which also sold coffee and cigarettes, had already attracted a few customers early in the morning, sitting in front, sipping coffee, smoking, and chatting.

When she saw Cô Năm Lài and her children enter, Mrs. Hai quickly greeted them:

"My goodness! It's been a long time since I've seen you, Mrs. Teacher Hai. You have three kids now, don't you?"

"Yes, two boys and one girl, Mrs. Hai!"

"Is teacher Hai still doing well?"

"Yes, my husband is doing fine, still teaching regularly. Oh, do you also sell coffee here?"

"Yes! This coffee is really fragrant, brought down from Ban Mê Thuột. You should buy some for your father to try! Just the other day, someone from Saigon stopped by, asked about Út Châu's family, and tried the coffee. He liked it so much he bought a whole kilo to take back to Saigon."

Cô Năm Lài was startled to hear Mrs. Hai mention the familiar name Út Châu from her past. After a moment of regaining her composure, she casually asked:

"Do you know who from Saigon came by?"

"He stopped by in the early afternoon, and we chatted for quite a while. He said he was born in Rạch Đào but moved to Saigon-Chợ Lớn for work. His nephew is a soldier in Ban Mê Thuột, and he was asked to check on Út Châu's family and let them know that he's still doing well."

Cô Năm Lài asked for clarification:

"Was it Út Châu from the Cầu Nổi neighborhood, Mrs. Hai?"

"Yes, that's him, Mrs. Teacher! I heard he's been in the army for a long time, and his family hadn't received any news from him. Now that they know he's still alive, that's a relief. It's a dangerous time to be a soldier."

Tèo, who had been waiting impatiently while his mother talked to Mrs. Hai, couldn't wait any longer. He tugged at his mother's dress, urging her:

"Mom! Buy candy!"

Carrying Tý, who was also starting to get fussy, wanting to be put down, Cô Năm took the items Mrs. Hai had wrapped in old newspaper for her to bring home, and as she left, Mrs. Hai called out:

"Come back and visit when you can, Mrs. Teacher!"

The early morning sun spread its light across the empty fields after the harvest, making the landscape look lonely and desolate. Memories of those old rendezvous suddenly flooded back to her—the passionate first kisses, the tight embrace that nearly took her breath away. For the first time, Út Châu had made her feel such intense emotions, a mixture of fear and desire, like someone sneaking a forbidden treat.

After more than two years of waiting, with no news from Út, she had to yield to her family's wishes and marry teacher Hai. Since then, she had kept her past love affair a secret from everyone. As time passed and her children were born, her days were filled with household duties, and she completely forgot about the events of her youth.

Now, she had a husband and a house full of children. If she were to meet Út Châu again, she wouldn't know what to say. They would probably just look at each other, then turn away as if they were strangers.

The Vietnam War had intensified dramatically. The U.S. military and its allies had found it too difficult to continue, so they had washed their hands of the situation and left over a year ago. They treated their previous promises with the same casualness as merchants haggling over goods, taking what was profitable and discarding the rest. They had promised to support the Army of the Republic of Vietnam (ARVN) with enough strength to protect the independence and democracy of South

Vietnam, but in the end, the U.S. Congress gradually cut that support for vague, unjustifiable reasons. In the years that followed, the news reports on the front pages and radio broadcasts from both national and international sources stated: Phước Long had fallen... the communists had taken Ban Mê Thuột... President Nguyễn Văn Thiệu had ordered a tactical withdrawal from Kontum, Pleiku, etc.

People in Nha Trang were abuzz these days with rumors that Tuy Hòa, Quy Nhơn, and the provinces and districts up to Huế had been abandoned, leaving Nha Trang as a frontier town. Many families who had the means had already fled southward to escape the impending disaster.

Châu and Phước were on 100% alert, ready for action at their unit. Phước was sitting in the café run by Sergeant Lộc, near the meeting room of the defense unit. Châu had just finished his overnight shift and was heading back to the barracks to sleep when he was called back by Phước:

"Hey, Châu! Come in here for a cup of coffee before you go back to the barracks!"

Hearing the call, Châu immediately recognized his lifelong friend who adhered to the bachelor lifestyle. He turned around and walked in:

"It's barely morning, and you're already on duty at the café?"

"I couldn't stand the chatter in the barracks, so I came out here for a 'starter' coffee to clear my head."

"Yeah! At least we don't have wives and kids, so it's one less thing to worry about. I heard some of our defense officers have already sent their families to Saigon. Are you planning to send your mom somewhere?"

"My mom's old and rooted up there with my older brother. She said that since I'm a soldier, I can go wherever I need to."

Just as Phước finished speaking, Châu pulled his chair closer and lowered his voice, just enough for Phước to hear:

"I have a close friend who's a flight engineer with the 215th Helicopter Squadron. He told me that if there's a tactical retreat, he'll let me know."

And what was bound to happen finally did. Today, Nha Trang was almost abandoned, with no law and order left. Soldiers from other places had flocked to the city in a chaotic manner, like headless snakes. People were saying that the province chief had secretly evacuated his family long ago. The police and military police had deserted their posts to take care of their own families. It was clear that Nha Trang was headed for the same fate as the other cities.

Châu was jolted awake by the sharp sounds of gunfire outside, as some unruly soldiers took advantage of the chaos to loot and rob. They fired their weapons to show off their thuggish strength.

Châu jumped out of bed, stepped over to Phước's cot, and called out urgently:

"Get up, Phước! Hurry up! I hear a lot of gunfire in the streets!"

Phước seemed to have already woken up but remained lying there, listening.

"I hear it too, but it's not the sound of AK-47s; it sounds like M16s from our own troops."

"Forget whether it's AKs or not! Both can kill! Get your shoes on and change quickly. Now's not the time for details!"

"So where do we go now?"

"Let's head to the 215th Squadron and see what's happening!"

"But our commanding officer hasn't said anything yet!"

Frustrated with Phước's slow response, Châu cursed:

"Damn it! I said we're just going there to check things out, not to make any decisions! Forget about the commanding officer!"

When Châu and Phước arrived at the 215th Helicopter Squadron's parking area, they saw Sergeant Hiền, the flight engineer, busy inspecting the equipment of the helicopter parked on the tarmac. Châu was excited to see his friend and called out loudly:

"Hey, Hiền! Are you getting ready to fly somewhere?"

Unlike previous times, Hiền didn't stop to chat for a few minutes. Instead, he answered briefly while continuing his work:

"If you want to go, get your stuff and get on board quickly."

Châu was puzzled:

"Where are we going?"

"Damn it! Nha Trang is in its final hours, and you're still asking where we're going? The lieutenant pilot is in the operations room getting his flight helmet right now. His wife and kids are already waiting outside by the airstrip."

Realizing the gravity of the situation, Châu turned to Phước and spoke quickly:

"Phước, park the motorbike by the airstrip, lock it up, and let's go with Hiền."

"Why are you in such a rush? If I leave my bike here, it'll be stolen."

"We're in the final hours, and you're worried about losing a bike? Better that than being caught by the Viet Cong!"

Phước stood frozen, seemingly paralyzed by the sudden turn of events. Sensing he might have been too harsh, Châu softened his tone:

"I understand that this is your hometown, and if you don't want to go, that's fine. I respect your decision. But as for me, I have to go back to Saigon with Hiền and figure things out later."

At that moment, the lieutenant pilot rushed out, telling his wife and children to get on the helicopter. He turned to Hiền and asked about the status of the aircraft:

"Is it ready to fly, Hiền?"

"The log shows an oil leak in the gearbox, but I've checked, and it's not too bad. The airstrip crew has already left."

"I know they took off early. Let's just get on board and see how far we can go."

"I have a friend who wants to come along, Lieutenant. Can he go with us to Saigon?"

"Tell him to get on."

The HU-1 helicopter's rotors began to spin, slicing through the air with a loud whirring sound as the aircraft lifted off the ground.

Phước stood below, watching, his hair tousled by the downdraft as the helicopter ascended. Just then, a group of four or five soldiers arrived on motorbikes. Phước heard them cursing under their breath, realizing they had missed the flight, before they sped off in search of another way out.

As the helicopter rose into the sky, heading out to sea, Châu looked down and saw Phước still standing by his motorbike, gazing up. Châu didn't have the heart to wave goodbye as he had done on previous trips home; he felt guilty for leaving his good friend behind to seek safety for himself.

The wind howled outside the helicopter, and Châu reached out to close the door. Looking down from above, he saw that the waters of Nha Trang were as blue as ever when he first arrived. The mountains stood tall and silent, watching the ebb and flow of events.

From this moment on, Châu had no idea what his future would hold or where he would end up. After nearly thirteen years as a soldier, he had never imagined that his life would take such a tragic turn as it had today. He suddenly thought of Tiến and Tấn, wondering what had happened to them after Ban Mê Thuột fell. The thought made him feel small and cowardly.

As the helicopter left Cam Ranh Bay, Châu noticed Hiền pointing to a blinking red light on the instrument panel. The lieutenant pilot saw it too and nodded as if understanding the situation. Châu didn't know what was happening, but his instincts told him that there was a technical problem with the helicopter. The aircraft maintained its altitude and course without any signs of an emergency landing. Hiền, sitting in the co-pilot seat, turned around and gave Châu a thumbs-up, indicating that everything was okay.

About five minutes later, the HU-1 helicopter landed safely on the steel grid of the refueling station at Phan Rang Airport, designated for helicopters. Hiền unbuckled his seatbelt, got out, and attached the fuel hose to the tank. Châu also stepped out, ready to help if needed. Hiền walked over and shouted loudly to be heard over the noise of the rotors:

"The helicopters parked at the airstrip were siphoned off, so we didn't have enough fuel to fly straight to Saigon."

Châu thought to himself and smiled wryly: "Ah, so that's what it was! The blinking red light on the instrument panel earlier was the low-fuel warning."

At Phan Rang Airport, soldiers were still going about their duties, seemingly unaware of any imminent change. The HU-1 helicopter took off again, and Châu leaned back in his seat, closing his eyes, trying to forget the swirling thoughts about what had happened and what lay ahead.

After the fiery summer of 1972, with fierce battles raging across the country and helicopters being shot down repeatedly, Châu would occasionally visit the squadron to check on Hiền. He still remembered Hiền telling him: "In our line of work, sometimes being foolish will get you killed, while being clever might land you in prison. But at least you'll live to come home and take care of your family." It was a harsh but insightful observation for ordinary people who didn't have grand ambitions like him.

Back in Nha Trang, Phước had once told him a story about a platoon leader who led his men on a mission to chase Viet Cong sappers outside the gate during the Tet Offensive in 1968. The officer wore a steel vest and urged his men forward: "Advance, men, I'll take responsibility if anyone dies..." Whether the platoon leader had accidentally misspoken while giving orders, it became a story that was whispered and passed around, exemplifying the saying "My life, your death"—a mentality attributed to a few cowardly and unscrupulous commanders who tarnished the image of the Army of the Republic of Vietnam (ARVN).

Châu was still lost in scattered and incoherent thoughts when he suddenly jolted upright as the aircraft plunged downward at a fairly rapid speed. The young children of the lieutenant pilot, who were sleeping in their mother's arms, suddenly woke up and began crying.

Looking outside, Châu noticed many houses crowded together near the road, but the surroundings didn't resemble an airfield. He wondered if there was some emergency that made the lieutenant land so quickly. When they were about a hundred meters from the ground, the HU-1 helicopter made a sharp turn to the left, and Châu saw below an old abandoned American base surrounded by metal mats. While he was still uncertain, the HU-1 touched down and skidded along the open ground. The jet engine mounted in the rear seemed to have completely shut down, and all that could be heard was the whirring of the rotor blades.

Hiền unbuckled his seatbelt, stepped outside, looked at Châu with a forced smile, and made the sign of the cross on his chest.

"What happened, Hiền? It seems like the helicopter is broken, right?"

"We almost died! What do you mean 'broken'? The hydraulic oil line burst, and the engine shut off. We're lucky the lieutenant is a good pilot; otherwise, we would all be dead!"

"Where are we now? What do we do next?"

"This is Long Bình Base. We'll have to wait for the lieutenant to decide."

As Hiền was saying this, the lieutenant pilot, after ensuring his wife and children were safe, walked over.

"Hiền, stay here and secure the rotor blades and look after my family for me. I'll go inside to get some soldiers to guard the helicopter and contact Biên Hòa for a repair crew."

"Yes, Lieutenant! You go ahead; Châu and I will take care of things out here."

After completing the safety procedures for the helicopter, Hiền motioned for Châu to sit with him on the wooden casing of an artillery shell and took out a cigarette.

"Things are chaotic in Nha Trang right now, but here it seems quiet, like there's no war at all. We're really lucky to have gotten out; if we had delayed, those undisciplined soldiers in the streets might have made it difficult for us to take off from the airfield."

"I've been confined to the base for over a week, so I have no idea what's been happening in the city."

"Things have been a mess these last few days. The city of Nha Trang is almost deserted; everyone is just looking out for themselves. Even in the airfield, people were sending their families to Saigon to escape. Like this morning, the entire squadron evacuated to the South. Our helicopter was leaking oil, so they left it behind, and that's how we were able to escape."

"Now that we're here, where do you plan to report?"

"I was going to ask the lieutenant what his plans are."

About half an hour later, the lieutenant pilot returned with an infantry soldier to the helicopter. Seeing Hiền and Châu sitting and smoking, he gestured to them.

"Close the door and do a final check. We're leaving."

"Where are we going, Lieutenant?"

"We'll take a three-wheeler taxi home. Where else?"

"And when do we report in?"

"Probably tomorrow afternoon! We'll report to the Helicopter Command at the Air Force Headquarters to see where we're assigned next."

After answering Hiền, the lieutenant turned to Châu.

"You belong to the defense unit, so you should report to the personnel office to get instructions."

Châu nodded, "Thank you, Lieutenant!"

The Saigon-Biên Hòa "lam" taxi stopped near the intersection at the highway to drop off passengers. The lieutenant called a taxi to take his family home, while Châu and Hiền, two young airmen, strolled to a nearby café. Feeling more at ease now, Hiền began recounting the incident when the helicopter nearly crashed:

"We're lucky we don't have wives and kids to worry about. The lieutenant was scared to death when he saw the engine losing pressure. He kept asking me to check if anything had happened to his wife and kids."

"And I was just sitting there, not knowing what was going on! It's lucky we were close to Biên Hòa; otherwise, if it had been in the jungle, we'd have no choice but to wait for the Viet Cong to capture us."

"So, Châu! Where are you staying tonight?"

A hint of sadness appeared on Châu's face, and he slowly replied, "I don't have a place to stay. I'll probably look for a cheap room to rent, or maybe rent a cot at the bus station like I used to."

"If you don't have a place tonight, come stay at my house for a few days until we figure something out."

"No, I don't want to trouble your family."

"My house in Phú Lâm is pretty empty—just my mom and my younger sister. Come stay with us; it'll be fun, and we can take our time figuring things out."

Two "honda ôm" motorbike taxis stopped in front of the house. Hiền's younger sister, who was inside sewing, looked out and saw her brother returning from Nha Trang. She quickly ran out, calling loudly:

"Mom! Brother's home!"

Their mother, who was sitting in the back, responded:

"Are you saying Hiền's home, Vân?"

"Yes, Mom! Hiền's home, and he brought a friend!"

After paying the "honda ôm" drivers, their mother and sister came out to the front door.

"You just got home, Hiền?"

"Yes, Mom, I just got back."

Hiền turned to Châu and introduced him:

"This is my mother, and this is my sister, Vân."

After the introduction, Châu politely bowed.

"Hello, ma'am, hello, Miss Vân!"

Hiền's mother quickly spoke up:

"Come inside, dear. Vân, take your brother's luggage inside."

Vân nodded politely and smiled at the stranger.

"Brother, let me help you with your luggage."

Hiền handed his luggage to his sister, who carried it inside. Their mother stayed back to share the news she had heard:

"These last few days, I've been hearing on the radio that the central provinces are in chaos, which made me worried about how things were in Nha Trang."

"Nha Trang is pretty much done for, Mom!"

Their mother looked surprised.

"You mean Nha Trang is gone too?"

"Nha Trang hasn't really fallen to the Viet Cong yet, but the government has left! The city is lawless now."

"So, are you moving here permanently?"

"Yes, we decided to move to Saigon because all the top officers have already come here."

As he spoke, Hiền turned to Châu.

"Mom, Châu is also from Nha Trang. He's come here looking for a place to rent. In the meantime, I invited him to stay here for a few days until he finds something."

Hearing this, their mother looked at Châu and said:

"Our house is pretty spacious; you're welcome to stay with us temporarily. Don't worry about anything."

Châu bowed his head gratefully.

"Thank you very much, ma'am."

Their mother then turned to Hiền with instructions:

"Hiền, you and Châu should go upstairs and clean up so there's more space to stay. I'll go to the market to buy some fish and condiments for dinner. Vân, you stay home and start cooking; I'll be back shortly."

Even though the Vietnam War fever was nearing its peak, the newspapers in Saigon were filled with large, bold headlines reporting the intense fighting in the central provinces like Tuy Hòa's evacuation, Nha Trang being abandoned, and so on.

However, the people of Saigon and Chợ Lớn seemed indifferent. They continued to do business and enjoy themselves, as if nothing was happening, or they would gossip over a cup of coffee at street vendors to pass the time while waiting for customers.

But if we look a little closer, we'll see that the people of Saigon are divided into two completely different groups. The first is the intellectuals, wealthy capitalists, and the powerful; the second is the small business owners, civil servants, and laborers.

The first group seemed to be more silent, calculating, and busy consolidating their assets, buying precious metals, exchanging dollars, finding ways to send their children abroad, and transferring money overseas. They constantly kept up with and monitored the daily war news.

The second group, living hand-to-mouth, saw each day as just like the last. Whether the war was won or lost, they had nothing to lose, so the war for them was sometimes just a topic to debate or to measure how accurate their predictions were during idle hours.

After reporting to the Air Force Headquarters, Hiền and Châu were assigned to Biên Hòa Air Base to reinforce the combat and defense units there. But given the current situation—after the fall of Đà Nẵng, followed by Pleiku, Phù Cát, and Nha Trang—all the air force support and combat personnel from these areas had evacuated to Saigon, creating an excess of personnel at the remaining units.

Moreover, this group of displaced soldiers, like chicks separated from their mother, caused great psychological distress and confusion among the staff at the units they reported to.

The Vietnam situation was becoming increasingly dire. The South Vietnamese government's loss of the central provinces emboldened the communist cadres and the North Vietnamese political party to press their advantage. The Southern military was fighting under extremely difficult and desperate conditions. They were not only fighting the enemy in front of them but also worried about the safety of their families behind them.

At that time, Vietnam was like a sinking ship, with half of it already submerged. Passengers were scrambling to the rear, but eventually, the entire ship would be submerged beneath the sea.

The darkest day for the people of South Vietnam was, ironically, the brightest day for the North Vietnamese communist party. They had achieved Hồ Chí Minh's dream after nearly twenty years of war.

Vietnam was in misery! Vietnam was torn and tattered! Vietnam was one of the poorest nations at the end of the 20th century, while the rest of the world lived in peace and prosperity. People focused on production, research, and scientific and technological discoveries, while we, the small Vietnamese people, were still digging the well of hatred. What else could we do to benefit future generations?

Hiền and Châu were ordered to evacuate from Biên Hòa to Tân Sơn Nhất Airport on April 28, 1975, during the final days of the country. When they arrived at Tân Sơn Nhất by road, the Phi Long base gate was completely closed, and no military personnel of any rank was allowed in or out due to martial law.

After a few minutes of standing at Lăng Cha Cả, listening to news updates, Hiền spoke up:

"Let's go home, Châu! I think it's over! The Viet Cong are bombing the airport with A-37s. What's left to talk about?"

"Yeah, I see the people looking bewildered. They're running around like headless chickens!"

When they heard the sound of bombs exploding in the airport and the neighborhood's excited chatter, Vân and her mother were extremely worried about Hiền. So when they saw the two men arrive on a motorbike, Vân ran out to greet them.

"Oh my God! From last night until this morning, Mom has been praying to all the gods for you two to come back safely. Yesterday, Aunt Tú from Bình Trị came to visit and said the Viet Cong were marching down the streets with guns, not afraid of anyone anymore."

Their mother, who had just come out, looked anxious.

"Hiền, you just got back? What Aunt Tư said made me really scared for you two!"

Hiền looked at Châu and shook his head.

"It's probably over, Châu! You know Bình Trị is right at the end of the runway at Tân Sơn Nhất. If they've gotten there and set up SA-7 missiles, the airport is as good as closed. Saigon and Chợ Lớn are in for a rough time. Sooner or later, they'll start shelling to intimidate and attack, just like during the Tết Offensive."

After a moment of thought, their mother leaned in and whispered to Hiền and Châu:

"Maybe we should close up the house and go to your grandmother's in Mỹ Tho for a few days to see how things turn out?"

She turned to Châu.

"Where are you from, dear?"

"I'm from Rạch Đào village, ma'am."

Hearing that Châu was from a village, she quickly interjected:

"It's better to be in a big town; it's safer. The countryside is dangerous right now. Why don't you come with Hiền and our family to Mỹ Tho for a few days to see how things unfold?"

"I'm afraid of being a burden and costing your family more. I can stay here in Chợ Lớn."

"In these times of war and chaos, helping each other is what's right. Don't think that way."

Hearing this, Hiền chimed in:

"We've been running together from Nha Trang, so if you don't come with me, I'll be sad. Besides, you don't have any close relatives here in Chợ Lớn, do you?"

Seeing that Châu remained silent and didn't refuse, their mother turned to her daughter.

"Vân, go pack a few clothes for yourself and your mother. Hiền and Châu, you also get ready and pack a few things. I'll go to the market to buy some supplies and be back soon."

The sun was nearly at its peak when Vân finished packing clothes for herself and her mother into a small bag. Châu, who had only recently bought a set of civilian clothes in Saigon, had nothing else to pack since all his personal belongings had been left behind in Nha Trang.

While waiting for their mother to return from the market, Hiền and Châu changed into civilian clothes. The two men then headed to a café at the end of the street to listen for any news.

Just as they finished their coffee, Vân came out and called them back.

"Mom's back; she's calling for you and Brother Châu!"

As soon as Hiền stepped inside, his mother quietly instructed:

"Listen, so the neighbors don't get suspicious, Vân and I will take the cyclo to the bus station first. You and Châu lock up the house and follow us. Oh, are you taking the bus or riding the motorbike?"

"Châu and I plan to ride the motorbike so we can move around more easily once we get there."

"That's fine, but you two need to drive carefully."

The road from Chợ Lớn to Mỹ Tho showed no signs of war. The water buffaloes were grazing peacefully by the roadside. Children were playing tag in front of their houses, and the buses were as crowded as usual.

Châu felt that the scenes of people fleeing from the war were only happening in the provinces from Biên Hòa to the central region, while from Saigon to the South, everything seemed calm.

By about three in the afternoon, the sun had begun to descend from its peak. Hiền stopped to refuel at a small roadside shop leading into the province.

"Hey, Châu! Have you ever heard of Mỹ Tho's famous hủ tiếu (rice noodle soup)?"

"I've heard of it, but I've never eaten it in Mỹ Tho, where it originated."

"Well, after we refuel, let's go grab some hủ tiếu so you can taste the difference between Mỹ Tho's hủ tiếu and the one in Nha Trang."

"Sure, I'm hungry too. Hearing you talk about it makes my stomach jump with joy. But is your grandmother's house far from here?"

"My grandmother's house is on the outskirts of the city, near Mỹ Tho Lake. We'll be there in about fifteen minutes. Back when I was in Squadron 211 at Tân Sơn Nhất, I would occasionally land there to deliver supplies from the province to the outposts in Đồng Tháp Mười."

"So you must know the area pretty well then?"

Hearing Châu hit the mark, Hiền burst out laughing.

"You know us helicopter guys are like locals, from the big cities to the small towns, district capitals, villages, even high mountains and dense forests—we've been everywhere. Wherever there's an infantry operation or a national outpost, you'll find us. We also keep a black book listing all the good food stalls, the pretty girls who are fun and reasonably priced."

"So you guys are living the high life, huh?"

"High life, my ass! We make the most of it when we can. Sometimes when we fly supplies to an outpost surrounded by the Viet Cong, we just calmly land, knowing we could get shot at any moment. I used to have two friends who rented a house with me. At six in the morning, they'd go to the squadron for coffee and gossip before flying out to support the infantry in their operation. That day, I was the sergeant on duty at the squadron. At exactly 6:30, they took off, and by 7:20, the infantry command reported to the squadron that their helicopter had been shot down by an SA-7 missile, and the entire crew was killed. You see? Sometimes you see someone one minute, and the next minute, they're gone. You can't predict anything. I've survived until now, probably thanks to my ancestors' blessings, not because I'm any smarter or quicker than anyone else."

"As for me, I haven't been in as much danger as you, but in over ten years, I haven't been home once. I don't even know if my parents and siblings are alive or dead. The war just keeps dragging on, and I'm so sick of it."

The two young friends finished eating but lingered to chat about everything from family to love, from the sky to the earth, almost forgetting the time. It wasn't until the shop owner, who had been listening to the Saigon radio, reported that the Viet Cong regular army was concentrating to move towards Saigon that they realized how long they'd been there.

"You know what? The radio just said that the South Vietnamese Army has repelled some of the Viet Cong regular troops who are heading towards Saigon."

Hiển checked his watch and nodded at the shop owner.

"This situation doesn't seem promising!"

Châu stood up from his chair.

"We've been sitting here for over an hour without realizing it. Your mom and sister are probably worried about us."

"Talking to you after so long made me lose track of time. Let's head back so my mom doesn't worry."

The two of them rode the motorbike back, and as they turned into the alley, Vân saw them and called out loudly:

"Mom! Brother's back!"

Their mother, who had been preparing food in the back, breathed a sigh of relief when she heard Vân.

"Yes, I heard. I've been so worried, not knowing if they'd make it back safely."

Their grandmother, who was sitting on the wooden platform, also looked out towards the alley when she heard Vân call out.

"It looks like two people. Is that Hiển and someone else, Vân?"

"It's Brother and his friend, Grandma."

Châu and Hiển walked through the door, and Hiển respectfully greeted his grandmother:

"Grandma, I'm back!"

He turned to Châu and introduced him.

"This is my friend, Châu, Grandma."

Châu respectfully bowed his head to greet her.

"Alright! You two go to the well, draw some water, and freshen up before dinner. Your mom and Vân are cooking dinner in the kitchen."

"And where are Uncle Út and Aunt Út, Grandma?"

"They went to their in-laws' in Long An for the father-in-law's death anniversary. They'll be back tomorrow or the day after."

At dinner, they enjoyed sour soup with mồng tơi leaves, steamed đậu bếp beans, and fried cá rô fish dipped in fish sauce—a truly delicious meal. For over ten years, Châu hadn't experienced the warmth of a family gathered around the dinner table. After joining the military, eating for him was just a necessity for survival. Sometimes it was just a loaf of bread, a packet of corn, or a plate of sticky rice, eaten quickly to fill the stomach rather than to savor the warm, familial atmosphere of a shared meal.

After serving the first bowl of rice to his grandmother, Hiền's mother asked:

"In these uncertain times, what are you two planning to do?"

After a moment of contemplation, Hiền glanced at Châu as if seeking his opinion. Châu remained silent and shook his head, so Hiền replied to his mother:

"We don't know what to do yet, Mom!"

His grandmother put down her bowl and spoke:

"Do what everyone else is doing. In these times, keeping your life is more important. When things settle down, then come back."

Hiền and Châu bowed their heads in agreement because, in reality, they didn't know what to do in such an unprecedented situation.

That night, perhaps because of the unfamiliar environment, combined with the anxiety over the rapidly changing war situation, Hiền and Châu couldn't sleep. So, at the break of dawn, they decided to go to a café near the lake for a cup of coffee. On the way, the locals were still busy going about their daily lives—some selling goods, others going to the market. There was no sign that anything had changed.

Châu lit a cigarette and took a puff.

"You see, Hiền? Being a civilian is the best; they don't have to worry about anything. But as soldiers, we have to worry about everything: getting imprisoned, getting captured by the Viet Cong..."

Before Châu could finish, Hiền interrupted:

"Imprisoned, my ass! We didn't desert; our unit was disbanded when we left Nha Trang. We're like orphaned chicks, and when we got to Biên Hòa, we were treated like outsiders. Now, I'm just worried about the Viet Cong capturing us to take revenge."

"Is today April 29th, Hiền?"

"Yep. Yesterday, we evacuated from Biên Hòa early in the morning on April 28th and came straight down here."

"It's only been a day, but it feels like it's been much longer."

As they were talking, they suddenly heard the sound of helicopters circling the city. Hiền signaled Châu to be quiet so he could listen more closely. After a minute, a group of three CH-47 helicopters appeared, landing on the other side of the lake. Like someone grabbing onto a life raft in the middle of the ocean, Hiền quickly stood up and urged his friend:

"Hurry up, Châu! Let's go there and meet our guys."

"Do you know anyone in that squadron?"

"I know almost all the crew chiefs in Squadron 237 at Biên Hòa."

The three CH-47 helicopters looked massive as they took up most of the landing area by the lake. Sergeant Hoàng, the crew chief of the last helicopter to land, was just lowering the rear door to step outside. At that moment, Hiền was parking his motorbike nearby. He recognized Hoàng and called out loudly:

"Hey! Hoàng, what are you guys doing down here?"

Hoàng looked surprised to see Hiền there so unexpectedly.

"When did you get down here? You look pretty relaxed."

"I ran away from the war and almost died, so I'm not relaxed at all! What are you guys doing here?"

"You don't know anything?"

"Know what? Yesterday morning, we evacuated from Biên Hòa to Saigon, but the Air Force MPs closed the gates and wouldn't let us into the airport, so Châu and I came down here to wait for news."

"Last night, we slept under the aircraft at Tân Sơn Nhất Airport. At four in the morning, the Viet Cong shelled the place, setting fires all over. The officers tried to contact the Operations Center but received no orders, so they decided on their own to move the helicopters to Cần Thơ. When we flew over here, the aircraft commander asked to land and pick up his family, so the whole squadron landed. What about you? Are you staying here or going somewhere else?"

"What's there to do here? Châu and I will probably go with you guys. How long are you staying here?"

"I don't know. You should ask the squadron commander to be sure."

After learning that the Chinook squadron would be staying temporarily while waiting for news from Saigon radio, Hiền and Châu hurried home to change into their military uniforms and join the flight crew.

Vân was sweeping the yard when she saw her brother return on the motorbike. She immediately informed him:

"Brother, Châu, I just saw three huge helicopters fly over. Did you see them?"

Hiền quickly replied to his sister:

"Yes, we saw them. They landed by the lake. Where are Mom and Grandma, Vân?"

"They're in the back, cooking breakfast. Mom's been asking about you!"

Hiền ignored his sister's words and ran straight to the back to find his mother washing rice.

"Mom, we might have to leave with the helicopter crew that landed by the lake."

His mother stopped washing the rice and stood up, looking at her son in surprise. Before she could react, his grandmother, who was sitting on a small stool nearby, chimed in:

"But you two should wait and have breakfast before you go."

"We're not hungry, Grandma."

Regaining her composure, his mother asked:

"So where are you two planning to go?"

"I don't know yet, Mom! We'll follow wherever they go."

Vân, who was standing nearby and listening, interrupted:

"Why don't you stay here until things settle down, Brother?"

Hiền shook his head at his sister:

"It's not as easy as you think."

Seeing that the conversation was going nowhere, his grandmother pulled the scarf from around her neck to wipe her face, then walked over to Hiền's mother.

"Listen, Hiền's mother, he's grown up and joined the military, so let him go with his comrades. He won't be able to do anything here, and it might even be more dangerous. That's my opinion, but you two should decide."

"I'm not stopping him from going! You two should change clothes and leave. When you get somewhere, send us a letter or a message."

From the start, Châu had been quietly listening without offering any opinions. Whether he wanted to or not, he still felt like someone living temporarily under someone else's roof. Now, before leaving, he wanted to express his gratitude to Hiền's family, so he spoke up:

"Ma'am, Grandma, I've been very fortunate to receive your family's help during this time. I don't know how to express my thanks, but I hope that one day in the future, I'll have the opportunity to meet your family again in better circumstances."

"In these times of war, helping each other is the right thing to do. When the country is at peace, no one will have to suffer like this. Now, you two should get ready and leave early so they don't have to wait outside."

By a little after one in the afternoon, the flight crew and their families were chatting under the shade of the helicopters, waiting.

At that moment, Châu noticed a lieutenant colonel with the insignia of the General Staff of the South Vietnamese Army, who appeared worried, driving up in a jeep to speak with the squadron commander. Châu couldn't guess what they were discussing. Afterward, the squadron commander called all the pilots together to discuss and gather opinions. Finally, he called all the flight crew members to a meeting to announce the final decision. He said:

"According to the news from the lieutenant colonel Nam of the General Staff who managed to escape here, we are now at the final hour of the war. To ensure the safety of all crew members and their families, and this is also the common opinion and desire of all crew members, given the current situation, we have no other option but to fly out to the Seventh Fleet, which is currently offshore in the Pacific Ocean. However, this is a voluntary decision. Anyone who wants to go should follow us, and anyone who wants to stay can feel free to return home. At this critical final moment, we only have our camaraderie left; there are no more orders. I wish those who choose to stay the best of luck."

The squadron commander's final words moved everyone. They all remained silent and one by one climbed aboard the helicopters. Châu noticed that only one crew member, a gunnery sergeant who was about to get married, volunteered to stay in Mỹ Tho.

The helicopters took off. A few minutes later, the city of Mỹ Tho, with its dense houses below, was just a blur. Clusters of white clouds drifted by the helicopter windows.

Hiền tapped Châu on the back and pointed downwards.

"That's Gò Công. We're heading out to sea."

Châu didn't reply; he just nodded. He looked down at the rice fields divided into neat squares, with a few thatched houses standing isolated in the middle of the fields, almost cut off from the outside world. Memories of home suddenly flooded his mind—those early season rains, the dry fields just beginning to fill with water, the nights spent catching fish from the village ponds, going frog hunting, picking young tamarind leaves, gathering sesbania flowers, pulling water spinach for his mother to boil or make soup.

Châu remembered one time when he was getting ready to go fishing for perch in the fields. His sister-in-law accidentally stepped over his fishing rod, which was lying on the ground, and he got so angry he yelled at her. His mother defended him, scolding the sister-in-law: "The youngest son has a knack for catching fish. How could you step over his rod like that and not expect him to complain?" Because in the countryside, it was believed that if a woman stepped over a fishing rod, it would bring bad luck, and the fish wouldn't bite.

As these mixed thoughts swirled in his mind, he suddenly remembered Miss Năm Lài. He figured she must be nearing thirty by now. Time had worn away the old memories. His love for Miss Năm in the early days had been like a strong wine, making him dizzy, but now she had probably forgotten him completely and was happily fulfilling her family duties. Now, everything had become a thing of the past.

The helicopter had been flying over the sea for quite some time. The coastline of Vietnam was just a faint gray line stretching along the horizon, gradually fading behind them. Ahead, there was only blue water and white clouds. Now, everyone on board seemed anxious.

Châu leaned close to Hiền's ear and whispered:

"They've been flying over the sea for a while. Where's the American fleet?"

"I don't know! They're probably searching for it."

Hiền's answer had barely left his lips when the helicopter made a sharp turn to the left. Sergeant Hoàng, the crew chief, after communicating with the pilots via the radio, looked at the passengers and gave them a thumbs-up, signaling that everything was okay. To be sure, Hiền got up and walked over to Hoàng.

"They've spotted the American fleet, right?"

"Not yet, but they've made radio contact with the Seventh Fleet on the emergency frequency. They're adjusting their course to head toward the fleet now."

About five minutes later, Sergeant Hoàng pointed out the left window. A large ship flying the American flag was moving across the sea. Seizing the opportunity, the Chinook helicopter circled the ship, searching for a place to land. Because the helicopter was too large to land on the deck reserved for HU-1 helicopters, the squadron commander (Lt. Colonel Nam) had to hover in place for everyone on board to jump down. After ensuring that everyone had safely jumped onto the ship's deck, Châu noticed that the helicopter began to fly away from its previous position. He was puzzled.

"Hey, Hiền! Is he flying back to Mỹ Tho?"

"I don't think so. He's probably going to 'ditch'."

"What's 'ditching'?"

"I've only heard that it means the pilot will crash the helicopter into the sea and then swim out, but I've never seen it happen."

Châu was utterly astonished and silently shook his head. He felt sympathy for the commander, who was ready to sacrifice himself, accepting to be the last one to leave the helicopter, like a conscientious ship captain.

What followed were long days drifting on the ship in the middle of the Pacific Ocean. Châu and Hiền saw that there were many compatriots on board. They were from all walks of life, having abandoned everything to escape. These people didn't even know where they were heading—some said the Philippines, some said the United States, and some believed they would only stay on the ship temporarily until it was safe to return to Vietnam.

Everyone was making assumptions based on their own opinions. Another day quickly passed, and the sun, now a bright yellow, began to dip below the horizon, casting the next sunset over the vast ocean.

Hiền, sitting at the bow of the ship, waved Châu over:

"Come over here, have a smoke, and let's chat!"

"You still have cigarettes?"

"Of course! Before we left, I made sure to stock up. I can go without food, but not without a smoke—it's too depressing."

"This afternoon, did you have a quarrel with the guys handing out the canned food?"

"Forget about it, those 'sneaky rats' are too greedy. I can't out-yell their shrill-voiced wives, so as soon as I hear that screeching tone, I'm out of there."

"Those guys volunteer to distribute canned goods to the refugees, but they keep all the good stuff for themselves. They give us only the things they don't want, like beets and hotdogs that are so awful you can hardly swallow them. When confronted, they deny everything, sometimes even cursing us out, saying that we, the 'political refugees,' have no right to expect better food."

Hiền laughed at Châu's complaints:

"I get a chill when I hear the term 'political refugee,' like you and I are famous politicians or revolutionaries from back home, now living abroad waiting for the day to return and restore the nation."

"I never said that! Those people who hoard the canned goods call themselves that and then go on to insult us. We're just two soldiers who fled because we were scared the Viet Cong would seek revenge, running from Nha Trang to Saigon, lost like motherless chicks. We just happened to run into some friends and got out here by chance, no planning or forethought at all! But let them say what they want. The important thing is, have you heard anything about where we're going?"

"I don't know any more than you do. I've heard rumors that we might be going to the Philippines or America. But whatever happens, we'll just have to deal with it as it comes. Us 'political refugees' have beets and hotdogs to eat daily, so we won't starve. No need to worry ourselves sick."

Châu couldn't help but laugh at Hiền's response:

"Yeah, I suppose you're right. We've hit rock bottom; there's nothing left to fear."

###

Almost three months had passed since they left Vietnam. Today was the morning of July 22, 1975, and a thick fog still blanketed Camp Pendleton. Châu woke up earlier than usual, and after rolling up his blanket and putting it in a nylon bag, he reached over and gave Hiền's army cot a hard shake.

"Hiền! Get up, it's morning. We need to get ready to leave the camp."

Groggily, Hiền mumbled, half-asleep:

"It's still early; what's the rush?"

"It's almost six o'clock. Not early anymore! Pack your stuff, and let's head to the dining hall for breakfast."

After spending time on Guam, then Wake Island, the refugee agency had finally moved Hiền and Châu to Camp Pendleton in San Diego, California, where they had been for over a month, waiting for a sponsor.

Most families with children were sponsored by church parishes and allowed to leave first. But soldiers and single men without specialized skills found it much harder to find sponsors.

Luckily, a few weeks ago, the governor of Washington State volunteered to sponsor several hundred refugees, including families and single men, to settle in the state. Tired of living in the refugee camp, where the only activity was waiting for the three daily meals, followed by naps, and with limited freedom of movement, the two men eagerly signed up when they heard the news on the radio.

There were plenty of mixed opinions and gossip about the pros and cons of leaving the camp. Still, for Hiền and Châu, getting out of the camp was the most important thing. They didn't care where they were going or what they'd be doing. As Châu often reminded himself after landing in this unfamiliar country: "Where there's life, there's hope." The way he saw it, they would manage, just like everyone else.

This morning, they were part of the first group leaving the camp, heading to the bus that would take them to the San Diego airport for a flight to Washington State. Châu sorted through the clothes he had gathered from the charity donations given to the refugees at the camp. Finally, he found a reasonably decent outfit that fit well and put it on. He turned to Hiền for his opinion:

"Hey, how do I look in this?"

178

"Looks good! You're lucky to have found pants that fit perfectly. My pants are all too short, and my belly sticks out like a pregnant toad."

"If the pants are too long, just tear off the excess and pin them up with a safety pin. If the waist is too big, wear a belt and leave your shirt untucked. It shouldn't look too bad."

"Yeah, I was thinking the same. Yesterday, they told us to dress nicely and cleanly for leaving the camp, but we don't have much to choose from, do we?"

"Forget it! They can say whatever they want, but we have to make do with what we've got."

The military bus began to move, heading toward the main road. The morning sun was already high in the eastern sky, casting its warm rays over the low brush to the left of the road. A few white rabbits, munching on wild grass, scampered into the nearby bushes as the bus approached.

Over ten minutes passed, and both Hiền and Châu remained silent, lost in their thoughts or pondering something deeply important. It wasn't until the bus suddenly braked hard at an intersection before turning onto the highway that Châu, jolted forward, snapped out of his daze. He turned to Hiền, who was sitting next to him:

"What's going on? Why did we stop?"

"There's a red light ahead. Didn't you see it?"

"Oh... I wasn't paying attention. I was just thinking about how funny it is that we've suddenly been discharged from the military without any paperwork or explanation."

Hiền, enjoying the chance to tease, replied:

"If you want discharge papers, go back to Vietnam. The Viet Cong will happily give you some. Or you could ask one of the high-ranking officers who made it here first to sign them for you so that the military police don't arrest you on the way."

"Let's drop that. What we need to focus on is what's coming next. With our English, we barely know ten words. What kind of job can we get?"

Hiền patted Châu on the shoulder:

"Hey! Didn't you say 'where there's life, there's hope'? Are you scared now?"

Then, leaning in close, Hiền whispered:

"Look at those old fishermen sitting in the back. They're not worried like you are. Back in Vietnam, even kids who only knew how to say 'OK Salem' managed to make some money. We're grown men; we won't starve, right?"

"I agree with you. If we were still back home, I wouldn't be so worried. But here, in a foreign land with everything so different, I'm a bit nervous."

Hiền stretched out, leaning back in his seat, and relaxed:

"Just rest. Whatever's coming will come. No point in worrying now."

After almost two weeks at Murray Camp in Washington State, Hiền and Châu were sponsored by a farming family in Fall City. According to the interpreter, the sponsoring family promised to provide food, lodging, clothes, and work on their farm. Like blind men suddenly regaining their sight, everything seemed new and beautiful to the two men, so they eagerly accepted the offer.

The next morning, after finishing breakfast, they returned to the camp, where the liaison officer told them to be ready to go to the office with their sponsors.

Through the interpreter's introduction, they learned that the sponsors were Bob and Linda Lenchou, a couple in their fifties.

After exchanging greetings and handshakes, the two men followed the couple to their car parked out front to head to their new home.

Bob was a tall man with a country look rather than a city demeanor. While driving, he asked Hiền and Châu a few questions, but neither understood, so they nodded repeatedly like "ducks listening to thunder." It ended up being a one-sided conversation. Linda, the wife, seemed cheerful and occasionally turned around to smile at them.

Their station wagon left the main highway and slowed down on a provincial road connecting the county towns. Sometimes, the car passed through dense pine forests that almost blocked out the sunlight. Then, suddenly, an expansive field bathed in sunlight appeared. The surrounding landscape had completely changed, and Hiền and Châu realized they had left behind the familiar atmosphere of a society where everyone spoke the same language. They were now in a new, empty world, far removed from everything they had known. A subtle, lonely sadness began to seep into their souls, like a bride leaving her family to move into her husband's house, only to sneak out to the back door in the evening, longing for her childhood home:

"Evening after evening, standing at the back door, Looking towards my mother's house, my heart aches."

The earlier excitement about leaving the camp and having a job was now replaced by a sense of emptiness. It wasn't that they were afraid of facing reality, but they felt like lost birds or animals trapped in the wrong cage.

The station wagon slowed down and then turned onto a small gravel road. Bob pointed out a large house on a piece of land surrounded by a white fence, not far ahead. Although they didn't fully understand what he said, the two men guessed it must be the couple's home.

Hiền nodded to Bob, signaling "OK." Châu leaned over and whispered:

"Do you know what he said?"

"I just nodded and said 'OK.' No need to understand everything."

Finally, the car stopped beside the house. The couple got out first, and Hiền and Châu followed. Bob signaled to a young man sitting on a lawnmower in the backyard, who Châu assumed was their son. Another round of introductions followed; the son's name was Joe, and he was about twenty-five years old, cheerful and energetic. After his mother said something to him, Joe eagerly opened the back of the car, grabbed the bags of clothes for Châu and Hiền, and motioned for them to follow him inside.

Joe showed them to a spacious room on the ground floor, complete with a bathroom, shower, and two neatly arranged beds in opposite corners.

After showing them their room, Joe left, closing the door behind him. Hiền looked at Châu, raising his eyebrows, indicating his satisfaction.

"Our American sponsors are really thoughtful, huh?"

"Yeah! Escaping like this is the best thing ever. I remember when the French brought soldiers into our village to hunt the Việt Minh, my family had to run for their lives. We didn't have enough food, and our clothes were so rough they were practically made of burlap. Running from the enemy back then was so tough. Let me take a shower first. Enjoy it while it lasts."

Around six o'clock in the evening, as Châu was sitting on a chair watching TV, he heard a soft knock at the door. He was about to get up when Joe opened the door slightly, gesturing to his mouth and following it up with a stream of English. Châu guessed Joe was calling them for dinner.

Bob and Linda were already seated at the ends of the dining table, with Joe on one side. Bob motioned for Châu and Hiền to take the two seats next to Joe. Once everyone was seated, Bob bowed his head to say grace. Hiền and Châu mimicked the family's gestures before eating. Dinner that evening consisted of boiled corn, bread, butter, and vermicelli with minced meat. Though the meal was simple, the table setting—with its elegant cups, plates, and utensils—was impressive and sophisticated.

Through the Vietnam War, Châu had witnessed the advanced military technology of the United States, which he had admired. Now, he was amazed at the lifestyle of a typical farming family. It was beyond what he could imagine.

In Vietnam, a farmer who could afford a pair of Japanese sandals or a new outfit would hide it away for special occasions like weddings or funerals. A large gourd or squash wouldn't be eaten but instead sold at the market to buy a few extra yards of fabric for new clothes. Vietnamese farmers worked hard to produce what they needed to survive, with no thought of luxuries or beautifying their homes.

Most Americans, on the other hand, worked hard to enjoy life. In contrast, many Vietnamese seemed to work just to survive, with little chance to indulge in the luxuries that people in more developed countries took for granted.

After dinner, Linda began clearing the dishes and loading them into the dishwasher. Bob, holding a cup of coffee, motioned for Hiền and Châu to join him in the living room.

Although they didn't fully understand the couple's intentions, Hiền and Châu had a sense of what would happen next.

Everyone gathered in the living room as if for a family meeting. Bob looked at Hiền and Châu, speaking slowly, almost word by word. Occasionally, he used hand gestures to ensure they understood. After each sentence, he nodded, as if to ask if they were following. Although they only grasped a few familiar words from his speech, they got the gist of what Bob was trying to convey.

Hiền turned to Châu to confirm: It seemed that Bob was telling them that tomorrow they would be working with him and Joe somewhere and that Linda would drive them to evening English classes afterward.

Châu agreed, thinking this interpretation was close enough, so they both nodded in agreement with Bob.

After a day of travel and the unfamiliarity of their new surroundings,

Hiền and Châu couldn't fall asleep easily. They stayed up talking late into the night, finally drifting off out of sheer exhaustion. When a "knock knock" at the door woke Châu, he sat up with a start. The light came on, and Joe pointed to the watch on his wrist. He repeated after Bob, slowly saying:

"Six... six o'clock... morning... eat breakfast... and go to work... OK?"

Châu understood Joe and nodded:

"OK."

After Joe left, Châu walked over to Hiền, who was sleeping soundly with his face buried in the pillow, and shook him hard:

"Wake up, Hiền! It's already six o'clock! Joe's calling us for breakfast so we can go to work!"

"Yeah, I heard."

"If you heard, then get up and wash your face. What are you waiting for?"

"You go first, then I'll go."

"What should we wear to work?"

"I have no idea what kind of work we'll be doing. Just put on the new clothes we wore yesterday."

After breakfast, Bob asked Joe to load some garden tools—rakes, hoes—onto the pickup truck.

Once everything was ready, Bob got in the driver's seat, and the three men climbed into the truck bed.

Châu seemed puzzled:

"I've heard that people come to the US to work in industrial factories. What are we doing with all these garden tools?"

Hiền chuckled:

"That interpreter at the camp told us that a farming family would sponsor us and give us work. So today is our first day of work."

"So, you're saying we're going to be farmhands?"

"I'm not sure. We'll see."

After about fifteen minutes, the pickup truck stopped in front of a large barn on a wide piece of land, probably a hundred acres. It looked like the crops were mostly corn, with the plants just about knee-high. Joe jumped out first, carrying the hoes and rakes into the barn. Bob got out of the car, walked to the back, and pointed out a large, empty field of a few acres to Châu and Hiền. After explaining what needed to be done, Bob seemed unsure if they understood, so he called Joe over to work with them for a while to make sure they got the hang of it.

When people think of American farms, they often imagine thousands or hundreds of acres of land, with most of the work done by machines for plowing, planting, harvesting, fertilizing, spraying pesticides, and so on.

But Bob's farm was only a few dozen acres in a large valley surrounded by other farms. His methods were somewhat old-fashioned, with machines used only when necessary, and most of the work was done by hand. On this farm, he mainly grew corn, strawberries, and cucumbers, with some secondary crops like tomatoes and vegetables. His market was local grocery stores.

Before Hiền and Châu arrived, Bob often hired students for part-time work, paying them minimum wage, which was about two dollars an hour at that time. But most students didn't stay long; they quit as soon as they had enough spending money. Because of this, Bob often faced interruptions due to a lack of workers. So, when he heard that the refugee camp was looking for sponsors to help and provide jobs, Bob and Linda volunteered immediately. They called the refugee camp and explained what they could offer.

After the necessary paperwork, the camp introduced Bob and Linda to two young, single, and healthy former soldiers of the South Vietnamese Army.

After giving Joe some final instructions, Bob drove to town to buy more fertilizer and insecticide. Joe handed a hoe to each of the men, who slung them over their shoulders and followed him out to the field that Bob had pointed out earlier. When they got there, they saw that the field had already been planted with cucumbers, the small plants just a few inches tall. Their main task was to weed out the grass growing among the cucumber plants and repair the rows of soil damaged by animals, among other things.

The work was straightforward, but given the large area planted with cucumbers, it required a lot of time and effort. Once Joe saw that Hiền and Châu understood the job, he left to water the corn and potato fields using a pump.

After working for a while, the white shirts and new shoes they had worn to leave the camp were now dirty with mud. Hiền couldn't help but voice his frustration:

"Man, I thought people came to America to study in fancy schools and live a luxurious life, but here we are, working as farmhands, and not even good ones!"

Hiền wiped the sweat from his forehead:

"Come on, man. Those people had government support for their education. We were running for our lives from Nha Trang to Biên Hòa and Mỹ Tho. We got lucky that someone took pity on us and brought us here. We're lucky not to have died on the way, so what's the point in complaining?"

"I know, you don't need to tell me. But the thing is, I've never done farm work in my life. I was lazy as a kid, so when I failed my baccalaureate exam, my mom scolded me so much that I joined the Air Force out of spite. I trained as a mechanic and became a flight crewman. I've never worked in fields!"

"Well, my story is different from yours. Your family was probably better off, but mine was poor, and there were many siblings. We had to work hard on the farm and fish since we were kids. Then, when I was old enough, I got drafted and sent to Ban Mê Thuột. Finally, fate led me to join the Air Force, where I met you in Nha Trang. My life has been full of hardships, so farm work like this isn't too bad for me, given what I went through as a kid. But after months of running from the enemy, eating and sleeping all the time made me lazy. I think we'll get used to it after a few weeks."

After hearing Châu's story, Hiền felt somewhat comforted. His tone was less bitter and more resigned:

"What do you think other people are going through?"

"We're in the dark, just like everyone else. Who knows anything for sure?"

That evening, after dinner, Linda told the two men to prepare for their evening English class at a community college for new immigrants.

Even though it was after six o'clock, the summer sunlight was still bright, casting a beautiful golden hue. Washington State, located in the far north of the United States, has long nights and short days during the winter. But in the summer, the situation reverses, with long days and short nights. Sometimes, even at nine o'clock in the evening, the sun is still visible on the horizon.

Hiền and Châu were ready. While waiting for Linda, they stepped outside to have a smoke:

"Hey Châu! What do you think?"

"Working in the fields on the first day was tiring because I'm not used to it."

Châu hadn't finished his sentence when Hiền interrupted:

"That's not what I meant. I'm asking if you think we're going to spend the rest of our lives working in the fields."

A faint sadness crossed Châu's face as he lowered his head, trying to hide a weary sigh:

"As I've told you before, I come from a poor farming family, so I understand this kind of work better than you do. Even though things are more modern here, with machines to help, the life of a farmer is the same everywhere. They're always tied to the land, waking up with the first rooster's crow and returning home when the sun sets. They toil all day, hoping for good weather to ensure a good harvest, and maybe save a little. Their children usually follow in their footsteps, generation after generation, just like Joe here.

To be honest, I've lived that life, so I understand it, and I'm tired of it. I want my children to go to school, get a good education, and have stable jobs in big factories. I don't want them wading through fields and streams like I did as a kid."

"So what's your plan?"

"We've only been here a day or two; we're still getting our bearings. I think we should wait a little longer, get familiar with everything, and then make a plan. Besides, Bob and Linda are treating us well, even though the language barrier is still a challenge."

They had just finished discussing this when Linda stepped out of the house, signaling for them to get in the car for their English class.

To Châu's surprise, as he stepped out of Linda's car, he saw a man with a familiar face and figure approaching from the parking lot. Châu stopped in his tracks, squinting to get a better look, and muttered to himself, "Could that be Captain Tiến from Ban Mê Thuột?"

Seeing Châu's odd behavior, Hiền asked:

"Do you know that guy?"

"Yeah, I think that's Captain Tiến!"

"Which Captain Tiến?"

"My old boss from Ban Mê Thuột!"

The man seemed equally surprised, staring at Châu before speaking:

"Are you Vietnamese?"

Châu's face lit up with recognition, and he raised his hand:

"Are you Captain Tiến from Ban Mê Thuột?"

"Yes, it's me! Is that you, Châu? My God, what a stroke of luck! I never thought I'd see you here."

The sudden, unexpected reunion was deeply emotional. Châu's words seemed to catch in his throat.

Sensing his friend's strong emotions, Hiền patted him on the back:

"We've just made a new friend!"

Regaining his composure, Châu introduced Hiền:

"Captain Tiến, this is Hiền, my friend from the Air Force in Nha Trang. We escaped together and came here at the same time."

Tiến nodded, shaking Hiền's hand:

"It's great you two have each other for company. I came here alone, and it's been lonely."

Hiền blurted out:

"Don't you have family with you, Captain?"

"My parents and siblings are still back in the countryside, like Châu. Luckily, I'm not married yet, so I didn't have any attachments. I've been working as a janitor at a winery in Woodinville for about a week now."

Châu added some context for Hiền:

"Captain Tiến is from the same village as me, Rạch Đào. He helped me with the paperwork so I could transfer to the Air Force."

Châu then turned back to Tiến, puzzled:

"I knew you were stationed at Ban Mê Thuột for years. How did you manage to escape?"

Tiến chuckled:

"It's a long story. I'll tell you all about it when we have time. In short, after Ban Mê Thuột fell, some of us made it to Nha Trang, where I ran into an old friend who was in the Navy. He got me on his ship, and we ended up here in the US, and now, by some miracle, I've found you."

That night, there were about eight Vietnamese students in the English class, all overjoyed to see each other as if they had reunited with long-lost family. They asked each other about everything, mostly about where they lived and their jobs. These were people who had fled their homeland because of war, their souls as torn and tattered as their lives, which made them feel a deep connection with one another.

The evening class was only two hours long, but for these Vietnamese refugees, it was a precious time to meet and share their burdens.

Despite the heavy hearts of those who had fled their homeland, their worries were eased somewhat by the presence of others with whom they could share their hopes. Though those hopes might seem distant, they were vital for their survival and their future. As one scholar once said, "People who are alike tend to stick together." Human conflicts and jealousy often stem from differences in beliefs and wide disparities in social status.

The needs of a person evolve over time. For these Vietnamese refugees, the first rung on their ladder was security and survival, and finding each other in this new land was crucial in this regard.

Since meeting Tiến, Hiền and Châu had a new friend to share their journey. They became more enthusiastic and optimistic. After long days of hard work on the farm and at the winery, the three friends eagerly met up in the evenings for their English classes, where they exchanged ideas and news.

Time seemed to pass quickly there. Summer quietly slipped away, taking its warmth and sunlight with it, leaving behind only gray skies and constant drizzle.

Autumn arrived slowly, like an old man, turning the green leaves to a faded yellow. The tree in front of the house had already shed most of its leaves, leaving its branches bare.

Hiền quickly downed his hot coffee, which he had just brewed. Châu had already dressed and was ready to head to the fields.

"I made you a cup of coffee, Châu. It's on the table. Drink it and let's go."

"Okay, but do you know what we'll be doing today?"

"Yesterday, Bob said we'd be checking the cucumber beds to pick any remaining fruit for pickling. Then, we'll cover the water pumps outside with tarps to keep them from rusting. He said Linda has a cold and will stay home to rest, so it'll just be the three of us with Joe."

"If Linda's sick, who will drive us to class this evening?"

"We'll ask Joe to take us."

"Yesterday, during the break, Tiến mentioned that he might call a friend in Seattle to look for a new job."

Hiền seemed surprised by this news and pressed for details:

"Are you saying Tiến is looking for a new job?"

"That's what he told me, but he wasn't sure yet. We should ask him again this evening."

The English classes for refugees had been running for over three months now, and the number of Vietnamese students had grown to over twenty. Châu noticed that some had even bought their own cars and drove themselves to class, looking more confident and proud. During breaks, they often gathered in small groups to discuss car-related topics, like which brands were good, which were bad, American cars, Japanese cars, and so on.

Hiền and Châu were smoking and chatting in front of the classroom when Tiến walked in. He seemed more excited than usual. As soon as he saw them, he called out:

"Hey, Hiền, Châu! I've got hot news!"

Châu stubbed out his cigarette in the ashtray:

"What's got you so excited?"

"I just got a call from a friend in Seattle. He signed up to go to Alaska to fish or catch crabs, something like that. He said the pay is pretty good, and they provide food and lodging on the boat. You can work as much overtime as you want, and they pay one and a half times the regular wage. I think us single guys should give it a shot. What do you say?"

Tiến's words seemed to light a fire in both Hiền and Châu.

"Hey, Châu! Let's sign up and give it a try with Tiến. If things go well, we could hit it big. If not, at least no one dies. We've got nothing to lose, so why not go for it? Working on Bob's farm, who knows when we'll ever get ahead?"

"I'm not scared, man. Let's ask Tiến to find out more about how to apply, and then we'll sign up together for some fun."

After more than two weeks of searching for workers, the canning company in Alaska had enough people signed up for the job. Tiến, Châu, and Hiền received notification from the company that they would be departing from Seattle on the following day, Friday at noon, on a regular flight with Alaska Airlines. Bob and Linda had known about this for over a week, as Hiền and Châu had informed them. The couple seemed a bit sad, but Bob still saw this as a good opportunity for the two men. He promised that if anything went wrong, they could return to the farm and resume their previous work.

In the group of people going to work, Châu noticed that most were young, with only two married couples, while the majority were single.

The "MD80" jet of Alaska Airlines landed at Anchorage Airport to drop off passengers. The entire group of Vietnamese workers was transferred to a local route by propeller plane to Dutch Harbor, located at the northernmost tip near the Soviet border, which was the final destination for their work.

About an hour later, the engine noise seemed to lessen. Hiền looked through the window and saw a vast, desolate landscape covered in snow, stretching out below. He turned to Châu and Tiến with wide eyes: "Brother Tiến, look down! Snow covers everything, all white. We must be flying over the Arctic region, right?"

After observing, both men shook their heads. Tiến, trying to keep the mood light, joked: "We're all in this together. We fought in the war against the Việt Cộng without flinching, so why fear a little Arctic snow?"

Châu forced a smile to chime in: "We're just going to be 'Eskimos' for a while; what's the harm in that?"

Not long after, the plane landed safely and taxied to a stop. The passengers gradually made their way into the waiting area. Tiến immediately recognized the representative from the canning company holding a large sign to help the new workers identify themselves. After collecting their luggage, the group was taken by bus to the company's office to complete the necessary paperwork before starting work.

While waiting, Tiến invited Châu and Hiền to a quiet corner to smoke a cigarette. As if recalling something amusing, Hiền grinned to himself: "Life moves so fast. In just about seven or eight months, we've already changed jobs three times—from aircraft maintenance soldiers to farm laborers, and now we're about to become crab sorters and fish processors. Who knows what job we'll take up after this contract is over?"

"It's good to be hired at all; otherwise, we'd all starve," Tiến added, sitting nearby.

Tiến chimed in: "Starve? Probably not, because the social services will give us money and food stamps to keep us afloat."

Châu, raising an eyebrow, looked at Tiến in surprise: "You say we won't starve? What is this social service or food stamps you're talking about? I've never heard of it before."

"You don't know because your sponsors didn't want you to apply for social benefits. In my case, the owner of the winery didn't let me file any paperwork either. I found out about it from a friend's family in Seattle. He has a wife and seven kids; they stay at home and collect social welfare for the whole family. With the extra time, he takes cash-in-hand jobs to save money, buy a car, and drive his family around."

Just then, Hiền blurted out: "Ah, no wonder I saw one or two guys driving cars at the evening English class, talking loudly as if they were showing off. I thought they were lucky to have landed good jobs despite being as bad at English as we are and bought cars just to flaunt them. But where does this welfare money come from, and how do they keep collecting it?"

"From our taxes! The government takes it out of our wages!"

"The government plays this game, making us work our asses off to feed these lazy bums. If we lose our jobs here, we'll go and claim social benefits like them to spite them."

Tiến laughed heartily: "Don't bother with social benefits; it's embarrassing. Instead, you can claim unemployment benefits from the unemployment office, which sounds much better."

"How do you know so much about this?" Hiền asked.

"I overheard it from people who know more about life in America than I do," Tiến replied.

Their conversation was interrupted as everyone in the group finished their paperwork. After receiving some final instructions from the company representative, the group was guided on a tour of the factory, their future workplace, and their accommodations during the time they had committed to working for the company.

Autumn here was lonely and desolate. Everything was draped in white snow, and the heavy gray clouds stretched endlessly, merging with the distant, shadowy mountains on the far-off horizon.

For months, the sun seemed to have disappeared. The lights on the ship were nearly always on, never going out. The work here was almost constant, 24 hours a day. Those who were healthy enough could work as much as they wanted. After eating and sleeping for a bit, they'd clock back in and continue working. Many people worked 16-hour days as if it were normal. Fishing, shrimping, and crabbing were seasonal, so canning companies maximized their labor force during this time to harvest as much seafood as possible.

After about six months of labor on the ship—the duration of their contract with the company—work was no longer as hectic as before, and Tiến, Châu, and Hiền had more time to sit together, drinking coffee and smoking, chatting about life.

In this isolated region near the Arctic, there was nothing to enjoy for all four seasons. For about six months, the sun never set, and for the rest of the year, the sun seemed to disappear on a southern vacation, leaving the North in an eerie, endless twilight.

Now, it was late winter. Outside, it was still snowing. The sea had frozen into a massive, solid expanse. The bitter cold felt like it could slice through a person's insides, making everyone dread having to go outside.

Taking a sip of hot coffee, Tiến leaned back in his chair, legs propped up on the armrest, and looked at Châu: "So, Châu? What do you think? We've been working non-stop for nearly six months. Should we head back or stay and work some more?"

Hiền, sitting next to him, quickly chimed in: "No, that's enough, brother! Let's go back to Seattle and find another job that's more fun. This place is boring as hell."

"He's right, boss! What Hiền says makes sense. This place is nothing but snow all day, bone-chilling cold, and as dull as a graveyard. When this contract ends, let's head back to Seattle and find something else to do."

"OK! I agree with you both. We've saved up a little capital now, so when we get back, we can pool our resources and start a business. Who knows, it might be better than this soul-crushing work."

Not long after, Tiến, Hiền, and Châu bid farewell to the Arctic and flew back to Seattle.

Today, spring had just begun. Châu felt elated as he watched the warm sunlight streaming through the airplane window on their return flight. Below was Mount Rainier, still covered in snow, standing tall amid the vast green forests, a breathtaking scene crafted by nature, starkly different from the endless white snow and dim, mixed light of the Arctic, which had cast a gloomy, lonely shadow over Châu during those long, cold months.

After collecting their luggage, Tiến stepped outside and immediately spotted Can waiting nearby.

"Tiến, my car is parked right up front," Can called out.

Tiến set his suitcase down to rest his hand and turned to Châu and Hiền, who were following behind, introducing them: "Châu, Hiền, this is Can, a friend of mine from our officer training days at Thủ Đức. Last week, I called him to help us find an apartment."

Can greeted Hiền and Châu warmly, shaking their hands: "Yeah, I've already rented an apartment with three rooms for you, Hiền, and Châu. It's in the south of Rainier Avenue, and the rent is pretty cheap—just one hundred and fifty dollars a month. I've already put down fifty bucks as a deposit."

"Great, we'll take care of the rest. No problem. But first, take us downtown for some pho, and we'll figure out the rest later."

"No need for pho. Come back to my place. My wife heard you guys were coming back and made a big pot of bún bò Huế. She's been waiting for you."

"Oh man, you're bothering her like that? We're bachelors; we can eat anywhere."

"Don't worry about it. We don't have gatherings like this every day. She insisted I invite Hiền and Châu too, to make it a proper gathering."

"Alright then, let's not disappoint her. We'll come over, and I can visit the kids too."

At nine o'clock on Saturday morning, the three men were still enjoying a lazy sleep-in. There was a knock at the door. Châu, whose room was closest to the front door, got up, peeked through the window, recognized the visitor, and went to open the door.

"Come on in, Can."

"Thanks, Châu. Is Tiến up yet?"

"Yeah, he should be. I'll go get him."

A minute later, Tiến emerged from his room, stretching and yawning lazily: "Aren't you supposed to be working today? What are you doing here so early?"

"My boss is sick, so I'm off this weekend. I thought I'd come by early and see if you guys want to grab some coffee and shoot the breeze."

"No need to go anywhere. I bought a bag of coffee yesterday that hasn't been ground yet. The store owner said it's really good. I'll grind it up, and we can try it."

"By the way, have you guys started looking for jobs yet?"

"We planned to, but we've been feeling lazy. We just took the driving test and got our licenses yesterday. We're planning to look for a used car this week. Waiting for the bus all the time is getting old."

"Hey, Tiến! My boss just bought two more old houses and needs some extra hands to fix them up, paint, and get them ready to sell. If you, Hiền, and Châu are interested, I can talk to him about hiring you guys. It'd be fun to work together!"

When Can offered this suggestion, Châu perked up: "Yeah, Can, ask him for us! Staying at home for too long gets boring. How long have you been working for this boss?"

"I've been working for Bill for about three months now."

Wanting to know more about the job, Châu continued: "Is Bill a contractor?"

"Well, calling him a contractor might be a bit much. He usually buys these small, run-down houses for about ten to fifteen thousand dollars in poor neighborhoods, fixes them up, and sells them at a profit."

Just then, Tiến returned from the kitchen, carrying four steaming cups of coffee. Hiền, who had just woken up, came out of his room. Seeing Can sitting there, he greeted him: "Morning, Can. The smell of Tiến's coffee was so good it wouldn't let me stay in bed any longer."

"Enough with the flattery. I've already made you a cup! And where's the smokes? Drinking coffee without a cigarette is like dying slowly."

"Right here. Cowboy smokes from Marlboro Country, just for you, my friend."

Châu, who was sitting nearby, lightly tapped his fingers on the table: "The smell of this coffee reminds me of the time we spent in Ban Mê Thuột with Tấn. Back then, Tiến was our platoon leader."

Before Châu could continue, Tiến interrupted: "Don't bring up the old days, Châu. You're talking about the coffee shop owned by Hằng's mother, Aunt Tư, right? I remember that whenever I wanted to find you two, I just had to drive to that shop, and there you'd be. Tấn was crazy about Hằng, and you had a girl named Năm back in your hometown, right?"

The memories of old times that Tiến accidentally brought up stirred up long-buried feelings in Châu. He fell silent, his gaze fixed on the coffee cup in front of him, watching the thin white steam rise and quickly dissipate.

Realizing he had gone too far and stirred up old emotions, Tiến quickly changed the subject to cover it up: "Let's put those memories to rest. Anyway, I heard Can mentioning a job earlier, but I was grinding coffee, so I didn't catch it all."

"I said my boss is looking to hire more people to help fix up some houses he just bought. If you guys are interested, I can talk to him about hiring you."

"Yeah, talk to him for us. We've been cooped up here for weeks, and it's getting old."

At six o'clock on the Monday morning of the following week, the three men were still deep in sleep when the phone rang loudly in the living room. The shrill ringing echoed in bursts. Annoyed, Châu muttered to himself, "Who the hell is calling this early? Can't they let people sleep?" He got out of bed, went to the living room, and picked up the phone, ready to give the caller a piece of his mind:

"Hello?"

"It's Can. Who's this?"

"It's Châu."

"Hey, Châu! My boss agreed to hire all three of you. Get ready and change your clothes. He'll be by to pick you up in about half an hour."

"We're starting work this morning?"

"Yes, he needs people right now. Tell the others to hurry and get dressed!"

A little over a month later, two dilapidated houses, which had been neglected and overgrown with weeds, had been meticulously repaired, painted, and landscaped. They looked so appealing that as soon as Bill put up a for-sale sign, the houses were sold immediately. But now, all four of them—Hiển, Châu, Tiến, and Can—were temporarily unemployed again, waiting for Bill to find new properties to buy so they could start working again.

On payday, as they discussed their last paychecks before being temporarily unemployed, Tiến suggested to the group: "I think this job is pretty straightforward. We could buy old houses, fix them up, and sell them for a profit. It's not rocket science. All you need is a bit of knowledge about paperwork, market prices, and some capital."

Can added: "Market prices and paperwork are things we can learn. The only problem is that we don't have the capital to buy a house because the banks or homeowners usually want cash before they'll transfer ownership."

Châu, wanting to understand more, asked: "It probably takes a lot of capital, right, Can?"

"I've been working with Bill for months now, so I've noticed that these houses are sold pretty cheaply—about fifteen or twenty thousand dollars at most. After fixing them up and paying for labor and materials, Bill still makes a profit of one or two thousand."

Hiền, who had been standing nearby smoking, chimed in: "Why don't we pool our money together and give it a try? Working for someone else like this, sometimes we have work, sometimes we don't, and we're not getting anywhere."

Sometimes being unemployed isn't such a bad thing; it can be an opportunity to move forward.

After dinner that evening, Châu brought up Hiền's suggestion to see what Tiến thought: "Tiến, this morning Hiền suggested we should pool our resources and start a business together. What do you think?"

Tiến didn't respond immediately. He rested his hand on his forehead, deep in thought. After a moment, he slowly replied: "You know, Châu, it's easy to talk about, but when it comes to putting it into practice, there are a lot of other factors to consider. In this country, everything requires paperwork and permits, unlike back home where you could just do whatever you wanted. I talked to Can about it, and he said that to do anything, we need licenses—like for plumbing, electrical work, carpentry, and so on. As for the capital, if we all chip in, we should be okay."

Hiền, who had been washing dishes in the back, heard Tiến's discussion and quickly added: "Those specialized licenses can be obtained by going to trade school. I know a few guys who are plumbers, and they say it's easy. You just need to go to vocational school and you're set."

Hiền enthusiastically continued: "So, we can each go to school, learn a trade, and then work together. We're still relatively young and single, so what do we have to lose? Even if we fail, we can just go back to Alaska and fish for crabs again. No big deal!"

Seeing that Hiền and Châu were serious and positive about the idea,

Tiến became more enthusiastic as well: "So, you two are ready to take off the gloves and get into the ring?"

Hiền responded quickly: "Exactly! Win or lose, we're going to give it a shot. We won't know if we'll win or lose until we try. I'm tired of being someone's lackey, taking orders all the time."

More than a year passed, and the four friends resolutely pursued their plan. During the day, they worked as construction laborers for Bill to maintain their daily living expenses and gain experience in the trade. In the evenings, after dinner, they attended vocational schools to learn specialized trades.

Today was graduation day for Hiền, Châu, and Can, who had completed their training in their respective trades: electrical work, plumbing, and carpentry in the home construction industry.

Everyone dressed more neatly and cleanly than usual. Can, in particular, was accompanied by his wife and children, who came to share in the joy of his big day.

Of the four friends, only Tiến would have to wait until the following summer to complete his studies because he was training to become an accountant, which required more time to finish the program.

For Châu, this was a significant milestone in his life. He now had the opportunity to return to school, not to pursue some grand ambition, but to find a "tool for survival" in a highly technical society. With his strong health and muscular build, he decided to pursue a career in plumbing and drainage, which suited his abilities. Hiền was drawn to industrial electrical and mechanical work, while Can, who had some prior experience in carpentry from his father's workshop in Biên Hòa before joining the army, naturally gravitated towards carpentry. Tiến was pushed by the other three to study accounting because they argued that if they were going to establish a company to bid on home repair contracts, they would need someone knowledgeable about legal matters, paperwork, and finance, so Tiến reluctantly took on this relatively challenging subject.

The "Tứ Hải" contracting company, specializing in home repairs, was officially established in the early summer of 1979. All the savings they had accumulated while working in Alaska were pooled together to form the company's initial capital.

The partnership agreements and shares were clearly outlined by Tiến, and all four friends agreed and signed the papers. The company's primary market was the Vietnamese community. At that time, many Vietnamese families who had immigrated to the U.S. had relatively stable jobs. They often adhered to the traditional belief that "a person must have a home in life and a grave in death," so they would buy houses.

After four years of leaving their homeland and starting anew, most of their monthly income was still relatively low, so they typically looked for small houses or those in need of some repairs that fit within their budget and allowed them to secure a bank loan.

The "Tứ Hải" contracting company was highly regarded by Vietnamese customers for its careful work and reasonable prices. In the first month alone, the company took on three house repair projects. Initially, the four friends divided the work among themselves, doing the labor to make a profit. After a few months, due to word of mouth from satisfied customers, many people started calling to request quotes. Sometimes, there were five or seven projects in a single month, and as the workload increased, the company decided to hire five to ten part-time workers.

In the first year, the company's financial statements showed that the total capital had grown to ninety thousand dollars, meaning that their profits and capital had doubled since they first started the business.

At the year-end meeting, Tiến presented his proposal: "As you all can see, in the past six months, our company has made nearly forty-five thousand dollars in profit, not including our monthly wages and the wages of the part-time workers we hired. This shows that the company is growing and has more work and issues to address. I propose that we hire a secretary to handle phone calls at the office and meet with customers while we're busy working outside. At the same time, we should hire three full-time employees to help with regular tasks, and keep part-time workers as needed for daily tasks."

Châu offered his opinion: "I agree with Tiến. To avoid losing customers, the company needs a secretary to handle phone calls. I suggest that we ask Can's wife to take on this role. I see she's already at home watching the kids, so she could manage the company's calls as well. Since the company is still small, we can operate in a home-grown style for now, and expand when the company grows."

Hiển also supported Châu's suggestion: "I think Châu's idea is very practical. If everyone agrees, Tiến can draft a clear agreement regarding salary, duties, and hours, and then Can can discuss it with his wife. It's better to be upfront and clear about everything now, rather than risk misunderstandings later."

After Hiền and Châu shared their thoughts, Tiến turned to Can: "What do you think, Can? Any objections?"

"For me and my family, this is a great opportunity. My wife can't work because of the kids, so our finances are a bit tight. Now that she has this job to do from home, I think she'll be very happy."

Tiến continued: "Okay, Can can handle the secretary issue. Now, what about hiring three more full-time workers to work with us? What do you and the others think?"

Based on what I've observed in the past six months, we've only taken on small repair jobs, painting, or adding small structures. We haven't moved on to phase two, which is buying houses, fixing them up, or buying land and building houses to sell. I fully agree that in the next phase, we should hire more people to help us with phase one, so we have more time to prepare and move into phase two."

Hiền, who had been listening carefully to Can's suggestion, said: "I agree with Can on some points. In the past, we've only focused on phase one, and even then, we've been overwhelmed with work. Moving on to phase two, as you suggest, might be too fast."

Châu, knowing that Hiền was cautious and feared failure if they didn't plan carefully, spoke up: "Hiền is right to be cautious, but if we stay stuck in one place, we won't get anywhere."

To strike a balance, Tiến proposed: "I think we should move on to phase two while still maintaining our current work in phase one. We now have enough capital to buy one or two small houses, fix them up like Bill did before, and sell them for a profit. At the same time, we'll continue working and overseeing the employees in phase one to maintain the company's good reputation. The main goal is to keep everyone working regularly. As phase one and two will support each other, when one job ends, we can immediately move on to the next."

In the end, everyone agreed with Tiến's proposal.

The "Tứ Hải" contracting company continued to thrive. Within four years, the company had around fifty full-time employees and about a hundred part-time workers. The company had two main offices in Oregon and Washington State. The total capital invested in real estate was about four million dollars.

Tiến, Châu, Can, and Hiền no longer had to do the physical labor themselves. They organized into a board of directors, managing and overseeing the business operations through paperwork, while all the specialized work was handled by managers and contractors. Occasionally, they would go out to inspect the work to ensure it met the company's standards or meet with clients to negotiate new contracts. The company's market expanded beyond the Vietnamese community to include local customers as well.

Today, being a Saturday afternoon, the "Tứ Hải" contracting company office closed earlier than usual because Hiền and Châu had an appointment to visit a friend who had recently arrived in the U.S. as a refugee.

Hiền got into the car first, started the engine, and waited. Châu was still closing the windows. Just then, the phone in the office rang loudly. Châu, a little annoyed, muttered to himself, "It's Saturday! Can't they let people rest for a bit? Just be quiet!" The phone kept ringing persistently. Running out of patience, Châu went to answer it:

"Hello, Tứ Hải Contracting Company, how can I help you?"

A man's voice responded on the other end: "We're calling from the International Red Cross. May I speak to Mr. Lê Văn Hiền?"

Hearing that it was the Red Cross, Châu realized it must be important and quickly put down the phone. He rushed to the door and called out to Hiền: "Hey, Hiền! The Red Cross is calling for you. Hurry up!"

Looking puzzled and worried, Hiền quickly jumped out of the car and ran into the office, grabbing the phone: "This is Lê Văn Hiền."

"Mr. Hiền, we're with the International Red Cross, calling to inform you that Mrs. Trấn Ánh Tuyết and Miss Lê Thị Thu Vân, your relatives, have successfully escaped by boat. They are currently staying at the Bulau-Tanga camp in Malaysia. They have asked us to find you and pass along this information. Here is the address for you to contact them... If you need any assistance, you can write to us through our organization. Thank you."

As soon as he hung up the phone, Hiền's face lit up with realization. He quickly ran to the front door and called out: "Châu! My mother and Vân have escaped to the camp in Malaysia!"

"Wow! Your mom and Vân made it out! Now you need to quickly get the paperwork done to bring them here!"

"It's Saturday, so nothing can be done today. I'll call the Red Cross on Monday morning and ask them how to proceed with the sponsorship paperwork."

###

More than six months later, with the help of a humanitarian organization, Hiền's mother and sister received official papers to leave the camp and reunite with their family in the U.S.

Today, the main office of the "Tứ Hải" contracting company was closed so that the entire board of directors could go to SeaTac International Airport to welcome Hiền's mother and sister from Malaysia, arriving on a regular flight with Singapore Airlines.

Hiền and Châu were both excited and anxious. They paced back and forth, occasionally checking their watches and looking out the window. Their nervous behavior made Tiến laugh: "You two are pacing around like hens laying eggs. They'll arrive soon enough!"

Can joined in the teasing: "Châu might be trying to score points with Hiền here. Remember to address him properly as 'brother,' not 'you' or 'me.' Otherwise, Vân might take offense and give you a piece of her mind."

A female airport attendant removed the rope barrier at the gate, allowing passengers to enter the customs and immigration area for baggage inspection.

Hiền carefully watched each person as they lined up. Those arriving from refugee camps were easy to identify, dressed relatively cleanly and neatly. At the front of the line was a woman holding a toddler, while leading a four or five-year-old girl by the hand. Her husband followed behind, burdened with luggage.

A little while later, Châu tapped Hiền on the shoulder: "Hey, Hiền! Look at the second row from the back; I think that's your mom and Vân, isn't it?"

"Yeah! That's them, my mom and Vân."

Tiến and Can, now excited too, began asking: "Where? Where...?"

"Point them out!"

Hiền pointed through the glass: "At the back of the second row. My mom is the one with her hair tied up and a scarf around her neck. Next to her is Vân, with long hair, carrying a nylon bag."

Tiến squinted to see where Hiền was pointing: "Ah, I see them now. Your mom looks like she's still in good health. And your sister Vân—how old is she? Her face looks so young!"

"She's the youngest in the family. She must be around twenty-three years old now."

Can couldn't resist teasing again: "No wonder Châu and Hiền are so close, like a brother-in-law and sister-in-law!"

Can's comment made everyone burst out laughing....

The house that Hiền and Châu had jointly purchased in the Redmond area was spacious and situated on a large piece of land next to a small, picturesque stream. In front of the house stood a "Cedar" tree that provided shade in the summer. Vân and her mother were picked up by Châu and Hiền and temporarily stayed there.

This morning, being the weekend, the two men slept in longer than usual. Hiền's mother, who was used to waking up early, went downstairs and busied herself cleaning up, putting things away, and preparing breakfast for Hiền and Châu before they went to work.

The English-style wall clock in the living room chimed eight times. The soft sound created a pleasant little melody.

The morning sun's rays had begun to filter through the window blinds.

Vân had been awake for more than half an hour. If there was a lot to do, she usually went downstairs to help her mother, but today there were only a few dishes left from Châu and Hiền's late-night snack, so she wasn't in a hurry.

Vân reached out and grabbed a body pillow, hugging it tightly. The warmth and comfort gave her a strange, pleasant sensation. It was more accurately described as the fluttering and anticipation of a flower in full bloom, waiting for the attention of butterflies and bees.

She remembered when she was sixteen, living in Chợ Lớn. Every morning, Vân would wear her white áo dài to school. Hiền's mother often marveled silently, noticing how her daughter was blossoming into a young woman. It seemed to her that Vân grew more beautiful with each passing day.

After her husband passed away, the aunt took on the responsibilities of both mother and father to ensure that Hiền and Vân were raised properly. Due to some bad luck with his education, Hiền had to join the military a bit earlier than usual, which made the aunt somewhat displeased. However, she complied with her son's wishes and did not stop him.

When Hiền left, the house became lonely, with only Vân staying by her mother's side. As a young girl just entering adolescence, Vân was growing rapidly. The white áo dài that her mother had tailored for her over six months ago now fit tightly around her waist, and her now well-developed chest had grown too large for the dress, necessitating a new one. "Girls grow like a weed when they hit puberty," her mother would often say to Vân when she asked for new clothes.

Sometimes, boys from the neighborhood would pass by and whistle as they looked into the house, hoping to catch a glimpse of Vân. But when they saw the aunt inside, they would quickly back away, embarrassed. Since Hiền rarely had the chance to visit home, the aunt didn't want Vân to marry too early. Yet, she was aware that the allure of a beautiful young woman was like sweet honey—sooner or later, bees and butterflies would come to her. Knowing this, the aunt sometimes felt a bit sad, fearing that one day Vân would get married and leave her.

...Then the South fell, and the aunt took Vân to her maternal grandparents' home to take refuge. Hiền, her eldest son, had flown away with his comrades, and there had been no news of him since, leaving them uncertain if he was dead or alive. The old house in Chợ Lớn, where the family had lived since her husband was young, was nearly emptied by thieves and looters when they returned after the war. With her son missing, the house ransacked, and all their savings gone, the mother and daughter could only sit and cry.

The aunt remembered that before, when the liberation forces entered Mỹ Tho, the local officials had told her, "The South has been completely liberated. Now the people are in charge, and in Sài Gòn and Chợ Lớn, people are celebrating with grand feasts to welcome the victory of the people and the Party."

Indeed, as the local officials said, now, standing before this scene, were she and her daughter happy? Because the Party had helped her become impoverished, and her son had disappeared, leaving her not just poor but also without a family. Wasn't that wonderful? So why should she be sad? She was crying tears of joy, wasn't she? Who would say she was crying in sorrow?

Knowing that there was no way to return to the house filled with memories from her youth, which she and her husband had bought when their first son was just a few months old, Vân advised her mother to sell off all the land and what remained of their property and move in with her grandparents. Seeing the sense in Vân's suggestion, the aunt sold the house to a neighboring family for about five gold pieces. The two returned to live with the grandparents in Mỹ Tho. About a year later, the grandmother, weakened by age and lacking medical treatment, passed away.

Since April 30, 1975, the day the communists called the liberation of the South, Vân no longer had the opportunity to return to school. The savings and assets accumulated over the years from the hard work of both parents had been nearly entirely confiscated through mandatory declarations and currency exchanges by the communist government. Vân had to stay home to help her mother, selling small goods in the neighborhood to make ends meet.

Nearly two years later, Vân and her mother received a letter from Hiền, sent from France, through an old friend of her father's. Hiền informed his mother and sister that everything was well. He and Châu were currently living in the United States, and he urged his mother to find a way to escape the country, promising to find someone to sponsor them to America.

After reading Hiền's letter, the aunt sighed in relief, knowing that her son was alive. However, she thought that escaping the country would be too difficult. Unlike her mother, Vân was very enthusiastic after hearing her brother's suggestion and continually urged her mother to find a way to leave.

More than two years later, an opportunity arose when a relative from her mother's side was arranging a boat for an escape in Gò Công. They only needed two more gold pieces to secure seats on the boat. Thinking of her children's future, the aunt took a gamble, gathered all their remaining money, and handed over the two gold pieces to the boat's owner to secure two seats on the hopeful voyage.

Fear affects people partly because they still have hope for the future. But when hope is lost, fear becomes meaningless.

This was the case for the aunt and Vân, as they had completely lost hope. Once a major rice-producing region in Southeast Asia, South Vietnam now faced food shortages. Even basic sustenance was no longer guaranteed, making the future seem like a dark, uncertain void.

###

Nearly a year later, mother and daughter lived on an island, supported by the United Nations and the International Red Cross, who had managed to locate Hiền.

The aunt's nightly prayers were finally answered, and today the family was reunited on this distant land of freedom.

There were footsteps outside the door. Vân knew her mother was about to come in and wake her up. The door, left slightly ajar, opened, and Vân heard her mother's voice urging her:

"Wake up now, dear! What kind of girl sleeps in so late? People will laugh at you!" Vân playfully protested:

"Mom, what are you saying? Who's laughing at me?" Her mother leaned closer and whispered in Vân's ear:

"Châu's laughing at you, of course!"

"Again! You always say that. If he hears, he'll laugh at me for real. Oh, aren't you going with brother to see the Chinese herbalist in town later?"

Yes, the weather here is different, a bit colder, so I often feel achy and have joint pains. Hiền said he'd take me to the herbalist today to get checked and pick up some medicine. And Châu will take you shopping to get some more groceries for the week.

Seattle was in late August. The summer sun was still shining, but the heat of the past few weeks seemed to be retreating south, giving way to the usual chill that flowed down from the far north.

Châu reached into the car to grab a jacket and handed it to Vân:

"It's a bit chilly today. Wear this jacket to stay warm."

"If I wear this jacket, what will you wear, Châu?"

I've been here for a few years, so I'm used to it. I'm just worried that you, having just arrived, might not handle the cold as well. Let's cross the street and grab some breakfast first, then we can shop afterward.

"Whatever you say, I'm not very hungry."

As the pedestrian light turned green, Châu extended his hand for Vân to hold as they quickly crossed the street. It was a natural gesture, but Vân, feeling the security of a man's protective hand for the first time, was a little embarrassed and quickly let go as soon as they reached the other side.

Châu noticed Vân's bashfulness and teased her:

"Vân's hand is soft and warm, while mine is rough and calloused, isn't it?" Vân blushed, covering her mouth as she smiled:

"That's not true at all! Don't accuse me of such things. I was just embarrassed, that's all!"

Châu led Vân into a nearby pho restaurant. He chose a table by the window so they could look outside:

"Let's see if the pho here tastes better than in Mỹ Tho, Vân."

Before '75, we would sometimes go out for pho, but in the years after, just having enough rice to eat was a blessing!

In the years before I came here, Hiền and I felt the same way, craving Vietnamese food terribly. We'd buy a chicken, chop it up, stir-fry it with soy sauce, and eat it with rice. It was delicious. Now that your mom and you are here, with all the dishes you cook, I want us to always live together. Vân smiled and gave Châu a sideways glance:

"Don't you plan on getting married and starting your own family?"

"I do! But I'm afraid people will think I'm old and unattractive." Vân looked down at the table, speaking softly enough for only Châu to hear:

"Who told you that? I think you're a good man!"

Thank you, Vân. This is the first time I've been complimented by a woman. Surprised by her own reaction, Vân looked Châu straight in the eye:

"You've been in the U.S. for nearly seven years, and you've never dated anyone?"

I have, but it was mostly for business, and after work, everyone went their own way. I never had the time to sit down and chat like we're doing now.

The construction company "Tứ Hải" was very busy during the summer, but when autumn ended and winter approached, business slowed down. The company's board members took the opportunity to rest or travel.

The rain outside had been steadily falling since yesterday. Heavy gray clouds blanketed the sky, with occasional gusts of wind blowing newly yellowed leaves from the "Alder" tree in front of the house, sending them fluttering like butterflies.

Hiền was organizing some paperwork and putting it into his briefcase when Châu, who was sitting nearby sipping coffee, complained:

This rain just keeps going on day after day, nearly drowning the earth. There's nowhere to go! Hearing Châu's grumbling, Hiền burst out laughing:

"You know, because of this weather, people are hesitant to come up here, so we don't have much competition in our business. If it were sunny and warm year-round like in California, it'd be much harder to compete. By the way, according to the schedule, you're supposed to go down to Portland today to check on the office there, right?"

"Yes! There are a few things Tiến asked me to look into to see if they're in order."

"Hey Châu! If you're going, why not take Vân along to show her Portland? She hasn't been anywhere since arriving here."

Châu didn't respond immediately. He looked around to see if Vân or the aunt were nearby before replying quietly to Hiền:

"Yes, Brother, I'd be honored." Hiền smiled, clearly pleased:

"Good! That sounds great. Do your best!"

As Châu's Toyota crossed the boundary of Olympia's city limits, one of the tires blew out. Châu had to pull over to change it.

The weather outside was still cold, with a light drizzle, and Châu grumbled:

"Another nail! On a rainy day, of all times—what a pain." Vân, sitting in the passenger seat, noticed Châu's frustration with the flat tire:

"Let me help you change the tire." Châu had just stepped out of the car when he heard Vân, and he turned back:

"It's okay, Vân! It's wet out here; I can change it by myself."

"I want to help, just for fun."

"Well, if you want, you could hold the umbrella for me to keep the rain off."

Vân reached for the black umbrella in the back seat and held it up to shield Châu as he bent down to work on the tire. When he paused to take a break, he turned and noticed that Vân was only covering him, while she was getting soaked. Châu dropped his tools and stood up, scolding her gently:

"My goodness, Vân, why aren't you covering yourself too? Your head and neck are getting all wet in the rain; you'll catch a cold!"

"I'm wearing a warm jacket, so it's not that bad. The umbrella is small, so it's just enough to cover you."

"If the umbrella is small, then you should sit closer to me so it can cover both of us."

"Okay, I'll do that. You keep working; I'll cover us."

Châu smiled at Vân's sincerity. He sat back down and continued changing the tire. About half an hour later, with the tire replaced, they both got back into the car to continue their journey. Châu turned to Vân with a teasing smile:

"Thanks, Vân. Thanks to the umbrella, I didn't get wet, and I stayed warm too."

Understanding the subtle message in Châu's words, Vân blushed and looked down, smiling:

"The umbrella was small, so I had to sit closer to you to keep you dry..." Around two o'clock in the afternoon, Châu and Vân arrived in Portland, Oregon. The weather there was much the same as in Seattle—drizzly rain and gray skies, giving the city a gloomy feel.

Châu parked the car in front of the office of "Tứ Hải." Vân glanced over as Châu prepared to get out:

"Should I wait in the car or come in with you?"

"Come in with me, Vân, then we'll go get some food together." The manager, who had been sitting inside, saw Châu open the car door and hurried out to greet him:

"Hello, Châu!"

"Hello, ma'am. I got a call this morning saying you'd be arriving, so I've been waiting for you." Châu turned to Vân to introduce her:

"This is Uncle Tư, our representative in Portland." Then he turned back to the manager:

This is Vân, Hiền's younger sister.

"If Châu hadn't introduced you, I would have thought you were married without inviting us!"

"When that happens, I won't forget you and your wife down here." They both laughed and walked into the office. After about an hour of exchanging ideas and gathering information from Uncle Tư to take back to Seattle, Châu shook hands with the manager and led Vân back to the car:

"Let's go downtown for lunch, Vân. You must be hungry, right?"

"Just a little, but you're the one who's probably starving." Châu grinned mischievously as he replied:

"If you hadn't come with me, I'd probably have fainted by now!" Knowing Châu was joking, Vân just gave him a playful side-eye without saying anything.

After a meal of sour soup and caramelized fish at a newly opened Vietnamese restaurant, Châu felt more relaxed. The rain was still falling outside as the restaurant owner brought over a pot of hot tea:

"Please enjoy some tea to warm up. It's been raining a lot today, so the restaurant is a bit quiet. You two can stay a bit longer and wait for the rain to stop before heading out..." Châu glanced at Vân, who was sitting across from him, and smiled at the way the restaurant owner addressed them. He whispered to Vân:

"She thinks we're really married, you know. Did you hear that?"

A blush spread across Vân's cheeks as she looked down at the table, playfully complaining:

"I heard her, but you didn't have to mention it again—you're making me embarrassed!" Châu pretended to explain:

Earlier at the office, Uncle Tư also thought we were a couple, and now the restaurant owner too. I didn't say anything to suggest that! Sensing a change in Châu's tone, Vân looked up at him:

"Are you upset with me, Châu? I was just saying I was embarrassed..."

"Not at all! I thought you were upset with me." Vân reached over and playfully pinched Châu's arm:

"You're always teasing me and making me feel awkward."

Châu stood up, grabbed a warm jacket, and handed it to Vân:

"Hiền told me to take you down here to see Portland, but with all this rain, it's not as fun as it could be. Now, put on this jacket, and let's head over there to get the tire fixed. Then we can take a walk around the city, okay?"

"I'll go wherever you take me, Châu." The City of Roses, in late autumn, was draped in a light drizzle. The last leaves had long since fallen, leaving the trees bare. Gentle gusts of wind carried the northern chill, encouraging couples to walk closer together.

Châu led Vân along the main street, past grand shops with large glass windows displaying luxurious and expensive items. The two walked closely together under one umbrella. Vân remained quiet as Châu pointed out the names of each shop they passed. At one corner, Vân tugged on Châu's arm, stopping him.

"Let's turn down this small street, Châu. Walking on the main street makes me feel out of place! Maybe I'm just not used to the grandeur of this country yet."

Châu reached over and lightly patted Vân's slim waist:

"Why didn't you say so earlier? I thought you liked these fancy places."

"I do, but maybe it just doesn't suit me right now. I want to walk on quieter streets to see if they remind me of Vietnam." Châu gently smoothed Vân's rain-dampened hair:

"You're right, Vân. Over the past seven or eight years, I've been so caught up in work that I forgot what it was like."

"I heard Hiền say your hometown is in Rạch Đào village, right?"

"Yes, my home is in Cầu Nổi hamlet, Rạch Đào village. Since joining the military, I've only been back once, after I graduated from Quang Trung training school nearly nineteen years ago. Now, I have no idea what's happened to my parents and siblings."

"You know, after you and Hiền left with the Air Force from the Mỹ Tho lakeside, my mother, grandmother, and I were so worried. We just waited and prayed for your safety. It wasn't until we got a letter from Hiền a few years later, saying you'd both arrived safely in the U.S., that we finally stopped worrying."

"I was just as sad when I saw the plane leave Vietnam. I thought we might never see each other again."

"The world is big, but it's round, so if you keep going, you're bound to meet again, right?"

"You're right, Vân. If it's meant to be, we'll surely meet again somewhere."

As they reached this point in their conversation, Vân suddenly stopped. She looked at Châu and smiled as if she wanted to say something important but then hesitated. She looked down at the ground and continued walking, her sudden pause causing Châu to feel curious.

"It seems like you want to tell me something, Vân. Is that right?" Vân remained silent for a long moment, carefully considering her response:

"Châu, when you walk with me, do you feel like I'm too country-like?" Châu stopped and turned to look directly into Vân's eyes:

"What? That's nonsense! Walking with you, I feel proud! You're young and beautiful, while I'm older. I was worried you might feel there was too much of an age gap and not like it."

"I like it very much! You're kind and wise, but there's one thing I don't like." Châu chuckled:

"Is it because I'm a bit old?" Vân gave Châu a playful look:

"No, not at all. You're good at putting words in my mouth. I never thought that."

"Then what is it that you don't like? Tell me, so I can fix it." Vân smiled and glanced sideways at Châu:

"You know what? I don't like it when you call me "cô" and use the word "tôi" with me anymore! That's all." Vân's words delighted and excited Châu. He gently placed his finger on Vân's delicate nose and repeated:

"Alright, I'll do as you say. I won't call you "cô" or use "tôi" anymore. How's that?" Vân nodded, her face down, answering softly:

"I prefer it when you speak to me like that..."

The aunt was cleaning and rearranging the flower vase on the altar. Hiền had just come in from outside and informed his mother:

"Châu and I have finished checking and setting up everything at the restaurant for tomorrow afternoon's event."

"You're back, but where's Châu?"

"He stopped by Tiến's house to remind him about the arrangements for the wedding tomorrow."

"Alright! When he gets back, remind him to take Vân to buy a few more necessary things so we don't forget anything with all the busy tasks tomorrow."

"Yes, I'll remind him for you, Mom."

The person who was the most excited and anxious about Vân and Châu's wedding was, of course, Vân's mother. Ever since the uncle passed away, leaving Hiền at ten years old and Vân around six, she had single-handedly cared for and raised her two young children.

After many ups and downs, Vân had grown into a mature woman. Tomorrow, she would officially marry and start a new family. And soon, the aunt would likely have grandchildren to hold and cherish, something that brought her much joy. But even so, like all mothers on the day their daughters marry, she couldn't help but feel a little sadness as she realized that Vân would no longer be under her care as she once was. It felt like losing her daughter forever.

The sound of a car stopping outside the house drew the aunt's attention. She pulled back the curtain to look out. Châu and Tiến were just stepping out of the car and heading toward the front door. Vân, who was cooking in the kitchen, heard the car and knew that Châu had returned. She hurried out to open the door, intending to ask him to take her to the market. But when she saw Tiến entering with Châu, she greeted him:

"Hello, Tiến!"

"Hello, Vân! Are your mother and brother home?"

"Yes, my mother and brother are in the back." Vân stepped aside to let Tiến enter and took Châu's hand:

"Later, can you take me to the market to buy a few more things?"

"Sure! I'll take you. Hiền was in the back, making coffee, when he heard Tiến enter." He called out:

"Tiến, come back here and have some coffee. I'm in the middle of making it." The aunt stepped out of the living room, and when Tiến saw her, he quickly bowed:

"Hello, Auntie!"

"Ah, Tiến, you're here. I've been arranging the altar for tomorrow's ceremony since this morning."

"Yes, I wanted to ask you about a few things for tomorrow so we don't have any problems."

"That's right. Planning ahead is good, so we don't have the drums beating in one direction and the trumpets beating in another, making people laugh at us." Vân and Châu, standing nearby, joined in:

"Tiến is really good at this, Mom! When I was in the army in Ban Mê Thuột, he officiated at many weddings!" After hearing Châu's praise, Hiền laughed heartily, teasing:

"Tiến is great at marrying people off, but he's still single himself. Maybe he hasn't found anyone to officiate his wedding yet!" Hiền's joke made everyone burst out laughing, and Tiến quickly responded:

"Hiền, you talk about me, but don't forget to consider your own lonely situation!" Vân, hearing Hiền and Tiến teasing each other, stepped in to mediate:

"I know a few single friends who are beautiful, kind, and sweet. I'll introduce them to you both when I have the chance!" Châu, hearing Vân's suggestion, immediately supported the idea:

"My future wife has a point! You two should keep up your vegetarian diet for a little while longer!" Châu hadn't even finished speaking when Vân gave him a sharp side-eye.

"Hiền and Tiến haven't said anything, but you're the one jumping the gun." Caught off guard by Vân's teasing, Châu weakly defended himself:

"I didn't say anything!" Tiến laughed loudly, stepped forward, and patted Châu on the shoulder, teasing him even more:

"You're about to get married, so from now on, you'd better watch your mouth and think before you speak. Don't blurt things out like you used to!"

After nearly ten years of operation, the construction company "Tứ Hải" had slowed down from its previous vigorous pace. This was largely due to the fatigue of the board members, as the four main pillars—Tiến, Can, Hiền, and Châu—had all started families and were busy with children, reducing their business interactions with clients.

Moreover, several new construction companies had emerged, aggressively competing for the construction market. Because of this, about a month ago, Tiến suggested to the board that they sell "Tứ Hải" to someone else so they could each pursue their own family businesses.

Today was the final meeting of the board to make a definitive decision about the company's future. After presenting a detailed report on the company's finances and its real estate holdings, Tiến also provided a list of two major construction companies in Bellevue interested in purchasing "Tứ Hải." The final price would be decided by the board during negotiations with the buyers.

A minute of silence enveloped the meeting room. There were a few sighs of regret. It seemed that no one wanted to say anything. Tiến understood this better than anyone.

Over twelve years ago, everyone had been struggling to make a living in a foreign land. Then, by chance, they met again, pooled their resources and labor, and built the company into what it is today. Over the past ten years, the company has provided them with livelihoods and prosperity. Now, they were forced to give it up due to circumstances beyond their control. From an emotional standpoint, who wouldn't feel a sense of loss? Even though the company was just an inanimate object.

To break the heavy atmosphere, Tiến spoke first:

"I fully understand the feelings you all have for "Tứ Hải." However, to stay on schedule for today's meeting, and as per my financial summary and ownership report, before negotiating with the buyer, I suggest we set a price of three million U.S. dollars, including our two main projects in Seattle and Portland. It's up to you all to decide whether to increase or decrease this proposed price." Can raised his hand and spoke:

"As you reported earlier, after settling the company's debts, there would still be nine hundred and fifty thousand U.S. dollars left in the account, correct?"

"That's correct. After we pay off our debts to the suppliers, the account will still have nine hundred and fifty thousand U.S. dollars." Can continued:

"So, if we sell the company smoothly, would the total cash assets amount to nearly four million U.S. dollars?"

"That's the price I'm proposing for negotiation, and I hope we'll achieve that." Châu raised his hand to ask a question:

"As you mentioned, we have two main projects in Seattle and Portland. The Seattle project is nearly complete, but the Portland one isn't finished yet. So, if we sell the company, the new owner will continue the unfinished work, right?"

"Yes, exactly as you asked. I've included all those details in the agreement documents for the new company. They'll have to sign off on completing the work within the specified time frame." After answering Châu's question, Tiến turned to Hiền:

"Hiền, do you have any other concerns besides what we've discussed?"

"It seems you've covered everything, but have you also considered taxes?"

"I'll take care of the taxes and other necessary procedures after we officially sell the company. Now, does anyone else have a new price suggestion or any other opinions?" Châu raised his hand to suggest:

"I propose we take a ten-minute break, then make our decision. If two-thirds of the votes agree, we'll go with it."

"Hiền and Can raise their hands in agreement with Châu's suggestion. After the break, the board held a secret ballot to decide on the final offer. Everyone unanimously agreed to accept the price Tiến had proposed.

A week later, they received the good news that "Tứ Hải" had been sold for the proposed price.

Tiến quickly called an emergency meeting that afternoon to report the sale and announced that he had signed all the necessary documents and received a check for one hundred thousand dollars as a deposit while they waited for the title transfer. The paperwork was expected to be completed in about four weeks.

With everything almost done, they only had to wait to distribute the shares from the company's joint account and sign the final documents to wrap things up. After that, the members would be free to pursue their own individual and family ventures.

###

Outside, tiny white snowflakes drifted down like delicate drizzle on a morning that felt more like winter than late autumn. The pine trees around the house had a few drooping branches, their deep green now sprinkled with freshly fallen white snow, creating a beautiful scene.

Châu sat quietly at the dining table by the window, watching two small birds, no bigger than the tip of his thumb, hopping around under the pine branches, searching for food. Vân was nearby, boiling water for coffee, a bit surprised by her husband's silence:

"What are you looking at so intently out there?"

"I'm watching two little birds, like the ones back home in Vietnam. They're hopping around looking for food—so cute. Oh, is Thu still asleep upstairs?"

"Yes, she stayed up late last night, so she's still sleeping."

Since "Tứ Hải" was sold, Châu's daily work routine was no longer as hectic as before. With the substantial capital, the company distributed to its members—around one million dollars each—Châu and Hiền pooled some of their money to open a store supplying interior decoration materials. They hired a manager and staff to run the store, so they only had to check in occasionally to review the store's finances, leaving them plenty of time for family and personal matters.

Now that their business was running smoothly and their family life was secure, Châu was considering talking to his wife about taking their daughter, Thu, back to visit her grandparents in Vietnam.

As Vân set a cup of coffee with milk in front of her husband, Châu pulled her into the chair next to him.

"Sit down for a bit, Vân. I have something to discuss with you." Vân smiled mischievously and wrapped her arms around her husband's neck:

"Didn't we discuss everything last night? What else is there to talk about, honey?"

"What we discussed last night was about giving Thu a sibling to play with, but this is something different."

"Alright, I'm sitting down. Go ahead."

"I want to talk to you about taking Thu back to Vietnam for a few weeks to learn about her roots. It's been over twenty years since I've heard anything about my parents and siblings back home. I don't even know if they're still alive, so I want to take you and Thu back to my hometown to see where I come from…"

Before Châu could finish, Vân, clearly excited, chimed in:

"Last week, I ran into an old friend at the market who had just returned from Vietnam. She said things over there are much better now than they were a few years ago. So, why don't you check out the flights and see what's available? If we go, let's take Mom with us. She'd love it."

"If we decide to go, we'll need to fill out passport applications at the post office. It might take about a month to get them, and then we can buy the plane tickets."

"That would be great since winter is almost here. We could go while it's warmer over there, stay for about a month, and then come back. Oh, I heard Tiến say that his hometown, Rạch Đào, is near your village. Maybe we could invite him to go with us."

"I asked him a few days ago, but he's still hesitant because he was an officer in the South Vietnamese Army and is worried that the local officials might give him trouble. He said if we go first and everything's fine, he'll visit later."

The Thai Airways Boeing 747 had just stopped at the international terminal of Tân Sơn Nhất Airport. The passengers on board were bustling, gathering their carry-on luggage in preparation to disembark and board the bus to the terminal.

Châu told his wife to hold their daughter, Thu, and stay seated until everyone else had disembarked. This was his first opportunity to return to his homeland since the national tragedy of 1975.

Châu came from a poor farming family that had been renting land for generations. Growing up, he was raised with the traditional rural values of Southern Vietnam, where the expectation was to follow in the footsteps of his ancestors, learn just enough to be literate, return home to farm, and when the time came, his parents would find him a rural girl who knew how to handle household tasks and farming. Then, they would marry, have children, and continue the cycle for generations.

Farming was a one-season affair each year, with three simple meals a day. If they were lucky enough to have a good harvest, they might build a three-room house with a red tile roof, buy an "An Sông" bicycle, and possibly even a three-band radio. If they were even more fortunate, they might afford a record player to listen to traditional Vietnamese opera in the evenings, becoming the pride of the neighborhood.

But by Châu's time, this cycle was broken due to the military draft under President Ngô Đình Diệm.

Whether this was fate or a turning point is debatable, but it was certainly a pivotal moment that redirected Châu's future. This turn of events coincided with the old saying, "No family stays rich for three generations, nor does any family stay poor for three generations."

His ancestors, going back to his great-grandfather, were tenant farmers, so they could hardly be called wealthy. If anything, they were among the poorest of the poor. By the fourth generation, represented by Châu, things had completely changed. He was no longer a farmer with mud on his hands and feet, working from dawn until dusk, eating hurried and simple meals of steamed fish and boiled vegetables.

Today, Châu's social standing has placed him among the wealthy, and he is one of the new millionaires in the United States, thanks to his business ventures and investments in major global corporations. Conversations with friends no longer revolved around farming and the weather but instead focused on the ups and downs of the stock markets in New York or Tokyo.

Châu was a symbol of the new generation. Although he didn't have a high level of formal education, his quick wit and intelligence allowed him to learn from real-life experiences and apply them accurately in his business dealings.

There was a time when Châu and Tiến were fishing for crabs in Alaska during the harsh winter months. Châu saw his former commanding officer enduring great hardship. Out of sympathy, Châu often volunteered to help with the heavy work. Tiến, noticing this, once patted Châu on the shoulder and confided, "I appreciate your help, and I understand what you're thinking about those old military memories. But let me tell you something: "no matter how glorious the past, it's only something to be quietly respected. Nowadays, you can't sell those memories for even a penny. Only the present and future matter in assessing a person's worth." From that day on, Châu felt more encouraged and confident in himself.

Now, as most of the passengers had disembarked, Vân nudged her husband:

"It's our turn to go now, dear!"

Châu snapped out of his brief daydream:

"Ah, yes... You take our daughter down first, and I'll grab the carry-ons." After navigating the bureaucratic paperwork and customs inspections at Tân Sơn Nhất Airport, they finally made it through smoothly, thanks to the "blessings" of President "Washington" and the motto "In God We Trust."

The crowd waiting to meet family members outside the airport was packed, filling almost the entire walkway. It took considerable effort for Châu to carry two suitcases outside. Vân, holding Thu, led the way and soon spotted Uncle Út and his wife waiting in the crowd.

Since Vân's grandmother passed away and Vân and her mother escaped by boat, about ten years had passed. Vân's mother occasionally sent letters and gifts to support the family, while Vân's younger brother, Út, stayed behind to maintain the ancestral home and take care of the family's graves.

Last month, Vân's mother wrote that Vân, her husband, and their daughter Thu would be visiting. Although Uncle and Aunt Út lived in the ancestral house with over two acres of family land, under the communist regime, the expensive and hard-to-get fertilizers made it difficult to grow crops.

If they managed to harvest anything, they had to sell it to the government at a paltry price, so the family barely scraped by. When they received the letter from Vân's mother in the United States, asking Uncle Út to go to Saigon to pick up Vân and her family, they were overjoyed but didn't know how to do it, as Uncle Út had no money left to rent a car to Saigon. While Uncle Út was scratching his head trying to figure out how to raise the money, Aunt Út took off her gold earrings and handed them to him:

"Dear, pawn these to get money to rent a car to Saigon and buy some fish and meat for them to eat while they're here." Seeing the gold earrings his mother had given to his wife on their wedding day, now in her hand with her generous and kind gesture, deeply moved Uncle Út.

"Oh, no, if you pawn them, what will you wear?"

"Just pawn them for now, and we can redeem them later when we have the money."

Uncle Út silently accepted, knowing full well that once pawned, it would be almost impossible to redeem these keepsakes. But given their dire situation, he had no other option but to follow his wife's suggestion.

With a little bit of money, he had a purple floral áo dài (traditional Vietnamese dress) and new black pants made for Aunt Út, something she had always dreamed of. For himself, he bought a white long-sleeved shirt to wear when they went to Saigon. The remaining money was used to rent a car and buy groceries.

According to the letter and instructions from his sister, the Thai Airways flight was scheduled to land at 11:30 AM, but Uncle Út urged the driver of the state-owned tourist car to leave Mỹ Tho at 7:00 AM. By around 9:00 AM, they were already in Saigon, waiting nervously among the crowd outside the main gate.

When Vân and Thu finally emerged from the crowd, Aunt Út, who was standing not far away, nudged her husband:

"Look, I think that's Vân with her child up front, and her husband is carrying the suitcases behind them." As they were discussing and identifying, Vân recognized her relatives waiting to greet them. She was overjoyed and led her daughter straight toward them, calling out:

"Uncle Út, Aunt Út, have you been waiting long for us?" Before Uncle Út could reply, Vân bent down and told Thu to bow and greet her great uncle and aunt, then turned to introduce Châu, who was standing behind her with the suitcases:

"Uncle, Aunt, this is my husband, Châu!"

"Oh! We saw you from over there, but we weren't sure because you've changed so much since you were little." Aunt Út, who was reaching out to carry Thu, added:

"You're so much fairer and prettier than before, so Uncle Út hesitated to recognize you. Let's go home now!"

It had been over fifteen years since the end of the Vietnam War, and this was Châu's first visit back to his homeland. Everything seemed completely different. Saigon's streets were more crowded and dirty, with small shops packed tightly together along the roads.

It seemed like every space was used for selling something. Châu noticed that there were more sellers than buyers, with the most crowded places being the simple eateries. Curious, Châu asked Uncle Út, who was sitting beside him:

"Where did all these people come from? I don't remember it being this crowded before."

I heard that after the liberation, people from the North came to the South, saw how wealthy and easy life was here, and brought their whole families down. That's why it's so crowded.

"How is life here now, Uncle?"

"It's better now, but we're still poor. At least we have rice to eat; we don't have to eat cornmeal mixed with worm-infested rice or cassava like in the years before. Fortunately, you and Hiền escaped in time, or else you'd have had to go to re-education camps, and when you were released, who knows what you'd do to survive."

"Yes, I heard that former officers, after being released from re-education, were discriminated against by the communists and couldn't find any work. They had to resort to pulling rickshaws, pushing carts, or patching tires on the side of the road to get by."

"That's true. I used to think that after reunification, there would be peace, happiness, and prosperity. But who would have thought that people would suffer a hundred times more than during the war? Can you believe that the South, known as the rice bowl of Southeast Asia, would run out of rice to eat after liberation? Oh, I was so caught up in talking that I forgot to ask, where is your hometown?"

"My hometown is in Cầu Nổi Hamlet, Rạch Đào Village, which used to be part of Cần Đước District, Long An Province. I don't know which district or province it belongs to now."

"Oh, I've heard of Cần Đước District, but I've never been there."

The car had just passed over Bình Điền Bridge, and the houses along the roadside were becoming more sparse. The green rice fields were neatly divided into squares, with the rice stalks spreading wide, almost ready to bloom. The scene before him suddenly took Châu back to his youth when, at this stage of the rice's growth, the rice paddy fish (cá rô) would have just grown to about three fingers' length. You could simply part the rice stalks, sprinkle some roasted rice bran, and within moments, the fish would come sniffing.

Rice paddy fish are greedy; they bite at anything. If you found a good spot, you could catch enough fish within an hour to fry or grill over a straw fire, then dip them in fish sauce mixed with basil for a meal you'd never forget. If you weren't lucky enough to find fish, you could snack on some "đòng đòng," the sweet and fragrant milk rice kernels, though you'd have to be careful not to get caught by the field owners, who would curse you for ruining their crop.

As Châu was lost in these memories, the car reached Bình Chánh Junction. Vân, sitting in the back with Aunt Út and Thu, reached forward and lightly shook her husband's shoulder:

"Honey, isn't this where we turn left to go to Rạch Đào Village like you mentioned before?" Her touch and question brought him back to the present:

"That's right, we'll turn here, and in about forty minutes, we'll pass Rạch Kiến Village, and then we'll be in Rạch Đào Village. But it's been so long since I've been back; I don't know what condition the roads are in. They used to be full of potholes, hard to drive on."

"Once we're at Uncle Út's house, tomorrow we can ask this driver to take us to your hometown?"

"That's a good plan...."

The crowing of a rooster in the chicken coop behind Uncle Út's house startled Thu awake. She quickly rolled over and clung tightly to her mother, crying softly.

"Mom, I'm scared! After spending the entire day on a plane and then traveling to Uncle Út's, Vân was exhausted." She had fallen into a deep sleep on the wooden bed and only woke up when Thu's crying roused her.

"What are you crying about, sweetie?"

"That loud noise scared me!"

"Oh! That was just Uncle Út's rooster crowing in the morning. Nothing to be scared of."

"What's a rooster, Mom?"

"It's a "rooster," honey. Tomorrow morning, I'll show you what it looks like. Now, go back to sleep."

Thu, who had just turned five, was born in Seattle, USA, where she grew up in a wealthy, modern society. When she got off the plane at Tân Sơn Nhất Airport, everything she saw was strange. She clung to her mother's hand, asking questions about everything—from the rickshaws roaming the streets to the vast, water-filled rice fields in the countryside.

To Thu's innocent mind, America was her real home because that's where her friends lived, where there were "McDonald's," "Burger King," and "Jack In The Box." Vietnam, on the other hand, was just some distant place her parents had taken her to visit, like a camping trip. That's why Châu had discussed with his wife the importance of bringing Thu back to visit her ancestral homeland, so she could understand where her grandparents and parents came from.

Outside, dawn was beginning to break. The clucking of a hen calling its chicks to eat in the yard woke Châu earlier than usual. The sky was just beginning to brighten, still shrouded in mist, with a slight chill in the air. Châu reached for his jacket draped over the chair, put it on, and stepped out onto the front porch. The surrounding neighborhood was still quiet. He lit a cigarette, the sharp taste of the early morning smoke waking him fully.

Nearly fifteen years had passed, and in this very house, he had once lived with Hiền's family. Back then, Vân, Hiền's younger sister, was just fifteen or sixteen years old, a young girl as innocent as a budding flower. Châu had looked at her like a sweet little sister.

But then the war had separated Châu and Hiền for many years, drifting through foreign lands, making it seem as though he would never see his young sister again. As he reflected on this, Châu suddenly remembered the fortune teller at the Hậu Giang bus station in Chợlớn, more than twenty-five years ago when he was just a new army recruit. The fortune teller had told him that he would have a benefactor, would become wealthy, marry a good wife, and have dutiful children, among other things.

The fortune teller's words had amused him at the time. But Châu knew that if a fortune teller only gave bad news, he wouldn't make any money and might even get beaten up. He thought that fortune telling was a profession that sold hope because people who sought out fortune tellers usually lacked confidence in themselves and needed someone to show them a way forward. Back then, Châu had nothing to believe in—how could he become wealthy with just an army kit and a few uniforms? His salary wasn't even enough to cover his meals at cheap diners, where he often had to settle for bread, let alone afford to marry a good wife.

Now, sitting in front of the reality he had never imagined, with Vân, his good-natured, capable wife, and their daughter sleeping warmly in her mother's arms on the bed inside, Châu couldn't help but think back to the fortune teller's predictions. Surprisingly, they had come true, even though the fortune teller hadn't charged him a single penny.

Châu chuckled to himself, "I really misunderstood him...." He wished he could meet that fortune teller again to praise his accurate predictions and reward him with a sum of money that the fortune teller could never have expected.

As his cigarette burned down to the end, Châu stubbed it out and flicked it into the yard. The hungry chicks, mistaking it for food, scurried over to peck at it, then moved closer to Châu, chirping expectantly for the leftover rice that Aunt Út usually fed them in the mornings.

There was a light sound from the door behind him. Châu turned to see Vân stepping out, her arms crossed tightly in front of her chest. Seeing

Châu standing with the chicks on the porch, she asked:

"Aren't you cold? Why are you out here so early?"

"I couldn't sleep, so I came out to watch the chicks. Is Thu still sleeping inside?" Vân walked over and grabbed Châu's arm, pressing it against her chest for warmth.

"You know, Thu was so scared of the rooster crowing that she woke up really early. Now she's wrapped in the blanket and sleeping again."

"Think about it—she's never heard a rooster crow in her life. Of course, she's scared, but she'll get used to it."

"I had to explain to her in both Vietnamese and English what a rooster was before she understood!" Châu laughed, looking lovingly at his wife:

"See, if we don't teach her Vietnamese, she won't know her mother tongue when she grows up, and that's our fault, not anyone else's."

"I do teach her, and I watch her closely, but when she's playing with the American kids in the neighborhood, they only speak English, so what can I do? By the way, what time did you tell the driver to pick us up and take us to your hometown?"

"I told them to be here around nine in the morning to take us to town for breakfast and then head straight there."

"How long do you plan to stay there?"

"I'm not sure yet. My house is deep in the countryside. We'll figure it out once we get there."

"It's been nearly twenty-seven years since you last visited your hometown. Do you still remember the way?"

"Of course, I do. But I don't know if my parents are still alive or if my siblings have moved somewhere else."

A trace of sadness appeared on Vân's face as she squeezed her husband's hand, sharing his anxiety:

"I'll pray to Buddha for their health and safety." Châu bent down and kissed his wife gently on the forehead:

"Thank you for thinking of them." Vân looked up at her husband, somewhat displeased by his comment:

"Why would you say that? Your family is my family too; there's no difference.'

Realizing that his polite remark was unnecessary, Châu remained silent, quietly thanking the heavens for blessing him with such a wonderful wife who understood and supported him through all their challenges.

The driver, Chú Hai, had just turned at Bình Chánh Junction, heading toward Cầu Tràm. Vân, sitting in the back, told him to find a spot to park so she could buy some more fruit to offer to her husband's ancestors.

Aunt Út took Thu along to help Vân haggle with the vendors, knowing full well that the local fruit sellers would likely overcharge a well-dressed, fair-skinned customer like Vân.

Meanwhile, Uncle Út, the driver, and Châu went into a nearby coffee shop to wait.

Knowing that Châu was a former South Vietnamese soldier who had moved to the U.S., the driver was curious about the benefits former soldiers received in America. He had heard rumors and asked:

"Châu, you probably know some of the former South Vietnamese soldiers in the U.S.? I heard the American government pays them back pay from the war. Is that true?" Châu was just about to sip his coffee when he heard the question. He quickly set the cup down and looked at the driver, surprised:

"What? I'm a former soldier, and I've never heard of any back pay! Who told you that?"

"I have a cousin who used to be a paratrooper in Saigon. He wrote me a letter saying that when he first got to the U.S., the American government sent him a few hundred dollars every month. He didn't even need to work!" Châu knew this was just a wildly exaggerated rumor and nodded with a smile, understanding the confusion. He slowly lit a cigarette and explained:

"Yes, that's true, but not in the way people think. The government sent those checks to help new arrivals who didn't have jobs yet. I received some help when I first got there too, but once I started working, the payments stopped. It's not like you can just stay home and get paid forever!" Uncle Út, sitting next to him, agreed:

"That's what I thought. Anywhere you go, you have to work to eat. No one can sit around waiting for charity all their life." The driver scratched his head:

"It's good to know the truth." People around here believe those stories and are all eager to move to America. As they were wrapping up their conversation, Vân and Aunt Út returned with the fruit, some bánh mì, and roasted pork. Vân called out to the group in the coffee shop:

"Let's go! It's getting late!"

The weather was clear today, with dry patches on the road. The car snaked around to avoid the large potholes filled with murky rainwater. Châu glanced at his watch, then turned to Vân:

"It's just after ten o'clock. We should reach Rạch Đào around eleven, and from there, it'll be another fifteen minutes to my house. We should arrive before noon." Vân, eager to learn more about her husband's hometown, asked:

"You mentioned that your house is in Cầu Nổi Hamlet. Does that mean we'll have to cross a floating bridge over the river?" Châu couldn't help but laugh:

"Cầu Nổi is just a name from long ago. I don't know why it's called that, just like Cần Đước and Rạch Kiến. I guess Rạch Kiến was named because the area had a lot of ants building nests in the canals, so they called it Rạch Kiến."

"So there's no actual bridge to cross between Rạch Đào and your house?"

"There used to be a small wooden bridge over a canal that carts and small cars could cross, but I'm not sure if it's still there or if it's fallen apart."

The car rocked gently like a cradle. The countryside breeze wafted through the open windows, lulling Thu back to sleep in Aunt Út's arms. The driver pointed to a concrete signpost on a mound by the roadside, with black letters etched deeply into the white background: "NGÃ BA TÂN LÂN." Châu nodded:

"We're almost at Rạch Đào, driver! By the way, have you ever driven passengers down here before?"

"I've only gone as far as Cầu Tràm, never this far."

"The road from Rạch Đào to my house is driveable, right?"

"It should be fine for this car, but I heard your mother passed away nearly ten years ago. The house was sold to a schoolteacher named Hai."

"Châu thanked the driver and told him to follow the road to Cầu Nổi Hamlet, about two kilometers away."

By now, the sun was almost directly overhead. The road ahead stretched straight, flanked by scrubby bushes, wild grasses, and rows of tall bamboo casting shadows across the road. Occasionally, a woman returning from the village market would rest under a tree, fanning herself with her conical hat.

Although nearly four decades had passed, Châu vividly remembered, as if it had happened just yesterday, a young schoolboy about ten years old, dressed in a short-sleeved white shirt and black shorts, barefoot, carrying a satchel made of woven pandanus leaves, clinging to the back of a horse-drawn cart on his way to the village school.

In the summer of that year, when the pomegranate flowers in front of his house had turned red, signaling the end of the school year, he could still hear his old teacher's voice echoing in his ears: "Today is the last day of school. Our village doesn't have a secondary school, so you will all go your separate ways. I wish you all the best in becoming useful citizens for our village and our country…"

After the final school bell rang that day, the wealthy children of the village gathered in small groups, discussing their plans to go to Saigon or the big cities to continue their education, dreaming of becoming doctors or engineers. They laughed and chatted as if it were a festival. But there was one poor schoolboy named Nguyễn Văn Châu, carrying his tattered pandanus-leaf satchel on his head to shield himself from the sun and rain, who walked home alone along the village road that his great-grandfather, grandfather, and father had walked many times before.

That road, though inanimate, silently lying there, had borne witness to the trials and tribulations of many generations of villagers. From that day on, the little schoolboy, content with his lot, grew up like a banana plant or a betel palm, nurtured by traditional rural methods in the hope that one day he would bloom and bear fruit, providing for his family.

Seasons of plowing and planting came and went, and the little schoolboy and his water buffalo became inseparable companions, wandering together along the village paths. By the time they returned the buffalo to its pen, the sun would have already set. The years passed, marked by the rising and setting of the sun. The mango tree in front of his house bloomed and bore fruit many times, and the little schoolboy grew into a strong young man. As he matured, he began to feel strange stirrings in his heart when he caught glimpses of the village girls bent over planting rice.

Like millions of other young men, one day he found himself in love, receiving the affection of a girl from the neighboring hamlet. Their simple, rural love blossomed in secret, and they promised to marry after the next harvest.

But as the old saying goes, "Man proposes, God disposes." The escalating war and the national conscription order tore their rural love apart, and from that day on, the young man from the lower hamlet disappeared without a trace, never returning to his old village. By now, the girl from the neighboring hamlet, his first love, was probably a grandmother with grandchildren of her own.

The old dirt road, lined with gnarled, ancient trâm bầu trees, had led Châu down memory lane. He was only brought back to the present by the distant, tired voice of a grandmother or great-grandmother singing a lullaby to her grandchild from a thatched house by the roadside.

Châu turned and told Vân:

"We're almost there, honey! Wake Thu up so she can see her grandparents' hometown." Aunt Út woke Thu, then passed her to Vân:

"Here, wipe her face with a damp cloth to wake her up."

"I'll take care of her." As Vân wiped their daughter's face, she asked her husband:

"We're almost there?" Châu pointed to a thatched house by the roadside:

"See that shop over there? That used to be Aunt Hai's place. We'll leave the car there and walk across a few rice fields to reach our house."

Thu was now fully awake. Vân stood her up so she could see more clearly:

"This is your grandparents' village. When your father was little, like you, he went to school here. Your grandparents, aunts, and uncles all lived here." Thu looked confused:

"But Dad lives in America! Not here."

"Yes, Dad lives in America now, but he used to live here."

Although Thu didn't fully understand this strange story, she quietly accepted her mother's explanation. The driver parked the car in front of the shop. A few children, curious about the strange car, gathered around to touch and inspect it with great interest. Châu got out first and walked into the shop. The children, seeing a customer, called out:

"Grandma, there's a customer!" From the back of the house, a woman's voice replied:

"Just tell them to take what they need, and I'll be out in a minute." Châu stood at the counter and said:

"I just need to leave the car here; I'm not buying anything." At that moment, the woman emerged from behind a bamboo screen that separated the front and back of the house. As she stepped out, she began to speak:

"You just need to leave the car...?" But then she suddenly stopped, staring at Châu in surprise. Her hands, as if acting on their own, reached out slightly, as if she couldn't believe what she was seeing.

"Is that you, Út Châu???"

Châu felt as if he had been struck by lightning. He recognized the woman, with her hair tied up and dressed in a black bà ba outfit, as none other than Năm Lài, his first love before he joined the army.

"Yes, I'm Út Châu, and you must be Năm Lài?" The woman didn't respond verbally, only nodding slightly. One of the children standing nearby asked:

"Grandma, do you know this man?" He's rich; he has a car! After the child's comment, the woman seemed to snap back to reality.

"Yes, this is Uncle Út. He used to live in the lower hamlet here and was friends with Grandma. We haven't seen each other in decades." Châu reached out to pat the head of the child standing closest to him.

"Are these your grandchildren, Aunt Năm?"

"Yes, they're the children of Hoa, my eldest daughter. Their mother went to the village market and hasn't returned yet." Châu glanced inside the house:

"Where's their grandfather?"

"My husband is teaching at the village school. He'll be back this afternoon." After answering, the woman stepped over to a table, lifted the lid of a coconut shell container, and poured tea into a cup to offer to her guest.

"It's been decades since the liberation. Where have you been, Út?" People around here thought you were dead!

"I've been busy working far away, so I haven't been able to visit. I don't know if my parents and siblings are still healthy." The woman's eyes widened in surprise:

"You don't know anything?"

"Know what? What are you talking about, Aunt Năm?"

"Both of your parents passed away within a year of each other, not long after the liberation. I thought you already knew!" Châu didn't respond. He lowered his head and set the cup of tea back on the table. Thu, who had been sitting in the car with her mother for quite a while, hadn't seen Châu come out yet, so she climbed down and ran into the shop, calling out:

"Dad! Mom's asking if you're finished yet. Shouldn't we go?" Châu quietly picked up his daughter.

"Aunt Năm, this is my daughter. Her name is Thu." The woman walked over and gently stroked the little girl's back.

"How old is she, Út? Do you have any more children?"

"She's five years old! She's our only child." As he finished speaking, Vân walked into the shop and asked:

"Are we ready to go? We've been here a while, and it's getting late!" Châu, holding Thu, introduced Vân:

"This is my wife, Aunt Năm." Then he turned to Vân:

"Aunt Năm owns this shop." The two women exchanged polite smiles and nods. Vân then urged:

"Let's go. Uncle and Aunt Út have been waiting outside for a long time. It's not polite." Châu signaled to the driver. He reached into his pocket and pulled out four fifty-thousand-dong notes, then instructed the driver:

"Here's two hundred thousand đồng. Take the kids to the market for some gas, then treat them to some hủ tiếu. And don't forget to bring some back for Aunt Năm! Then park and wait for me here."

The green rice fields in front of him lay exposed under the scorching midday sun. Occasionally, a soft rustling sound was heard, caused by the rice leaves brushing against each other when a gentle breeze unexpectedly passed by.

The dirt road that separated the fields leading back home, where Chau used to walk thousands of times, still lay there quietly. On either side of the road, wild grasses, bitter herbs, and deep buffalo footprints remained, all images seemingly unchanged from the time when he was just a child.

Suddenly, Chau remembered what Ms. Nam Lai, the shop owner, had just mentioned—that people in the village thought he was dead. If death had truly come to him, what difference would it have made? The wide road he walked on remained as it was; nothing had changed. The rice was still green, the wild grasses and bitter herbs still grew as abundantly as ever.

"The only difference today," Chau thought, "is that I'm no longer the poor farmer, the tenant farmer, like my grandfather and father who had to beg to borrow rice from the landowner during bad harvests." Thinking of this, he quietly smiled at the changes.

Van, walking behind with little Thu, called out to Chau:

"Slow down a bit, wait for me and Thu. The wild grasses are pricking my pants too much; let me remove them first!" Chau, carrying the suitcase ahead, quickly turned around and laughed heartily, teasing his wife:

"Why worry about those wild grasses! From here to home, there's a lot more of them, remove one, and another will stick in just the same." Aunt Ut also stopped and told Van:

"Let me carry little Thu for you, it'll be easier to walk. These buffalo footprints are deep, if she steps into one, she might sprain her ankle." The group walked past nearly ten acres of fields before they reached the bridge over the canal, where water palms grew thickly. Chau set down the suitcase to rest his hands.

"Once we cross this palm bridge, we'll be home." Van took the opportunity to sit down and rest, removing the wild grasses stuck to her pants. Little Thu quickly slid down from Aunt Ut's arms and shouted:

"Dragonfly! Dragonfly!" Aunt and Uncle Ut were startled, not knowing what had happened to Thu, so they asked Van:

"What is she shouting about?"

"Oh, nothing, Aunt Ut. She just saw a dragonfly sitting on the top of that rush plant over there, it's just new to her."

"I thought something happened, scared me! Aren't there dragonflies where you live?"

"Where we live, we don't see them, but in the southern states, I think there are dragonflies like here."

After resting for a few minutes, Chau picked up the suitcase again. Since the bridge over the canal was made of two palm trunks placed parallel to each other, Chau warned everyone:

"When crossing the bridge, hold on to the poles and go slowly to be safe." Everyone safely crossed to the other side of the canal. Chau approached little Thu and pointed to a shabby three-room thatched house, located before a straw stack behind a bamboo grove, about two acres away.

"That's your grandparents' house." Little Thu looked puzzled and asked:

"Which grandparents' house are you talking about, Dad?"

"Your paternal grandparents, of course!" Still confused, she asked:

"What about the 'grandmother' who's living with us?"

"Oh! That 'grandmother' is your mother's mom. But here, these are my parents, your grandparents. When I was little like you, I lived here." It seemed she understood a little:

"Why does the house look so funny?" Chau knew his daughter didn't understand anything about the place, so he simply replied:

"It's our house! What's so funny about it?"

As Chau turned off the main road to head home, a brindle dog, seeing a stranger, ran out to the gate and barked loudly. Near the water moat, an elderly woman, dressed in black clothes with one pant leg rolled up and the other down, was carrying a woven bamboo basket. She raised her right hand to shield her eyes from the glaring sun as she looked out toward the entrance.

After noticing the group of city-dressed women and children entering the house, she thought they might be asking for directions to a nearby acquaintance's home.

She called out to the dog as she walked straight to the bamboo fence to guide the guests. At that moment, Chau, carrying the suitcase, walked directly toward her. He carefully observed the woman coming out, thinking to himself, "It looks like Sister Hai..." Sister Hai was Chau's sister-in-law, the wife of the eldest brother in the family. Chau quickly set down the two suitcases and spoke first:

"Is that you, Sister Hai?" The woman, walking toward him, was surprised to hear someone calling her Sister Hai. She hadn't recognized her brother-in-law yet, so she asked:

"Oh! Who are you, and how do you know me?"

"It's me, Ut Chau, your brother...."

As soon as she heard the name Ut Chau, she didn't wait for him to finish his sentence. She exclaimed with joy and surprise:

"Oh my God! Is it really you, Ut Chau? You've changed so much I couldn't recognize you. We all thought you were dead...." She suddenly stopped mid-sentence, and Chau understood, so he continued:

"I heard from the neighbors that they spread rumors about me dying before the liberation."

"Why haven't you come back here for so long? Our parents have passed away!" Chau bowed his head, sighing:

"I was busy making a living far away, so I couldn't come back." As if suddenly remembering the others waiting behind him, he turned around to introduce them:

"Sister Hai, this is Uncle and Aunt Ut, my wife's side, and this is my wife, Van. The little one is Thu, my daughter." After greeting everyone, the sister-in-law bent down to look at little Thu;

"Oh my, she looks just like you when you were a child! Has she started school yet?"

Seeing the unfamiliar woman, dressed differently and with her mouth stained red from chewing betel, little Thu shyly stepped back to stand behind her mother. Knowing that Thu was a bit scared, Van replied:

"Yes, she's already attending kindergarten, Sister Hai."

Chau reached for the two suitcases and urged:

"Let's go inside, we can talk later. By the way, where's Brother Hai? I haven't seen him around."

"The rice is just starting to sprout, so he went out to the fields early this morning to spread more fertilizer. He'll be back soon."

"And how are Brother Tu and Sister Ba doing, Sister Hai?"

"Well, Brother Tu and his wife moved out on their own after our parents passed away, but they still live nearby. As for Sister Ba, she just married off her youngest daughter a little over a month ago in the upper village. Everyone is in good health; only you, Ut Chau, have been missing, leaving the place ever since you went off to join the army."

"Yes, it's all because of the times, Sister Hai."

After everyone put their belongings on the wooden bed in the middle of the house, Chau asked Van to go to the back to get some plates to arrange fruits and snacks on the altar for the ancestors. Uncle Ut, feeling a bit tired from the long journey, lay down on a straw hammock strung between two bamboo posts on the left side. Aunt Ut sat on the bed with little Thu to rest. The sister-in-law busily moved around, lighting incense and preparing the offerings for Chau to present to the ancestors.

Van searched all over the kitchen but couldn't find any dishes or bowls, so she returned and whispered to her husband:

"I couldn't find any bowls or plates, dear!" Chau smiled sympathetically:

"Go outside to the front by the bridge near the moat; you'll find a dish rack out there. In this area, people wash their dishes at the moat and then leave them on the rack to dry."

Van nodded, smiling:

"You remember everything so well!" The brindle dog, lying quietly on the porch, suddenly stood up, barked a few times, and then ran out to the gate. The sister-in-law, who was pouring tea for the ancestors inside, called out to Chau:

"Brother Hai is back!"

Brother Hai was the eldest son in the family. The house he was living in was inherited from their ancestors. When their parents were still alive, Ut Chau and his siblings all lived together in that house. When they got married, they moved out on their own. Sister Hai had married into the family, living with her in-laws at this house when Ut Chau was about ten years old. Now that the parents had passed away, Brother and Sister Hai were naturally the ones staying here to take care of the ancestral worship and the family's traditions.

Brother Hai had just placed his hoe down near a banana bush in the front yard and walked over to the bridge to wash the mud off his hands and feet from the fields. His wife hurried out to the front and urged him:

"Hey, Ut Chau has come back to visit us!" The husband, bending over to wash his hands, was a bit puzzled when he heard that Ut Chau was back. He straightened up and asked:

"Which Ut are you talking about?"

"It's Ut Chau, your brother, who else could it be?" At that moment, Chau also stepped out of the house to greet his elder brother.

"Hello, Brother Hai!"

The elder brother, with surprise evident in his nearly wide-eyed stare, looked straight at Chau:

"Is that really you, Ut? My God, where have you been all these years without coming home? Even when our parents passed away, you didn't return?"

"Yes, I was very far away, Brother Hai…"

Ut Chau could only say that much before his voice choked up. He lowered his head, looking down at the ground to hide the two tears that were welling up in his eyes. Seeing his younger brother so emotional, the elder brother softened his tone, no longer reprimanding him:

"I was angry, so I said that, but I'm just glad you're alive and back here. The neighbors were all saying that you had died before the liberation. Come inside and rest a bit; I'll go and call Sister Ba and Brother Tu. They'll be so happy to see you. Oh, I almost forgot, do you have a wife and kids yet?" Chau turned back into the house and called out to Van:

"Honey, Brother Hai is home." Van, who was using some firewood to stoke the stove for making more tea, heard Chau calling from outside the bridge, so she stood up and went out.

"What is it, dear?"

Chau pointed to the man standing in shorts, shirtless, washing his hands at the bridge.

"That's my Brother Hai." Van politely bowed her head:

"Hello, Brother Hai, did you just get back from the fields?"

"Yes! And you must be Sister Ut? Do you and your husband have any children?"

"Yes, we have a five-year-old daughter, Brother Hai."

"Did you bring her here with you?"

"Yes, she's inside with Aunt Ut." Brother Hai's wife spoke up again, urging her husband:

"Hurry up and wash your hands and feet, then come inside to talk. We also have Uncle and Aunt Ut from Ut Chau's wife's side visiting us." After greeting the guests, Brother Hai told his wife;

"Go to the back and catch two capon roosters to make salad and porridge for the offering to our parents. I'll run over to Sister Ba's and Brother Tu's house to let them know that Ut Chau is back so they can come over."

The three-room thatched house of Brother and Sister Hai felt crowded and joyful today, like a wedding day, with the presence of all the siblings, nieces, nephews, and distant relatives from the village who had come over after hearing that Ut Chau was alive and back.

By late afternoon, as the sunlight dimmed, most of the neighbors had left. The rice fields in front of the house, now lacking the midday sun, had turned a dark green. A few frogs here and there had begun to croak, calling their mates.

Chau was sitting on the porch smoking a cigarette when Van came over and quietly said:

"Little Thu wants to go home, dear. This afternoon, when I took her to the banana bush for a bathroom break, she got so scared she held it in until now. She says she doesn't like it here and keeps begging to go back to America."

"Well, just try to comfort her, she'll get used to it, others live here just fine, don't they?"

"But what about Uncle and Aunt Ut? We should take them back to My Tho, it's not right for them to stay here."

"Let me talk to Brother Hai, and we'll figure something out." Just then, Brother Hai and Brother Tu also brought out a teapot and sat down on stools next to Chau, who was still smoking. Van excused herself and went back inside to help little Thu wash her face. Ut Chau took out a pack of cigarettes and offered one to Brother Tu:

"This cigarette is lighter than the rolled ones."

"Down here, Brother Hai and I mostly smoke rolled tobacco bought in bulk from the market because it's cheaper. We don't have the money for these. So, how long are you planning to stay here?"

"About four weeks, Brother Tu."

"If Brother Hai's house doesn't have enough room, you can stay at my place."

"Our daughter isn't used to new places; she's scared and keeps crying to go home. We'll probably rent a room in Can Duoc district to stay overnight, then come back here during the day to spend time with you all and take care of our parents' graves before heading back to America."

Brother Hai, sitting next to them, seemed to agree with Chau's idea:

"Ut's right, Tu. When we were kids, it was different. Now that he has a family and kids, they're not used to living in the countryside. Besides, if people know he's from America and has money, I'm worried it might not be safe for him." Sister Ba, who had just finished cleaning up inside, stepped out and joined the conversation:

"Yes, I think Brother Hai is right. Down here, at night, if they know you have money or jewelry, it could be dangerous. Let Ut stay in the district, it's safer." After finishing his cigarette, Chau went back inside to discuss with Uncle and Aunt Ut, finding that they both wanted to return to My Tho that evening. Aunt Ut suggested:

"When we get back to My Tho, you could leave Van and little Thu at our house so you can focus on taking care of the graves. That way, it'll be more convenient."

Chau found Aunt Ut's suggestion very reasonable because having Van and little Thu with him wouldn't be of much help and would only add to Brother and Sister Hai's burden, so he nodded in agreement.

The next morning, the driver, Uncle Hai, arrived at Uncle Ut's house very early. Van opened the door for Chau, who then drove back to Cau Noi village to handle the graves.

"Remember to come back as soon as you finish!" Van reminded him.

"Yes, I'll be back in two days. Then we'll have some free time to take our daughter sightseeing in Vietnam."

As they drove near Rach Dao, Chau told Uncle Hai to head straight to Can Duoc district to rent a hotel room, and then return to Cau Noi village to park at Ms. Hai's shop as before.

Everything proceeded as planned. Chau's brothers, Hai and Tu, found a contractor who specialized in grave construction in the village. He agreed on the construction method, provided all the necessary materials, and promised to complete the work in about two weeks. On the second day, Chau felt more relaxed and at ease. During lunch at Brother Hai's house, they had boiled okra, snakehead fish fermented sauce in a clay pot with shrimp fat and green onions, and sour soup with shrimp and white Sesbania flowers. These "homegrown" dishes reminded Chau of the last meal Brother and Sister Hai prepared for him before he left for his new military unit.

"Sister Hai, do you still remember the pot of fragrant rice you cooked about twenty-seven years ago?" Chau asked, making his sister-in-law laugh heartily:

"Oh my, Ut, you have a good memory!"

"How could I forget, Sister? Eating that kind of rice here is something you remember for a long time!"

Brother Hai, after taking a small shot of home-brewed rice wine, made a loud sound like a snake hissing. He then poured another full shot and handed it to Chau:

"Here, Ut, take a shot to warm up. This is the best rice wine from Ms. Hai's shop." Chau took the glass, sipped half, and then set it back on the table.

"This wine is too strong and spicy, Brother Hai."

"Well, good wine has to be like that!"

After a delicious lunch with those special hometown dishes, Chau thought about how the snakehead fish sauce was made by Brother Hai after draining the pond last dry season. He couldn't finish eating it, so he salted it and stored it in the jar. Okra was easy to grow; Sister Hai just had to till the soil, plant the seeds, and water them, and a few months later, they would have okra. The Sesbania plant was almost like a wild tree; when the fruits dried and the seeds fell, they would sprout on their own near the water moat. When they bloomed, you just needed a bamboo pole to hook them down. If you wanted to cook sour soup with shrimp, you only had to bring a net to the water moat, sprinkle some bran, wait for a while, then scoop up enough shrimp for a sweet soup for the family.

Chau still believed that the daily life of the farmers here hadn't changed much from the past, so he complimented his sister-in-law:

"Sister Hai, you seasoned the snakehead fish sauce perfectly. It's been a long time since I've had it with shrimp fat and green onions like today." His sister-in-law was carrying the rice pot to the kitchen when she heard Chau's compliment, so she turned and smiled:

"I'm glad you like it, Ut, but that snakehead fish sauce wasn't made by me. Knowing that you liked it, Brother Hai told me to buy the best kind from the village market for you." Chau looked surprised at his sister-in-law:

"Oh, I thought Brother Hai drained the pond and brought home the fish for you to make the sauce?"

"Yes, that was before the liberation, but in recent years, they sprayed defoliants that killed everything, even the leeches couldn't survive, let alone the shrimp and fish. The people here are so poor they go out to the canals and fish day and night, catching everything, even the smallest shrimp, lizard fish, and anything else, leaving nothing."

As his sister-in-law spoke, Chau suddenly remembered the two old buffaloes and the calf that used to be his companions during the afternoons in the fields. So he asked:

"I've been so busy lately that I didn't notice, where are our buffaloes?" Brother Hai, sitting opposite, licked the cigarette paper as he rolled a cigarette. After exhaling a long puff of smoke, he spoke in a sad tone:

"After you left, many of them came here. They would sometimes call the villagers for meetings at night, telling them that once they defeated the Americans and their puppet regime, our country would be rich and strong, with plenty of land and buffaloes for all the farmers. The villagers were all happy and excited, regularly providing them with food and rice. But after the liberation, they started holding endless meetings about the new communist policies. They urged everyone with buffaloes to join the cooperative or something... I didn't really understand it. And the rice we harvested had to be sold to the government at a very low price."

As Brother Hai recounted, he shook his head and picked up his tea cup to drink.

"If I were to tell you everything, it would be a long and tiring story. Let me just tell you about our buffaloes so you'll understand. A little over a year after the liberation, our parents were constantly ill. We didn't have money for medicine, so I sold one buffalo to get some money to buy Chinese medicine from the village for them. A few months later, our father passed away. At that time, our mother was still lingering, and with food shortages and the constant pressure from the local cadres to join the cooperative, I decided to sell the remaining two buffaloes to relieve the burden. I thought our mother would live a few more years, but four months later, she also passed away. Within a year, we had to endure the loss of both parents, which was truly difficult."

Hearing Brother Hai recount the family's hardships made Chau deeply emotional. He realized how selfish he had been, avoiding responsibility while leaving his siblings to bear the burden of the family's extreme poverty.

Chau spoke up to console his brother:

"I never imagined such painful events would happen to our family. I'm deeply grateful to you and Sister Hai for taking care of our parents during that time. Now that they've passed away, all I can do is build a sturdy grave for them as a lasting memorial. My wife and I will also provide some capital for you to save for old age. As for me, being so far away, it will be a long time before I can visit you again."

"Ut, your thoughtfulness means a lot to us. All we can do is thank you and your wife for thinking of us."

As if remembering that Ut Chau had planned to visit the neighbors that afternoon, Brother Hai prompted him:

"By the way, I heard from Sister Hai that you planned to visit someone, didn't you?"

"Yes, I was thinking of taking a walk to visit some old acquaintances in the village after lunch."

"Well, you should go while it's still early, and then you can return to My Tho in the evening. I need to check on the fields too, to make sure the buffaloes from the upper village don't trample the rice."

Chau walked to the front, scooped some water from the jar with a coconut shell, and washed his face. He was thinking of visiting Ms. Hai's shop to see how his old friends were doing and also asking the driver, Uncle Hai, to go to the village market and buy a few cases of beer to give to his siblings before returning to My Tho.

The cool rainwater collected from the thatched roof refreshed him, washing away the afternoon drowsiness. He went back inside to the altar and took out a horn comb to groom his hair in front of the small mirror hanging on the wall. As he looked at his reflection, he thought to himself: "Back then, when I lived here, my hair was jet black and stiff, hard to comb, unlike now..." Then he smiled to himself, "Ms. Nam Lai already has three grandchildren... Am I still so young...?"

Outside, the sun was still shining brightly. The green grasshoppers, with bodies the color of the grass, hopped away at the sound of footsteps on the dirt path, making "snap" sounds as they jumped.

As he crossed the coconut bridge over the canal, Chau suddenly remembered the Vo bridge from years ago. Curious, he wanted to see how it had changed, so he decided to take a left turn and head toward the Vo bridge, which was only about three acres away. The old path was now overgrown with knee-high grass and sensitive plants. Thanks to his thick "jeans" pants, Chau wasn't worried about the grass blades cutting him or the sensitive plants scratching his skin, but he was concerned about stepping off the path and falling into the water, soaking his clothes and causing trouble.

Finally, Chau reached a small mound near the water, where he thought the main entrance to the old Vo bridge used to be. He looked around carefully but couldn't recognize any traces of the past.

The Vo bridge had been removed at some point, and now only mangrove apple trees, spiny amaranth, and sesbania trees had grown thickly along the canal banks. Chau suddenly laughed at himself for being so sentimental: "Trying to rebuild a mirror to find an old reflection..." Ms. Nam Lai already had three grandchildren; she's no longer the young woman she once was!

Chau turned around and headed back to the shop. Ms. Nam Lai's grandchildren, seeing him walking up from the fields, recognized him as a family friend and shouted:

"Grandma! Uncle Ut is here!" Today, Ms. Hai, the shop owner, was dressed neatly and elegantly, wearing black silk pants and a floral blouse with a heart-shaped neckline, her hair tied up high and slicked with oil. She stepped out from behind the counter to greet Ut Chau:

"Uncle Ut, you've been here for a while but haven't stopped by the shop."

"I've been so busy with all sorts of things, Ms. Nam!"

"I heard you're building your parents' graves. Are they finished?"

"It'll take another two weeks."

"So are you staying here to wait or going back to My Tho?"

"I'm planning to return to My Tho this evening, and when the graves are finished, I'll come back to pay my respects and then return to America."

Hearing Chau say that he would be returning to America, Ms. Nam's face showed a hint of sadness. She knew that everyone had their own circumstances, but how could she hide the emotions and memories of the past?

The sound of children playing outside woke Uncle Hai, the driver, from his afternoon nap. Recognizing Chau's voice, he got up, opened the door, and stepped outside to ask:

"Mr. Chau, when are we heading back to My Tho?"

"Probably this afternoon. Could you take this money and go to the village market to buy me four cases of canned beer to give to my siblings?" After Uncle Hai drove off, Chau returned to the table, pulled out a stool, and sat down. Ms. Nam poured tea into a cup and offered it to her guest.

"Life in America must be fun, right, Uncle Ut?"

"Over there, it's lively, with lots of cars and factories, but I think life here is more enjoyable, Ms. Nam!"

"You must be saying that to comfort us country folks!"

"I'm serious, Ms. Nam, not just saying it to make you happy! Over there, work is hard, but at least you get paid for your efforts, unlike here, where people work all their lives and still struggle to make ends meet."

"How long have you been married, Uncle Ut?"

"Over six years now, Ms. Nam, and my eldest daughter is only five years old, as you saw the other day."

"So you waited until you were over forty to get married?"

"When I was single, without a steady job, I drifted like a water hyacinth, moving from place to place in a foreign land. I didn't have the money to think about marriage and starting a family."

After Ut Chau's response, Ms. Nam only sighed quietly, not saying another word. She thought that marriage was a matter of fate, something beyond one's control, no matter how many promises or oaths were made. After a moment of silence, Ms. Nam continued:

"Your wife is still young and beautiful, and your daughter is also very pretty and sweet!"

"Thank you, Ms. Nam. By the way, earlier, I went to see the Vo bridge, but it seems like it's been removed for a long time, hasn't it?"

When Chau mentioned the old meeting place from their youth, Ms. Nam suddenly felt a bit shy. She glanced at him and then looked away.

"It's been decades, not just recently. Even if no one had removed it, the rain and sun would have rotted it by now; it wouldn't have remained intact like before. Why did you go there, bringing back sad memories?" Chau scratched his head:

"Oh, I'm getting more sentimental as I get older. Maybe because I haven't been back here for so long, I wanted to revisit old memories to see how they've changed."

"Do you find it all very different now, Uncle Ut?"

Chau slowly took out a cigarette, placed it in his mouth, and lit it:

"The scenery, houses, and fields around here haven't changed much, but people seem thinner and more withered than before. Do you remember when I lived here, my hair was black and stood up like a post, and there was so much of it. Now look, I barely have any left, and what's there is mostly gray." In a melancholic tone, Ms. Nam looked at Chau:

"When you first arrived, I didn't recognize you, but your voice is just like it used to be. Over there, you still have everything, but here... you see I've gotten thinner too. You say your hair is gray, but look at mine, is it still black?"

Chau lifted the cup of tea, took a small sip, then slowly set it back on the table:

"We're all getting old, aren't we, Ms. Nam?"

"You're the only one with young children. All my friends, like Huong, Nhan, and Mr. and Mrs. Tu Dom from the inner village, are already grandparents." After Ms. Nam's words, Chau burst out laughing:

"If I had stayed here, I'd probably be a grandfather several times over by now!" Ms. Nam gave him a half-smile and said:

"That's likely the case!"

After pausing for a moment, Ms. Nam continued:

"You know, Uncle Ut, after the liberation, I saw many of the local soldiers and militiamen returning home to work. I kept looking out for you, but you never came back. Then people in the village started spreading rumors that you had died." As Ms. Nam spoke, Chau interrupted:

"If I had died, it would have been like any other single soldier; no one would have cared or noticed." Ms. Nam seemed to understand the subtle meaning behind Chau's words, so she lowered her head and spoke softly, just loud enough for him to hear:

"How do you know no one would have cared? After all, we knew each other from before..."

Just then, the children playing outside ran into the shop, shouting:

"Uncle Ut, the car is back!" The nostalgic conversation between the two came to an end. Chau stood up and said goodbye:

"Well, I'll take my leave, Ms. Nam. I need to get back to Brother Hai's place so we can return to My Tho this evening." Ms. Nam, now back to her cheerful self, called out after him:

"Don't forget to give my regards to your wife, Uncle Ut!" Chau didn't reply; he just turned back, smiled, and waved goodbye.

Today was the final day. After the offerings to the ancestors and a feast with the neighbors, everyone left, except for Brother Tu's and Sister Ba's families, who stayed behind to bid farewell to Ut Chau's family as they prepared to return to My Tho and get ready to go back to the United States. Chau turned to his wife and said:

"Hand me the handbag." Chau pulled out four bundles of Vietnamese currency, each worth fifty thousand dongs, and placed them on the table.

"A few days ago, I promised that my wife and I would help you all with some capital for your farming. Before we leave, I want to give each of you a little something. For Brother Hai, we'll also provide some funds for repairing the house to honor our ancestors and have some left for raising pigs and chickens to sell for profit."

After Chau's words and the sight of the large bundles of money, everyone was stunned, unable to speak.

Chau understood their surprise because he had experienced the daily hardships of farmers in the countryside. They only hoped for good weather and a successful harvest each year to have enough left over to buy a few new outfits for weddings and ceremonies or to buy leaves to repair the roof before it started leaking.

Such a life of poverty and deprivation stretched on from generation to generation. Now, with such a large sum of money in front of them, something they had never seen before, their astonishment was understandable.

After a moment, Brother Hai slowly lifted one of the bundles and then placed it back on the table, speaking slowly:

"Our family has always made just enough from farming to get by; we've never had much left over. I wish our parents were still alive to see how successful you've become, they would be so happy. With this money, we'll buy tiles to rebuild the house so it's clean and comfortable, a proper place for worshiping our ancestors and for the children to visit."

The last incense stick on the altar had burned out, and the shadow of the house extended all the way to the front yard. The old bamboo groves at the gate swayed gently in the breeze. Van checked her watch and then turned to her husband:

"It's past four in the afternoon, dear, let's start getting ready to leave." Chau nodded and stepped out onto the porch. He took out a cigarette, lit it, and watched the white smoke drift and dissipate into the distant sky...

Chau gazed out into the distance. In these final moments, he tried to recall his childhood, running in the dry fields with kites, catching crickets, or gathering resin along the bamboo hedges around the village. When the rains came, he fished in the ditches where water spinach was planted behind the house. Life was simple and carefree, without complaints or regrets.

However, he knew that the events of his childhood were a natural outcome of fate, shaped by the circumstances of each family. He remembered them fondly, not with regret, as his current life and the life his daughter, little Thu, was enjoying were far better.

As the afternoon sun softened, the car started to move. On the dirt road, Ms. Hai and her three grandchildren stood by the shop, waving goodbye, the last ones to see them off. Chau leaned his head out the window. He wanted to take one last look at Ms. Hai because she had once been Ms. Nam Lai from the upper village, who often wore tight-fitting black ba ba dresses that showed off her round hips, slim waist, and full chest. But Ms. Hai now was no longer the Ms. Nam Lai of his youth! With her black pants rolled up, surrounded by her grandchildren, she now symbolized a virtuous, hardworking Vietnamese woman.

Time passed, measured by the rising and setting of the sun, year after year, with no one able to hold it back.

Ut Chau and Ms. Nam Lai had aged with the passage of time. It was no longer a question of whether Ms. Nam Lai was beautiful or young, but rather why she had always been the last to see him off?

In those moments of nostalgic reflection, Chau saw the figure of Ms. Nam Lai and the image of the Cau Noi village of his childhood, faintly visible in the eyes of Ms. Hai, the shop owner, as she stood by the roadside, waving goodbye to him for the last time, just as she had done nearly thirty years ago when he left for military service.

The car continued to move... Time continued to flow... And Ut Chau still sat there, lost in his thoughts.

CHAPTER 6

Sorrowful Past

Original Title: "Dĩ Vãng Buồn"

Nam was born in early spring in a spacious three-room tile-roofed house situated in a garden full of fruit trees. The garden was surrounded by four deep water-filled moats and dense bamboo groves, leaving only a small gateway at the front, just wide enough for people to come and go.

Nam's village had about a few dozen houses, but they were divided into several small clusters, such as the "Hai Quoc" pig slaughterhouse area, the "Quan Nam" area, the "Nam Nang" area. Nam lived in the "Doi Cu" area. There were four houses in total on a piece of land about five acres in size. "Doi Cu" was Nam's maternal grandfather. The other three houses belonged to Mr. Ba, Mrs. Hai, and Mrs. Tam, who were all close relatives of Nam's grandfather.

The term "Doi Cu" was used because Nam's grandfather had worked at the Saigon Post Office under the French colonial government. People in the community eventually used this title as a nickname for him to easily recognize him.

Nam's grandfather was the only son of Nam's great-grandmother, who was over eighty years old at the time. Due to his work in Saigon, and having a home there, he entrusted the large ancestral house to Nam's mother, the eldest daughter, to take care of and look after his mother.

Nam's paternal hometown was in Long-Thanh village, about ten kilometers from Long-Hoa village where Nam lived. When Nam's father married, due to the circumstances of his wife's family, he agreed to stay with his wife's family temporarily so that Nam's mother could be close to her elderly mother and care for her in her old age.

Nam still remembers that, at that time, his village was relatively prosperous, with houses built from three-room red tiles, supported by large ebony pillars as wide as an adult's arms, and with shiny thick wooden planks over a foot wide. His grandfather liked to collect antique dishes, expensive ones with gold or silver rims, and exquisitely crafted brass incense burners. Every year, around the 23rd of December, during the kitchen god's day, Nam had the task of cleaning and polishing these antiques meticulously. Nam was quite lazy and disliked these chores, so he was often scolded for being so sluggish.

Nam's father worked in Cholon and only occasionally came home to visit, so Nam was the only son in the family. Although his elder sister, who was eight years old, and his younger brother, who was three, were both frail, all the heavy tasks like climbing trees to pick oranges, plucking star apples, and shaking mangoes down fell to Nam. Even though Nam was only about five years old at the time, he was chubby and larger than other kids in the village.

After Japan's unconditional surrender, French forces returned to Vietnam, specifically taking control of Cochinchina under the governance of the French High Commissioner. During this time, a youth movement emerged in the South, gathering Vietnamese patriots to rise against colonial rule. Nam vaguely remembers that his father also joined for a time, but due to a lack of modern weapons, relying only on sharpened bamboo sticks and local "ba-nha" knives, they were unable to resist the French army's cannons and tanks. Consequently, the youth movement was soon disbanded.

The French expeditionary forces gradually brought in mercenary soldiers from African and Moroccan countries to Vietnam. The villagers referred to these mercenaries as "black-faced demons" because their faces were deeply etched with frightening lines. Wherever they went, they looted, raped women, and burned down houses indiscriminately. When the villagers heard they were coming, they hurriedly gathered their belongings and fled deeper into the swamps and marshes to hide, following the policy of "scorched earth resistance." During these escapes, Nam's responsibility was to carry his younger brother, who was too weak to run fast, while his parents took care of the clothes and other necessities. Sometimes, they would be away for several days or even a month before daring to return home. During these escapes, in the early days, they feasted on whatever poultry they had to avoid the burden of caring for them, but eventually, food supplies dwindled, leading to very frugal meals.

Nam still remembers that they used pig fat to light lamps at night since kerosene was unavailable. During dinner, Nam would secretly pour the lamp oil into his rice, mixing it for a deliciously rich flavor, or occasionally, his mother would give him a piece of palm sugar wrapped in leaves to eat with his meal. As a growing boy, everything tasted good to him. Clothes and fabric for making clothes were expensive and hard to find during that time. Nam loved to play in the mud all day, causing his clothes to wear out quickly. There was a time when he had to wear shorts made from rough cloth, which he disliked because they were scratchy, heavy, and slow to dry when wet.

Nam's family, which consisted of six people, had to take shelter in a small house owned by Mrs. Bay in "Long-Duc," a close relative of Nam's mother, who allowed them to stay temporarily while fleeing from the war. During the day, Nam's parents worked in the fields, harvesting rice for a living, while his elder sister stayed home, keeping the house clean and cooking meals. Nam enjoyed fishing and spent his days by the fields. Fearing that he might fall into the moat and drown, his mother would tie his leg to a bamboo fence at the entrance before leaving for work each morning to prevent him from wandering off. Nam would pretend to sit quietly, waiting for his parents to disappear behind the rice fields before untying the fence and sneaking off to play. He would often go to the canal in front of the house to fish for small catfish, then return the gate to its place before his parents came back, acting as if nothing had happened. After a few times, his elder sister told on him, but Nam's mother just shook her head, defeated.

On full moon days, Nam would often run to the pagoda nearby, where his great-grandfather, the abbot, lived, to ring the big bell. Afterward, he would receive fruits or sweet porridge from the offerings brought by visitors.

After about a year, the situation began to calm down. The French had consolidated their power in the village, building strongholds and appointing local officials to manage administrative affairs. They forced the village's young men to join the local militia, using the strategy of "using a stick to hit itself."

During this time, the resistance movement against the French was secretly gaining momentum. Vietnamese patriots fled into the jungles to establish bases against the French, known as the "Viet Minh." Those who collaborated with the colonial government were labeled "Viet Gian" (traitors).

Nam's parents gradually brought the family back home to resume their normal lives. Young men in the village, if they did not join the resistance, were often imprisoned for various alleged crimes.

Due to the difficult circumstances of the time and the need to make a living for the family, Nam's father had to leave the village and his family behind to go to Cholon to find work. Nam also began attending the first grade at the village school despite being almost eight years old. Nam's life, from birth until then, had been marked by constant warfare and turmoil. His childhood was spent within the confines of bamboo groves, cut off from the outside world. He grew up waking in the night to the sound of distant cannon fire from the French army or the drums and gongs of the Viet Minh, who would disrupt the village's strongholds at night.

Nam's childhood was filled with bitter memories of a subjugated nation. Every morning, he had to stand at attention to salute the "Tri-color" flag of the colonial motherland and bow his head whenever he saw a Frenchman or a Frenchwoman pass by.

It had been a long time, but Nam still vividly remembers a shameful, humiliating incident involving a Vietnamese policeman. The policeman grabbed a cane from a horse-drawn carriage driver and brutally beat the old, frail driver on the head and neck while many Vietnamese and foreign passersby watched. The reason was that the horse had bolted and almost knocked over a Frenchwoman crossing the street. This despicable act of groveling to foreigners left a deep scar on Nam's young mind, instilling in him a deep sense of inferiority about his nation's subjugation.

In 1951, when Nam's great-grandmother passed away, his grandfather retired. Nam's mother, determined to find a better future for her children, left the ancestral home to her uncle and moved to the city for the sake of her children's education and future, preferring the opportunities of urban life over the simple rural existence of constant rain and sun.

When they moved to Cholon, Nam's family settled in a relatively decent house in Phu-Lam, which Nam's grandparents had sold to them. Although there was no spacious garden or moats like in the countryside, every afternoon after school, Nam and his elder sister had to fetch water from a public tap to have enough for the family's use. Though it was a bit tiring, it was a far cry from walking barefoot across cracked, painful, dry fields or wading through knee-deep mud in the countryside. The streets here were wide and paved, with horse-drawn carriages, rickshaws, and bicycles for transportation, no longer the long walks of the countryside. The bustling streets were filled with people coming and going, and the noise of traffic was constant.

The hurried, materialistic life of the city gradually made Nam forget the lonely afternoons of his village, where he would sit quietly by the bamboo grove, watching the sunset. Nam had always been fascinated by the colorful sunset, imagining it to be the palaces of the heavenly court, where a thousand fairies were singing and dancing. On misty mornings, he and his sister would watch the early-blooming orchids in the garden. Nam also fondly remembered the summer afternoons with the creaking of the hammock and the sound of rice being pounded by the neighbors, all forming a rhythmic melody of rural life. Now, these familiar sounds were replaced by the rush of city life and the noisy traffic of the streets during rush hour, occasionally interrupted by the quarrels of neighbors over trivial matters.

When Nam moved to the city, he was admitted to the second grade at Phu-Lam Primary School. In the beginning, he was shy and nervous, climbing the stairs to the third floor of his classroom. But after a few weeks, he became accustomed to the new environment. Nam was in Class 2B, taught by Mr. Ba Tam, who was very strict and disciplined with his students. Every Friday morning, he would make the students stand and show their hands for inspection. Those with dirty or long nails would get five painful slaps with a ruler. All the students were terrified, and every Thursday night, they would thoroughly wash and scrub themselves before going to school.

Especially during the dry season, the boys were often punished for kneeling because it was cricket-fighting season. They would bring crickets in matchboxes to school and store them in their desks to fight during recess. If one cricket chirped, the whole class would join in, chirping "Ren-ret," causing a commotion. The teacher would then have them bring their crickets to the front of the class and kneel for half an hour as punishment.

It's worth mentioning some of the experiences of playing with crickets, a common pastime among Vietnamese children. Crickets were often kept in small woven baskets made from thin bamboo strips, brought from the countryside, and sold to students. After choosing a cricket, the seller would set the price based on its size and color and then place it in a matchbox to deliver.

There were different types of crickets, but the two most popular among students were the fire cricket and the thang cricket. Female crickets didn't fight, so they were sometimes used as bait for fishing. Fire crickets were a bright orange color, while thang crickets were jet black. The male crickets looked very imposing, especially when they raised their wings to chirp "Ren-ret," challenging their opponents. In battle, they would spread their large front pincers and fight fiercely. If one was too timid and ran away, the owner would tie a thread around its head, spin it around to make it dizzy, and then place it back in the arena, where it would fight immediately.

Playing with crickets was an art form, much like cockfighting, requiring strategy and cunning to defeat opponents. Catching crickets also required patience. After the last harvest, when the fields had dried up, and the ground had cracked, children from the countryside would start catching crickets. Before going out, they would prepare a few small bamboo sticks about the size of a finger. After hearing a cricket chirp, they would determine its location and approach it carefully. If the crickets were fighting, they would chirp loudly, but if one ran away, it would only chirp a few times and then fall silent, making it easy to lose track. The easiest to catch was the male cricket luring the female, as he would chirp in a soft, drawn-out "Tak-tak-tak" to attract her. If they were in a cracked section of the ground, the children would use their bamboo sticks to block the cracks and then slowly dig them up. The female cricket, being chivalrous, would usually come out first, while the male cricket would stubbornly stay behind, only being pulled out once they dug deep enough. Sometimes, after a whole morning of searching, they would only catch a few crickets.

The educational system at that time was very challenging, with many children of school age. Government public schools were scarce, so student enrollment was very limited. Sometimes, a whole village would only have one primary school with three or four classes going from kindergarten to third grade. To study beyond that, children had to go to the district school, which offered higher grades. Due to the French colonial policy of keeping the population ignorant, many poor people in rural areas remained illiterate for life.

After completing the primary school curriculum, students had to pass an exam to receive a primary school diploma, called the "Higher Primary Certificate." In a village, it was fortunate to have one or two children pass this exam. Their parents would usually throw a party to celebrate and show off their pride to the neighbors.

After two years at Phu-Lam School, Nam passed the primary school exam and was preparing to take the entrance exam for the seventh grade at Petrus Truong Vinh Ky High School. Nam vaguely remembers that at that time, Saigon only had one public high school, "Petrus Ky." The "Chu Van An" school was established later and was prioritized for students who had migrated from the north.

Taking the entrance exam for the seventh grade at a public school was a very challenging task because most of the students were from Saigon, Cholon, and even neighboring areas and provinces. Many wealthy families from rural areas wanted to send their children to the city for education. The entrance exam was extremely difficult, beyond the capabilities of an ordinary child. The school's capacity was only a few hundred students each year, while thousands of candidates applied. Poor children who failed the exam faced the end of their education because their families couldn't afford the tuition fees to continue studying at private schools in Saigon.

Due to the rote-learning educational system and the challenging exams, the country remained impoverished, lacking practical talents. Students had to memorize lessons in subjects like physics, chemistry, and history. When a teacher called on them, they had to recite the lesson verbatim without stumbling or omitting a single word to receive high marks. Those who were lazy and didn't memorize their lessons or stumbled while reciting would be caned. Repeated offenses would lead to expulsion from school. Students had to study like "parrots," learning everything as it was written without questioning or understanding why.

The primary goal of education was to obtain a diploma as a mark of pride and then seek a position in society to make money and enjoy life, compensating for the long years spent studying. Only a few people had the passion to use their education and knowledge to contribute to the prosperity of the nation. Perhaps due to this general mindset and many outdated, bureaucratic attitudes that blindly revered academic achievements without practical application, Vietnam remains one of the poorest countries in the world today.

After failing the entrance exam to Petrus Ky High School, Nam returned to study at the continuation class in Binh-Tay. A few months later, the Cao Thang Technical School opened an entrance exam for the seventh grade, and Nam passed.

It seems that the Cao Thang Technical School in Saigon was initially a school specializing in training mechanics for the French Navy. Later, it was handed over to the Ministry of National Education and transformed into a National Technical High School. The school received funding and state-of-the-art equipment from the Federal Republic of Germany and the United States to train future technicians for Vietnam's developing industries.

The school's education program was divided into two parallel parts, training students from the seventh to the fourth grade. At the end of the fourth year, all students had to pass an exam to obtain the lower secondary diploma. After passing this stage, students can choose their path. If they wanted to continue their studies, they would pursue three more years, taking exams for the first and second parts of the baccalaureate. Afterward, they could take the entrance exam to the Phu Tho Higher Education School to study engineering.

For those who wanted to graduate earlier, after passing the lower secondary diploma exam, they could study for two more years at Phu Tho Technical College to obtain a technician's diploma, called a "cán sự," which was highly needed for the newly developing industry in our country.

Currently, Nam knows a few friends from the past who graduated from Cao Thang Technical School and are now running important industrial tool manufacturing businesses.

In the first year of seventh grade, Nam studied at "48" Phan Dinh Phung Street in Saigon. Every morning, Nam rode his bike from home to school, taking nearly an hour. At noon, he stayed at school, eating sticky rice or bread with sugar and drinking cold water. In the schoolyard, there were two large tamarind trees, and during the tamarind season, students would gather during lunch to throw rocks at the trees to knock down the fruit. Often, they would break windows or hit the roof of the teachers' quarters, causing them to run for cover. Sometimes, they would sneak out to the zoo and swim in the Thi Nghe River. Nam and his friends would challenge each other to swim across the river or jump from the bridge, despite the strong current, feeling invincible.

In 1992, when Nam returned to Vietnam to visit home, he sat drinking coconut water by the Thi Nghe Bridge and recounted these old stories to his mother. She shook her head, never imagining that her son had once played so recklessly.

In the following year, while in the sixth grade, Nam and his friends returned to the main school on Huynh Thuc Khang Street. Here, there was no more river swimming or tamarind throwing; instead, they would go to the Vinh-Loi cinema for double-feature films at five dongs for two movies, or they would go to support the school's soccer team. During that time, the Cao Thang School soccer team played very well, and they were also famous for fighting in Saigon. Every time there was a match, many students from Cao Thang would show up to support their team. At that time, the Tan-Thinh School team was their equal, and whenever they played in the finals, there was almost always a brawl, sometimes so intense that the police couldn't intervene.

###

It is worth briefly mentioning the historical context of Vietnam during this period. After their defeat at Dien Bien Phu, the French expeditionary forces were forced to surrender and sit down to sign the Geneva Accords, which divided the country. The area north of the 17th parallel was controlled by the communist forces under President Ho Chi Minh, while the area south of the 17th parallel was governed by the southern forces under the former emperor Bao Dai, who was known as the national chief. Bao Dai was the last puppet king of the Nguyen dynasty, protected by the French.

At that time, Mr. Ngo Dinh Diem was invited by the United States government to return to Vietnam as Prime Minister to establish the southern government. With the support and trust of the southern people, it wasn't long before former Emperor Bao Dai was deposed, and the people elected Mr. Diem as President, establishing the First Republic of South Vietnam, based on a democratic system like the Western capitalist nations.

After a difficult period of quelling the regional warlords such as Nam Lua, Ba Cut, and Bay Vien, the southern region of the country returned to relative peace and autonomy.

Nam remembers that around 1955, during summer break, he accompanied a classmate to visit his hometown. When the bus stopped at the crossroads of Cai-Lay on the national highway, they transferred to a horse-drawn carriage to enter the village. On the dirt road, which ran alongside a canal that brought water from the main river into the fields as part of the government's irrigation program, two rows of "trambau" trees stood tall and green along the canal banks. To the left, the rice fields were heavy with grain, a golden hue stretching across the sky, signaling the approaching harvest season. The midday breeze from the fields barely rustled the leaves at the tops of the trambau trees, carrying a faint scent of the countryside, hard to describe. The rhythmic clanging of the bell on the horse's neck kept time with the carriage's pace. Sitting in the carriage, Nam felt like a creature created within the brushstrokes of a rural Vietnamese landscape.

The homeland was truly at peace. Nam no longer saw the black-faced, ink-faced invaders or the arrogant, insolent foreigners strutting through the markets. The Roberts, Henrys, and Simones of mixed Annamite descent still imagined themselves as citizens of the motherland, struggling to speak their native language while working as interpreters or informants for the invading forces, digging up the graves of their ancestors' land for a share of leftover liquor and half-eaten butter from their foreign masters.

During his time in Cai-Lay, Nam truly experienced the most peaceful, joyful moments of his fifteen-year-old life. Friends would row boats to see outdoor performances late into the night, return to eat sweet egg soup, and listen to traditional Southern folk songs under the moonlight in front of their houses. At noon, they would lead buffaloes to the fields to catch prawns, grilling them with salt and pepper, and in the evening, they would climb trees to pick fruits. The fruits here were plentiful and came in all varieties with the seasons. Thanks to the fertile alluvial soil, the trees thrived without the need for fertilizer or watering.

A very special trait of southern rural people was their hospitality; the saying "Sacrifice your appetite to treat your guests" was very common. Sometimes, the best food in the house wasn't eaten by the family but was reserved for elders, guests, or offerings at village festivals or temples. Even now, Nam's mother still holds to that tradition. Whenever friends of her children or grandchildren visit, she bustles around, going to the market to buy all sorts of food to cook for the guests. Nam's mother often said, "When guests leave full and happy, the house is blessed." She took great pride and believed deeply in these actions, and as a result, all of her children and grandchildren were prosperous and well-off.

A few years later, Nam passed his lower secondary exam. Some friends suggested studying electrical engineering, but Nam's mother wanted him to continue studying to attend Phu Tho Technical College.

During his time studying for the upper secondary curriculum at Cao Thang School, Nam followed his friends and started taking the bus to school instead of riding his bicycle. His personality began to change from a mischievous boy who loved to tease to a more calm and gentle young man.

Nam started keeping a journal, writing poetry, and meeting with friends. He began to imitate the style of ancient philosophers, growing his hair long and dressing in a thoughtful, artistic manner. He dreamed of becoming a figure like Tu Uyen or Kim Trong from the timeless literature of Vietnam.

In the adolescent years of a young boy, a time when the heart beats differently upon catching a glimpse of the opposite sex, Nam began to experience the dreams of youth, yearning to "build a house by the stream" and drink cool water while singing the praises of love.

Nam was just starting to take his first steps, learning to speak, polishing his writing, and carefully grooming his hair, often clumsy and awkward in his newfound emotions.

Love is always new. It is an exploration, a discovery of something unfamiliar and enchanting between two individuals. It is the harmonious blending of the minds and hearts of two people of the opposite sex. It is a natural law of creation that has existed since the time of Adam and Eve.

Then, one day, by chance, on a bus, Nam caught the eye of a young girl, her face partially hidden under a tilted conical hat. She occasionally glanced up at him, then pretended to look away. Each time she did this, Nam felt nervous and unsure of what to do or how to respond. He discreetly discussed this with an older friend, knowing that the friend already had a girlfriend and would likely have some experience in matters of love. After receiving some basic advice, Nam felt more confident and daring.

Indeed, from ancient times and surely for millions of years to come, beauty and love will always possess a boundless power, surpassing all the forces created by humans. It exists in all ages and in every social class. Beauty can elevate people to the highest peaks of glory, but it can also cause them to suffer and die slowly in loneliness.

Nam's first love, as a young boy, was not as intense as a raging storm; instead, it was gentle like the clouds in the sky, delicate like silk. She wasn't as beautiful as Xi Shi, the legendary beauty of the state of Yue, or as striking as Yang Guifei of Tang Emperor Ming. But to Nam, her eyes were like a poem with no ending.

Nam knew her name and the school she attended. They took the same bus every day, and in the evenings, they would walk home together. Nam fondly remembers her school uniform, a pristine white ao dai. Whenever the evening breeze blew, her long ao dai would flutter, sometimes wrapping around Nam's legs, making him stumble slightly. She would shyly tilt her hat and smile, saying, "You're walking too close to me."

Nam became a poet. He wrote many schoolboy poems for her. He wasn't sure how deep his feelings were for her, but whenever she was absent or late, he felt inexplicably lonely and sad. Nam loved her in the simplicity of youth, with ink-stained white shirts and blue pens. It was just that, nothing more. They never held hands or shared a kiss.

###

Time passed, and Nam grew older, influenced by the circumstances around him. After graduating from high school with a baccalaureate degree, Nam couldn't enter Phu Tho Technical College due to technical reasons, so he enrolled in Saigon University's Faculty of Science instead. Gradually, his encounters with her became less frequent as their paths diverged, and he became busier with other responsibilities, such as running an evening school for children preparing for their lower secondary exams with a group of friends.

The North Vietnamese communists were determined to carry out Ho Chi Minh's vision of turning the Indochina Peninsula red, so the war grew increasingly intense. President Ngo Dinh Diem issued a general mobilization order, calling on students to volunteer for military training in the reserve officer classes at Thu Duc, or in the specialized military academies of the Navy, Army, and Air Force.

During this time, students had many different opinions and were often caught up in debates. Nam and his friend from school, Nguyen Van T. (a former squadron leader who died in 1970), discussed their future paths many times. Finally, they agreed to apply to the Air Force in the 63A flying class at the Nha Trang training center.

Nam's life had truly entered a new phase, a crucial chapter in his personal history. Although Nam's mother wasn't entirely happy with his choice, she didn't oppose it either, understanding that he had chosen this military career.

On January 1, 1963, Nam bid farewell to his family, slinging a heavy backpack over his shoulder, and together with his friend, Nguyen Van T., he set off for military service. The VNCH Air Force's C47 transport plane had just landed at Nha Trang Airport and was taxiing to the parking area. Nam and his companions were still hesitating, taking in the mountains and surroundings like tourists, when suddenly, a loud voice boomed from behind.

"Hey, you there, what are you waiting for? Put your backpacks on and get in line immediately!"

The whole group was taken aback, not knowing what to make of the situation. Nevertheless, everyone remained silent, jostling each other to line up, like elementary school kids following the teacher's orders before entering class. At that moment, Nam noticed an old classmate standing with the group that had just issued the orders. Nam was about to call out to him when another shout came right next to his ear.

"Get in line! Here, there are no friends or brothers, only seniors, got it?!"

Nam obediently tucked his tail between his legs and silently fell back in line. Once everyone was in place, the senior continued to give orders.

"All of you, raise your hands, forward march... One, two... One, two..."

A group of newly recruited young men, still rough around the edges, their rebellious, cowboy, student bravado quickly disappeared, and they became meek and docile, obediently following the commands of the senior cadets with their unmistakably military haircuts.

In the early days, referred to as the "humiliation training period," it was extremely exhausting and continuously busy from morning until night. It seemed that any action or response to the senior cadets was considered insubordinate or incorrect. Doing ten push-ups, twenty jump squats, or running a few laps around the camp were common occurrences, happening several times a day.

There was one particularly funny and unforgettable memory. Every night before going to bed, the cadets had to line up properly for roll call. Nam was in room six, which housed about forty cadets. After counting from one to forty, the last cadet had to report, "Room six reporting, forty present." If the cadet was from the south or north, there usually wouldn't be any issues. However, this particular cadet was from Quang Tri, and his accent was quite heavy. Instead of saying "present" with the correct tone, he used a falling tone, which made the whole room burst out laughing. Consequently, the "Carrot" senior officer ordered everyone to change into their combat uniforms, put on their backpacks, and report to the Cadet Office within one minute.

After being scolded, everyone had to do fifty push-ups, then put on their backpacks for a "night flight" with ten laps around the training field. More than half an hour later, everyone was exhausted and out of breath before being allowed to return to their rooms to change and sleep. The next day, we requested that the cadet change his standing position during roll call to avoid further issues.

After about three months of basic military training, the new cadets participated in the Air Force Training Center's "Alpha" ceremony. The flight cadets were awarded half wings. In class 63A, only Nguyen Van Ve-Ch. received full wings, which made him look very impressive, and everyone else dreamed of having the same silver wings. Ve-Ch. had previously been a pilot during the French colonial period, but now he was undergoing basic military training to adjust his rank. In addition to the flight cadets, there were many technical cadets who would become future specialists and leaders in the Air Force's technical fields.

After receiving the "Alpha," Nam officially became a full-fledged officer cadet of the Republic of Vietnam Air Force. No longer did he have to crawl and roll outside on the scorching runways or spend all day with "Rifles on shoulder... Forward march..." From Monday to Friday, every day was spent studying English or completing security and background paperwork while waiting for the call to go abroad for flight training in the United States.

Every Saturday morning, the cadets' faces would light up with joy. Shoes were polished to a shine, and uniforms were crisp and wrinkle-free. After the flag-raising ceremony, the new cadets were inspected by the senior cadets one last time for proper dress and grooming before being granted leave. Any cadet who was lazy, with unpolished shoes or slightly wrinkled clothes, would be confined to the camp immediately. This was particularly troublesome for those who had girlfriends and had promised to take them out on Saturday but were instead punished and had to stay home reading newspapers. Their girlfriends might think they were the type to "run down the runway and take off for good!" or that they had joined the "Frogmen squad!"

Nha Trang is known as the land of coconuts, of pine trees, and white sands—a place where the characteristics of both the south and central regions blend together in a natural, open-hearted way, like the vast sky and sea breeze.

Nha Trang was unfamiliar to Nam, but also very close to his heart. He loved the city from his first days of leaving the camp. Here, he didn't have a lover, but the blue sea, vast beaches, and rolling mountains, with their picturesque natural scenery, deeply touched the romantic soul of a young man easily moved by the mysterious and miraculous arrangements of nature.

Nam had long lost contact with the girl he used to walk home with after school. Her image gradually faded from his mind. The schoolboy poems and diary pages were tucked away in his thoughts, replaced by the soldier's backpack, combat boots, and short hair.

After nearly six months of military training, filled with mixed emotions and unforgettable experiences, Nam was about to leave Nha Trang and return to Saigon for a medical examination and procedures to go to the United States for flight training. While waiting in Saigon, Nam searched for and inquired about the girl from his school days, only to learn that she had quit school to marry someone new. After a moment of melancholy and nostalgia for the past, Nam told himself, "I'm just a guy with nothing, and the future is uncertain, so I can't expect anyone to wait for me!" Nam quickly turned and walked back to his friends waiting outside the Phi Long camp gate.

###

Early the next morning, around the first week of July 1963, a Boeing 707 jet from Pan Am Airlines took off from Saigon, carrying the cadets to flight training school in the United States. Nam was scheduled to train on helicopters that the U.S. military would provide to Vietnam, replacing the old aircraft left by the French military.

This was Nam's first trip abroad in his military career, so everyone looked relaxed and happy, making up for the tough days at the Nha Trang training camp under the scorching tropical sun. The training fields, where the sand glittered like diamond dust under the sunlight, were places where sweat soaked through their uniforms as they crawled and rolled during exercises. The three months of rigorous basic military training had completely changed the personalities and appearances of the "bookish boys." Their speech and movements became more disciplined. It seemed that the military had brought the man out of the boy. Nam thought that if he were to praise them, he would say that those who truly fought in the military were the real men among men.

After about fourteen hours on the plane crossing the Pacific Ocean, they finally landed at Travis Air Force Base near San Francisco, California. After collecting their luggage and being guided to the Officer's Transit Quarters for the night, Nam was exhausted from sitting still on the plane for too long and from the time change as the plane crossed several time zones. After a shower, he went straight to sleep. It wasn't until nearly 6 PM local time that his friend Th. from the next room knocked on the door and urged him to wake up.

"Hey! Get up and change. Let's go find something to eat. Did you come here just to sleep or what?!"

"Okay, wait a minute."

Nam slowly got up, looked out the window, and saw the soft evening light shining on the green lawn outside. Nam suddenly realized, "Oh! I'm in the United States now..." The beautiful land he had imagined through vivid and colorful images in magazines about the United States, which he had seen while still in school in his rural village, had come to life. He had been fascinated by the progress and achievements he saw, even though he didn't know exactly where the United States was at the time. But he secretly hoped that one day he would see it with his own eyes and touch the greatness of this civilization.

As he grew older and attended high school, Nam developed a different impression from watching American Western movies. He imagined that American men were constantly fighting and shooting each other with no regard for the law, and when they encountered women, they would jump on them and make love freely, like wild animals in the forest. These violent and sexual images had given the thirteen-year-old Nam a negative and immoral view of American society.

After putting on his sky-blue Air Force dress uniform and standing in front of the mirror, Nam felt more mature. Although he tried to look serious like a real soldier, he couldn't hide the youthful and innocent face of a young man just beginning his military career.

The wind from the vast golden fields outside carried a strange scent, different from the fragrance of ripe rice fields back home. Nam felt comfortable with the cool weather and the clear blue sky, with not a cloud in sight. The group of about ten cadets walked to the Officer's Club for dinner. This was also Nam's first time entering an American Officer's Club. The luxurious decor, expensive furniture, and artistic presentation, along with the respectful service, made Nam and his friends a bit self-conscious, even though they had all been briefed on proper etiquette and dress before leaving Vietnam. After being seated at a large table with enough space for everyone, they looked at the menu, which was filled with dishes they had never tried before. In the end, Nam only recognized one dish: "beef steak." To keep things simple and avoid any confusion, he ordered it. His friends weren't much better off than Nam.

Nam remembered a funny story about a rich man who was completely ignorant of French. Trying to appear sophisticated like the French, he went to a French restaurant to dine. After looking at the menu, he was overwhelmed by the foreign words. But trying to stay calm, he randomly pointed to a dish on the menu. A few minutes later, the waiter brought him a plate of bún (vermicelli) mixed with minced meat, which he tasted and found unappealing. Meanwhile, the man at the next table was enjoying steamed mussels with butter and a glass of fine wine. Frustrated, the rich man decided to wait for the man to order another round of mussels. When the opportunity came, he imitated the man's order exactly. After signaling to the waiter and saying "Encore," the waiter respectfully responded, "Oui, monsieur." Once again, the waiter brought him a plate of bún mixed with minced meat. Shaking his head, the rich man called for the bill and left (this story was often told to poke fun at those who tried to appear sophisticated but were actually ignorant).

Nam wasn't quite as bad off as that man since everyone in the group had a basic grasp of English to get by, unlike the rich man. In Vietnamese restaurants, they usually bring vegetables with beef for customers to eat together. However, in the United States, they typically serve salad first or bread with butter while waiting for the main dish to be prepared. When the waiter brought out Nam's beef steak, he also removed the salad plate, assuming Nam didn't like vegetables. Realizing his mistake, Nam quietly let it go to avoid further embarrassment.

After paying the bill and stepping outside, the group noticed a sign on the wall indicating that the area was reserved for senior U.S. and allied officers. No wonder Nam only saw colonels and generals dining inside! Perhaps the waitstaff assumed the group was composed of young, high-ranking allied officers, possibly even royalty or nobility, rather than new recruits!

The next morning, the group was directed to the train station to take a train to San Antonio, Texas, from where they would transfer to a bus to Lake Lane Air Force Base for English language training. Since the journey would take them through several states, the group purchased sleeper tickets. Nam lay on the top bunk, while Nguyen Van Th. was on the bottom, enjoying the warmth.

In the first few hours, Nam was captivated by the scenery outside the window, observing the surrounding hills and farms. He tried to find similarities between the countryside here and the fields back home but was disappointed not to see golden rice paddies or green bamboo groves. Instead, he saw rolling hills and vast, barren fields stretching endlessly into the horizon, occasionally interrupted by farms with neatly planted rows of corn or dense clusters of trees covering hundreds or even thousands of acres.

The United States is an enormous country, consisting of fifty states, each sometimes larger than an entire country like those in Europe or even Vietnam, Laos, and Cambodia. It spans from near the Arctic to the tropical regions, so the climate across the nation encompasses all seasons if you enjoy cold weather or want to experience short days and long nights, head to Alaska, near Siberia in Russia. If you prefer warm weather year-round, head to Florida or Hawaii in the Pacific.

The rhythmic "clunk-clunk" of the train running along the tracks began to give Nam a headache, so he returned to his bunk to rest. For the next two days, he lay in bed, suffering from a splitting headache, unable to eat, and surviving only on soft drinks that his friend Th. bought for him.

Finally, the train reached its final stop at San Antonio, Texas. Nam struggled to step out and sit on a bench in the station. He felt much better, no longer motion sickness from the train ride, but he was hungry after two days without food. He made his way to a vending machine nearby and was thrilled to see what appeared to be a sausage sandwich inside. He inserted his money, pulled the lever, and retrieved the sandwich. But after taking a big bite, Nam was shocked—it didn't taste like the pickled cucumber and sausage he was expecting. Realizing it was not sausage but a hot dog, Nam immediately spit it out into the trash can next to him. Another lesson learned in this new land. He continued to survive on chocolate bars.

While waiting for the bus to take them to Lake Lane Base, one of Nam's friends, frustrated, cursed out loud.

"Dammit! Even the water fountain is racist! Why does the water come out when the Americans drink, but when I open my mouth, nothing happens?!"

The whole group burst into laughter as they realized that the water fountain operated with a hidden foot pedal. When a black man stepped on the small button beneath the fountain, the water flowed. The group found the simplicity of the mechanism amusing, as they had overlooked the button hidden underfoot.

After World War II, electronic technology advanced rapidly, leading to many new inventions, not just for national defense but also for everyday use. For example, there are handheld electronic devices that can pinpoint your location anywhere in the world; with just a few button presses, satellites will locate you and provide nearly accurate coordinates. A few years ago, Nam encountered one such innovation in a hotel bathroom. He noticed that the water faucet had no knobs or buttons. Unsure what to do, he hesitated until he saw the person next to him simply place their hand under the faucet, and the water automatically flowed. This device was designed to conserve water.

Finally, Nam and his friends arrived at Lake Lane Base. According to the list, each pair of cadets was assigned to a room with daily housekeeping services. Meals were provided at the mess hall within the base. The only downside was the ban on cooking in the rooms, so when they craved Vietnamese food, Nam and his friends would improvise with makeshift meals, reminiscing about the flavors of home.

On weekends, they would often go to town to buy records or wander around the Alamo or along the river, watching people enjoying the scenery.

Nam learned that in previous years, Vietnamese cadets were allowed to buy and drive cars. However, after causing public disturbances, this privilege was revoked. Occasionally, when they wanted to go on a trip or deal with "homesickness," they would ask Turkish friends to drive them to Mexico. On Friday evenings, they would head to nightclubs and relax with

Mexican girls, whose petite, curvaceous figures and charming personalities made the Vietnamese bachelors forget their way home. The next morning, they would return to base, bleary-eyed and stumbling, struggling to find their way to the bus.

After six months of English training, the cadets were transferred to flight school in Mineral Wells, Texas. There, they joined other international cadets in class 64w.

Mineral Wells is a small county town near Fort Worth. It is situated in a vast flatland surrounded by numerous farms. The U.S. Army selected this location as a training base for helicopter pilots to supplement the forces in Vietnam and simultaneously train pilots from allied nations worldwide. In the early days, cadets wore their caps backward, with specific colors to distinguish each class. Once they completed their first solo flight, they were allowed to wear their caps correctly.

During his first training flight, Nam was not accustomed to the tight-fitting flight helmet and being strapped tightly into the cockpit seat. About half an hour into the flight, Nam became dizzy and nauseous and asked the instructor to land at a nearby landing zone to rest. By the second and third days, he began to feel better. After a week or two, he started to enjoy the sensation of taking off.

After some time, Nam's instructor allowed him to solo around the airfield. The helicopter used for training at that time was the OH-23. It resembled a dragonfly, with only two seats at the front and a long tail at the back. When the instructor allowed him to solo, they placed a few sandbags in the instructor's seat to balance the helicopter during takeoff.

Flying a helicopter may look easy, but it is quite challenging. The pilot must use both feet, both hands and their mind in a coordinated effort. Imagine standing still on a round ball, maintaining balance without falling—that's what it's like to hover a helicopter in place.

Additionally, a helicopter's movement—forward, backward, up, down, or hovering—is governed by vector dynamics, which depend on the tilt angle and direction of the rotating blades.

After his first solo flight, Nam felt more confident in his abilities. When the instructor was in the seat next to him, cadets tended to think, "The instructor is here, so it's okay," and didn't fully grasp the importance of making their own decisions when facing challenges.

That afternoon, when the bus brought them back to the base from the training field, it stopped near a farm's pond. Nam and a few others who had soloed that day were grabbed by the group and thrown into the pond as part of a "baptism ceremony."

Now Nam was officially allowed to wear his cap forward, just like those who had soloed before him. The next morning, his instructor told him to take the helicopter from the parking area to the training field about ten miles away on his own, while the instructor would drive and meet him there later. Nam was both excited and nervous because this was the first time he had to perform all the procedures like a real pilot. He had to conduct a pre-flight check, contact the control tower for taxi instructions, and line up for takeoff. Everything had to be done in the correct order. Eventually, everything went smoothly. About three months later, Nam and his classmates graduated and received their wings. However, some of the cadets were disqualified due to health issues, lack of ability, or repeated flight errors and were sent back to Vietnam to be reassigned to non-flight duties or other military branches.

After a few days of rest to pack their belongings, Nam and his friends left the base to continue their training at Fort Rucker in Alabama. There, they trained on slightly larger helicopters, the H-19 and H-34. Both were piston-engine helicopters developed during World War II, but they were still widely used in the early stages of the Vietnam War. Due to their effectiveness and speed, the U.S. military chose the H-34 to equip their elite forces, such as the Marine Corps and Special Forces, in Vietnam and around the world.

In the first few weeks, Nam trained on the H-19. Coming from the small OH-23, where you could easily step into the cockpit like getting into a car, the H-19 was much larger. To conduct a pre-flight inspection, Nam had to climb onto the rotor blades to check each bolt and the engine for oil leaks before climbing up to the cockpit. The startup procedure was also more complex, requiring strict adherence to the manual. Vietnamese cadets, being smaller than their American counterparts, often had to adjust the seat fully forward and upward to reach the pedals.

The H-19 was an older model with a heavy frame and a relatively weak engine, making it less capable of carrying heavy loads. During hot afternoons when temperatures exceeded 90°F, attempting to take off from a small clearing in the forest with full power over tall trees could often result in having to shut down the engine and wait for cooler temperatures. One key factor was that the pilot needed to be aware of the wind direction to take off into the wind. At airfields, pilots could contact the control tower for wind direction or look at the windsock. But in the forest, they had to figure it out on their own. There were various ways to determine wind direction, such as observing the movement of treetops or tossing grass into the air. If the wind was light, you could wet a finger and hold it up, and whichever side felt cooler indicated the wind direction.

After a few weeks of training on the H-19, the group transitioned to the H-34, which was slightly larger and equipped with a more powerful engine, allowing for greater payloads. The U.S. had equipped the VNAF with the H-34 to suit the current battlefield conditions.

During their time at Fort Rucker, despite being officer cadets, they were treated like other allied officers. Each cadet had a private room, and some rooms even had kitchens for cooking. It was here that many renowned chefs like Nguyen Van C., Tran Tan Th., and Huynh Van B. emerged. They would divide into small teams of about three or four cadets each. Nam was part of Nguyen Van C.'s team—C. was the head chef at the "Cây Mù U" restaurant in Bien Hoa. Nam was the dishwasher and Pham Van Tr. (former Lieutenant Colonel and squadron leader of the CH-47 squadron in Can Tho, who was killed during a combat mission in the Dong Thap Muoi region) was the errand boy. Since C. was skilled in cooking, he was the "boss," and whether the food turned out delicious or not was entirely in his hands. If anyone was lazy or didn't fulfill their responsibilities properly, C. would scold them immediately or go on strike, leaving the whole group to eat bread.

As we all know, when investing in a restaurant, the most crucial figure is the head chef. The number of customers a restaurant attracts depends on the chef's culinary skills. Sometimes diners will wait for hours or drive long distances just to enjoy a dish prepared by a skilled chef.

Even after nearly forty years, Nam still remembers C.'s braised chicken. In the beginning, Nam would often add water to the pot when braising chicken. But when it was done, the dish turned out to be boiled chicken, not braised, and everyone would shake their heads. The entire pot would then be thrown into the trash. Chef C. would scold them before giving instructions.

"If you want to braise chicken to make it flavorful, you need to add chopped onions, then season with pepper, soy sauce, sugar, and salt, and simmer. Never add water."

To this day, whenever Nam's wife refuses to cook, Nam resorts to his specialty: braised chicken. His children love it and praise how delicious it is. Nam also learned how to cook rice from his aunt when he was about nine years old back in the countryside. His aunt taught him to measure the water by inserting a finger into the rice. The water should be one finger joint deep over the rice, and you increase or decrease it depending on the amount of rice. Nam still uses this method today, so every time he cooks rice, it turns out exactly the same. However, there have been a few mishaps. Usually, his wife prepares the food, and Nam continues his old routine of wiping tables and washing dishes. One day, his wife came home late from the market, and they had guests. He had to cook the rice beforehand. Normally, they used two cups of rice, but with Mr. and Mrs. Ngoc Anh visiting, Nam used four full cups. After rinsing the rice, Nam applied the same old measuring method: half rice, half water, and turned on the electric cooker. Normally, the cooker would turn off automatically when the rice was done. But this time, when Nam lifted the lid, the top was still watery, and the rice was undercooked. He wasn't sure what to do. Ngoc Anh advised him to move the pot to the gas stove to finish cooking since the electric cooker wasn't accurate with large quantities. Mrs. "Hia Tỷ" was amused and laughed heartily. Nam learned a lesson that day.

###

After about a month, Nam was considered capable of handling the controls of an H-34 helicopter. This meant that, besides being able to skillfully maneuver the aircraft during takeoff, landing, and emergency procedures, the pilot also needed to anticipate the weather and understand the impact of different cloud types to avoid them. Additionally, the pilot had to be proficient in reading flight maps, familiar with air traffic control frequencies, and knowledgeable about basic aviation regulations when flying over major cities or restricted areas. Before graduating, each cadet pilot had to plan a cross-country flight through several cities designated by the instructor to familiarize themselves with real-world aviation procedures at unfamiliar airports. They practiced using and reading intercity maps, communication devices, and navigation instruments accurately on the aircraft to avoid getting lost when flying at night or in bad weather. The cadets also had to accurately calculate their estimated time of arrival at each waypoint on their planned route. These long-distance flights usually took off early in the morning, with stops only for refueling, lunch, or overnight stays at airports or provincial capitals according to the preplanned itinerary.

The flight instructors at Fort Rucker were all U.S. military officers with extensive flight hours and combat experience. Most of them had served in Vietnam for several years, flying combat missions for both the U.S. and South Vietnamese forces, so they were somewhat familiar with the customs, geography, climate, and guerrilla tactics of the communists at that time. This was in contrast to the OH-23 instructors at Mineral Wells, who were mostly civilian pilots hired by the U.S. government on long-term contracts to train cadets from allied nations.

After the cross-country flights, Nam and his classmates moved on to the next phase, which involved survival and escape training in the forest. The goal of this training was to teach pilots how to survive if shot down in the jungle. The exercise lasted about two days. The cadets were divided into small groups, each consisting of two or three people, and were dropped off at various locations at night in a wooded area on the side of a mountain. The cadets then had to use maps to navigate their way to a designated base camp located more than ten miles away. The afternoon before boarding the bus to the drop-off point, all cadets were thoroughly searched. They were not allowed to bring any food or drink and had to find their own sustenance for forty-eight hours.

In the first few hours, Nam and his group tried to hack their way through the forest, crossing dry creek beds, but they soon became exhausted. The group decided to move closer to the road to make the journey easier. They were instructed that only helicopters were considered friendly, while all vehicles and military patrols on the road were enemies. They had to avoid being captured while on the move. So whenever they saw headlights approaching, the cadets would jump into nearby ditches to hide.

The difficult part was that to reach the designated base camp, they had to cross several streams, some of which were waist-deep. If they tried to cross a bridge, they risked running into enemy patrols guarding it. In the end, Nam's group was captured while crossing a small bridge around 1 AM, thinking the enemy had gone to sleep in their vehicles. As the group reached the middle of the bridge, they were discovered by the enemy, who shone spotlights on them, revealing their position. The enemy then fired blanks at them to simulate pressure.

They were then escorted back to camp and each placed in an oil drum, where the enemy pounded on the outside of the drum with sticks to simulate an interrogation. After about ten minutes, another group of cadets who had also been captured arrived in a jeep. Nam's group was then released, crawling out of the drums so the next group could take their place. After being released, each cadet was given two oranges to sustain them on the way back to the main base. Everyone in the group was exhausted and sleep-deprived and decided to find a place to hide and rest for a few hours, waiting for daylight to continue. Nam's group consisted of three cadets. They found some bushes to sit against for warmth. One American cadet in the group pulled a chocolate bar the size of a thumb from his pocket, which he had hidden since the previous afternoon, and shared it with Nam, advising him, "Eat a little now and save some for when you're really hungry."

Nam had prepared by wearing two layers of thick combat uniforms. On top of that, he wore an Air Force jacket, pulled his combat helmet down to cover his ears, and wore gloves and tightly fitted combat boots. Despite the cold mountain air late at night, Nam felt warm and fell into a deep sleep.

The sound of machine guns firing and the shouts and laughter of the enemy, who seemed to be chasing another group along a trail, woke Nam's group. The sun hadn't risen high yet, and the forest leaves were still wet with dew. The group leader spread a map on the ground to determine their current position. Using the bridge where they had been captured the previous night as a reference, they estimated that the distance remaining was about five miles as the crow flies to reach the safe base. While this distance wouldn't take long if they followed a trail, crossing through the forest using a compass heading meant they had to wade through two more streams, which would take considerably more time.

After a few hours of traversing up hills and down valleys, following the stream, the group leader decided to take a short break to check the map and ensure they were heading in the right direction. Nam had already finished the chocolate bar his American friend had kindly shared with him, but he was still very hungry and uncomfortable. The leader decided they needed to reach their destination before nightfall, so the group could not rest longer than planned. Everyone, despite being exhausted and hungry, continued to push forward, maneuvering through bushes and climbing over rocks.

By late afternoon, as the shadows of the forest trees stretched eastward, indicating it was around three or four o'clock, one member of the group noticed a small plume of smoke not far away, towards the northeast. The group maintained silence and cautiously approached to observe. Once they confirmed it wasn't the enemy, the group leader identified himself and met with the friendly soldiers. They were roasting a rabbit they had just caught. Nam was given a small leg, and although the smell was quite unpleasant, the demands of his empty stomach forced him to swallow every bite.

With some food in their bellies, everyone felt a bit more at ease, and the two groups decided to join forces and continue together. By around six o'clock that evening, the entire group reached the designated camp.

They ate well and spent the night at the camp, waiting for the other groups to arrive. The next morning, Nam boarded the bus back to Fort Rucker. Everyone looked like they had just emerged from the wild, with tired, scratched faces covered in dust and grime. As soon as the bus pulled up in front of the barracks, Nam jumped off and ran straight to his room, quickly stripping off his clothes, throwing them into the bathroom, and immediately took a shower. After cleaning up, he locked the door and fell straight into bed, falling asleep almost instantly.

Today was the last day at Fort Rucker. Nam and his friends were waiting for a taxi to take them to the Greyhound bus station, where they would catch a bus to Travis Air Force Base in California to board a plane back to Vietnam. He took one last look at the barracks where the allied pilots had been housed, located next to a small forest. Every morning, Nam would cross a large grassy field to catch the bus to flight school. Today, the same scenery filled him with a strange sense of melancholy and longing. Those rooms had provided him warmth during the cold, rainy nights. He remembered the familiar sounds of the old black man adding wood to the stove in the basement, heating water. He thought of the small squirrels with their long, beautiful tails running around to eat bread crumbs and the black crows perched on the open lawn in front of the house, searching for food. Nam mused that he would probably never have the chance to relive these memories.

Today, the weather was beautiful, with warm, golden sunshine spreading across the yellowing grass on the gentle hills in front of him. California was famous for its nearly year-round warm weather, a homeland for women proud of their tan skin, captivating eyes, and full, youthful figures.

To Nam, California evoked a sense of romance, with its slender, alluring legs and outfits barely covering the curves of women's bodies, as if unintentionally inviting innocent glances.

The Boeing 707 was moving onto the main runway for takeoff. Nam was about to say goodbye to America, farewell to the flight school, to the warm California sun, and return all the bittersweet memories of two years spent there.

Through restless sleep, disrupted by the changing time zones, they were finally informed that they were about to enter the airspace of Saigon. Many young American soldiers, around nineteen or twenty years old, were peeking out of the windows. This was probably their first time in Vietnam. Nam couldn't guess what was going through their minds as he observed their faces. They were so young. Nam wondered, when they volunteered to come here, what roles they had envisioned for themselves. Did they volunteer to fight for the freedom of another nation? Or were they fighting to protect the interests of capitalists in their homeland? Nam was certain that these soldiers probably never considered either of the reasons he was thinking about, as their thoughts had likely been shaped by their country's politicians and leaders, who had instilled in them noble ideals like "hero" and "freedom defender."

Although Nam wasn't much older than these American soldiers, he believed he had a more concrete reason to fight than they did. Nam was Vietnamese, and he felt it was his duty to fight for his people to resist the expansion of communism from Hanoi.

Nam still remembered that President Ngô Đình Diệm of the First Republic of Vietnam had not invited these soldiers to defend the autonomy of his nation.

At this point, Nam recalled a story from long ago, involving an old friend. Nam wasn't sure whether to laugh proudly or feel nauseated by the overwhelming sense of subservience in the idea. Around 1976, after the communists from North Vietnam had taken over the South, Nam fled to the United States. At a gathering of fellow countrymen, a newly acquainted friend, who seemed to be older than Nam by about ten years, excitedly told him:

"Hey! Do you know? I'm writing a letter to the U.S. President, asking him to establish an expeditionary force made up entirely of Vietnamese soldiers to fight in battles anywhere in the world. I'm sure the President will listen to this idea..."

Nam felt a bit uncomfortable with this strange notion.

"So we came here to become mercenaries?" he asked.

The man seemed displeased with Nam's question.

"No, that's not it. We're helping the American military to protect freedom for other nations, and at the same time, repaying them for helping us come here..."

Nam thought that further discussion was pointless, so he found an excuse to leave. He later learned that this man had once served as a "Khố xanh, khố đỏ" soldier, a colonial auxiliary for the French army, and his mentality of servitude to the motherland still clung to him!

The plane gradually descended through layers of white clouds, revealing green fields and gardens in Biên Hòa and Lái Thiêu, to the north of Tân Sơn Nhất airport. The rooftops, weathered over many years, had turned dark brown, nestled in patches of green foliage. A few black buffalo were leisurely grazing by the roadside. These were some of the first images of Vietnam that entered the minds of foreign visitors as they looked out from the plane's windows.

After clearing customs and immigration, Nam took his luggage and caught a taxi home. It had been nearly three years since he first entered the military academy in Nha Trang. Saigon was still the same—its streets were noisy and bustling with taxis, pedicabs, motorbikes, bicycles, and pedestrians jostling each other during rush hour.

The first thing that struck Nam as he observed the reality of life in Vietnam was the sight of emaciated, poor laborers in tattered clothing, bending under the weight of their burdens as they walked along the roadside. This sight moved Nam deeply. Perhaps he had forgotten the harsh realities of a poor country riddled with conflict and personal ambition. Over the past two years in the United States, Nam had been surrounded by the beauty, civility, and modernity of another land, sometimes forgetting that he was also a Vietnamese, born into a people struggling in poverty. Suddenly, he asked himself, "What should I do for my people?" This question continued to haunt him, as Nam hadn't yet found a practical answer. Perhaps it was because he was still burdened by the greed, anger, and selfishness that are so hard to shed in this human life.

After a few days of leave in Saigon to sort out personal matters, Nam and his comrades received orders from the Air Force Command to report to the newly established 215th Squadron in Đà Nẵng. In 1965, the squadron's command staff consisted of Captain Trần Minh Th. (Colonel Th. later passed away in the United States after 1975) as the Commander, Lieutenant Đặng Trần Dj. as the Deputy Commander, and Second Lieutenant Nguyễn Văn Tr. as the Operations Officer. All the pilots were freshly minted second lieutenants.

Since the squadron was newly established and still in the process of training, it didn't have enough personnel or equipment. They had only received one or two aircraft, so Nam was sent to train with the U.S. Marine Corps squadron stationed on the other side of Đà Nẵng airfield.

A month later, the 215th Squadron received five or six CH-34 helicopters from the U.S. military. Nam and his comrades returned to train with their squadron instructors. Captain Th., Lieutenant Dj., and Second Lieutenant Tr. were all experienced helicopter instructors with the VNAF. Additionally, two more senior pilots, Lieutenant Vĩnh Qu. from the 213th Squadron and Second Lieutenant Phạm B. from the 211th Squadron, were brought in to reinforce the 215th Squadron.

During their time in Đà Nẵng, all the pilots, from the commanding captain to the junior pilots, lived in the single officers' quarters on the airbase. Every day, the new pilots would report to the squadron to divide into training groups or fly with instructors for about two hours.

It's worth mentioning why training a pilot is so costly for the nation. As we know, the Air Force's combat operations are fundamentally carried out by squadrons. The pilots who actually fight on the front lines not only carry out the responsibilities of soldiers but also must be highly skilled technicians. They operate sophisticated war machinery worth millions of U.S. dollars, sometimes carrying dozens of military personnel on board. They must thoroughly understand the capabilities of the aircraft they fly, their hands and feet must coordinate precisely and timely on the controls, and they must quickly recognize and respond to the numerous signals from the instruments and the myriad of switches in the cockpit. Additionally, external factors such as weather, wind, storms, and terrain significantly affect their missions, requiring swift and accurate decisions. Therefore, when selecting pilots, in addition to the mandatory basic educational qualifications, candidates must meet specific physical standards, including height, weight, and overall health.

During training flights, often over the Sơn Trà peninsula or the Non Nước area southeast of Đà Nẵng, because these areas had many open spaces and few civilians, new pilots had the opportunity to practice emergency landing techniques, such as engine failure or a 360-degree landing, where the aircraft had to spin sharply to land on mountain outposts or in dense forests to drop off special forces teams, etc.

On weekends, everyone had the day off except for the rescue crew, who had to stay on standby in the operations room. Occasionally, when Nam was stuck on standby duty, he would take the helicopter for a test flight north of the airfield to catch wild ducks swimming in the waterholes near National Route 1 on the way to Huế. All it took was lowering the helicopter close to the ducks, and the wind from the rotor would pin them to the ground. The captain would jump down, grab a few by the neck, and bring them back to the base for the cookhouse to prepare. That would be the afternoon's drinking snack.

Most of the new pilots were still single, except for the senior officers who were older, married, and had children, who carried themselves with more dignity and decorum. Outside of daily flying duties, the young officers enjoyed themselves in the city, seeking out cafés with good music and charming, sweet girls—the kind of places where young pilots liked to congregate.

After more than half a year in Đà Nẵng, Nam moved with the 215th Squadron to Nha Trang, under the command of Air Group 62, led by Major Anh (Brigadier General Anh, who later died in Cần Thơ before 1975). During this time, all the new pilots in the VNAF were officially promoted to Second Lieutenant. Nam was very proud of the two shiny new gold bars on his flight suit's epaulets.

Nha Trang was a famous tourist city in Indochina, clean and luxurious, with tall palm trees and green pines lining the beach and golden sands shimmering under the sunlight. Nha Trang welcomed Nam with open arms. He returned to this city as a young officer and pilot, not as a rookie cadet enduring the rigors of training at the military academy as he had three or four years earlier. The sound of the waves crashing on the shore no longer surprised Nam as it had during those early morning runs, back when he thought the thunderous noise was the sky about to burst into a storm!

The Vietnam War was also intensifying, escalating from sporadic guerrilla attacks at the squad or platoon level in remote villages to battalion- or regiment-level battles involving North Vietnamese regular forces in bloody engagements like Pleime, Đức Cơ, and Ben Hét.

The young "Elephants" of the 215th Squadron, known as "THẦN TƯỢNG" (THE IDOL), heroically soared across various battlefields, supporting friendly units in operations throughout the Second Military Region. Every month, they were deployed for about two weeks on missions to provinces like Quy Nhơn, Pleiku, and Buôn Ma Thuột, supplying outposts and district capitals cut off by the disruption of roadways. During these deployments, the squadron issued each member a ration of rice and canned meat for lunch. Occasionally, they were invited by friendly province chiefs or district commanders to dine in town for dinner, often on credit, before riding around the city in jeeps, looking for dance halls to enjoy the evening.

But it wasn't always so comfortable and carefree. Nam often found himself flying, hungry, and exhausted. On one occasion, while on a forward command post mission at the An Lão front in Bình Định province, where they were sweeping the North Vietnamese regular army's Sao Vàng Division, the crew had to eat dry rice and salty canned meat for lunch, with no water to drink or soak the rice in. The villagers pointed out a water jar with milky white liquid, like the water from washing rice. The crew politely declined and continued, waiting until the evening when they returned to Quy Nhơn to quench their thirst. In reality, if one were to compare the hardships faced by the aircrew during the war to those endured by the infantry, there was no comparison at all! Despite the challenges, the pilots were fortunate to have the means to return to a large city after their missions, where they could rest and eat properly.

As a newly established squadron, most of the pilots were around twenty-five years old, full of heroic spirit and the romanticism of young, single men. Sometimes, they act without thinking of the consequences for their lives or their emotions.

Nam still remembered a time in 1966 when Second Lieutenant Thái Văn Â. was on deployment in Quy Nhơn. His girlfriend wanted to visit Nha Trang to see her family, but she didn't have the necessary paperwork to board a military aircraft, so the military police at the gate wouldn't let her into the airfield. Second Lieutenant Â. told the situation to the group leader, Lieutenant DZ. The latter, being the deputy commander of the 215th Squadron, decided to show off his authority. He flew the helicopter outside the airfield fence to pick up the lady and then flew straight to Nha Trang. A few weeks later, the squadron received a telegram from the General Staff, docking thirty days of pay from Lieutenant DZ. for unauthorized use of military equipment!

Nam often sat in the co-pilot's seat during operations with Second Lieutenant Phan Chi H. Every time they landed at a forward base northeast of Bồng Sơn, he would ask the infantry soldiers below to pick a bunch of purple wildflowers growing abundantly on the hills near the landing zone. After completing the mission, as they flew back to Quy Nhơn in the afternoon, he would fly low over the roof of the women's teachers' college to scatter the flowers. The female students in their white áo dài would hear the helicopter and run out to the yard to watch him fly a circle before landing. On warm weekends, Nha Trang's beach was lively, with many attractive women basking in the sun, making it difficult for the young pilots to focus on their scouting missions, especially when faced with the distracting beauty along the shore. Such dangerous missions often resulted in a few weeks of confinement for flying too low and compromising safety!

As the months and years passed, the 215th Squadron grew along with the times. Nam and his comrades were no longer the young, carefree bachelors they once were. Captain Th., the commanding officer, was promoted to Colonel and assigned as an inspector at Air Force Headquarters. Lieutenant DZ., the deputy commander, was promoted to Lieutenant Colonel and transferred to the Political Warfare Department. Second Lieutenant Tr., the operations officer, was promoted to Major and became the commanding officer of the 215th Squadron. The heavy CH-34 helicopters were replaced with newer, more agile models, such as the UH-1, powered by jet engines instead of the older piston engines of the CH-34.

The "THẦN TƯỢNG" Squadron earned a proud reputation through its numerous heroic achievements, with medals for valor and gallantry adorning the chests of its pilots.

The Vietnam War, like a fierce storm, brought immense suffering and destruction to the homeland and its people. At times, Nam felt confused and questioned the purpose of his service, especially when he saw so many foreign troops in his country under the noble guise of protecting freedom. It was under this very banner of freedom that the nationalist leader Ngô Đình Diệm was executed because he refused to accept a freedom given by others and instead wanted his people to achieve it on their own. Nam couldn't help but think that in this world, no nation would sacrifice its resources, manpower, and wealth to help another country without expecting something in return.

Nam, as an officer in the RVNAF, was acutely aware of this. But if, by unfortunate circumstance, his vehicle broke down near a foreign military base, he knew he would be quickly shooed away as if he were a beggar, or worse, he might even be shot on his own soil. Nam despised the oppressive communist regime, but he also loathed the idea of being a servant to foreign powers. Sometimes, Nam dreamt, in a rather foolish way, of being a warrior like Lý Thường Kiệt, Hưng Đạo Vương, or Quang Trung, so that he could at least take pride in being a citizen of a nation with the right to self-determination and control over its destiny. Nam believed that many young soldiers and leaders in his country shared his thoughts. But sadly, the modern world was dominated by superpowers, with their grand, seemingly noble, but ultimately deceptive agendas.

The intensity of the war continued to escalate. After the Tet Offensive of 1968, when the North Vietnamese communists launched a major attack on key cities in the South, the internal political situation in the United States faced significant challenges. Protests against the Vietnam War grew increasingly fierce, contributing to the U.S. policy of "Vietnamization" of the war. Nam personally felt that the term "Vietnamization" itself indicated a loss of moral justification for the South Vietnamese fight for freedom. It implied that up until then, only the U.S. military and their allied forces had been the ones truly fighting against the communists to protect freedom for the people of South Vietnam, while the RVNAF was merely a puppet force. The entire concept contradicted the idea of fighting for the freedom and independence of a people. Nam thought that the U.S. government should have said, "Due to our own national interests, we can no longer continue to support you..." Psychologically, this would have been easier to accept and might have preserved some pride for the RVNAF.

Around 1970, Nam and Captain Trần Tấn Th. and Captain Phan Chí H. left the 215th Squadron and returned to Saigon to report to Air Force Headquarters, preparing to go to the United States for training on a new transport aircraft, the CH-47, which the U.S. military was about to hand over to the VNAF.

This time, Nam left the city of Nha Trang not only with his belongings but also with his young wife and their two-year-old son. After years of being stationed there, the city held many memories for him. His footsteps seemed to still be imprinted on the beaches and the moss-covered rocks of Hòn Chồng Xóm Bóng, but now he had to leave it all behind as if parting with a lover.

At the study abroad office, Nam met with Major Hồ Bảo Đ. and about ten other Captains from various squadrons across the four military regions who were also preparing for training. Most of these pilots had extensive experience flying combat missions in CH-34 and UH-1 helicopters.

The reason Air Force Command selected such experienced and high-ranking pilots was that they would eventually hold key positions in the new CH-47 squadrons that would be formed in the coming years as the U.S. military handed them over.

Major Đ. was the group leader and would eventually become the commander of the first CH-47 transport squadron in the VNAF. After completing the necessary paperwork and making sure their families were well-settled, they returned to flight school at Fort Rucker.

Most of the VNAF helicopter pilots had been trained at Fort Worth, Texas, on smaller aircraft like the OH-23 and T55. After graduating, they were transferred to Fort Rucker for further training on the CH-19, CH-34, UH-1, and the massive CH-47 transport helicopters. At Fort Rucker, Nam met Captain Nguyễn Văn H. and Captain Nguyễn Văn T., who had previously served as VNAF liaison officers at U.S. training bases, along with a group of new VNAF officers who had recently graduated from Fort Worth. The first training course had about twenty-two people, half of whom were veteran pilots, while the others were new officers. Each U.S. instructor was assigned to train one veteran and one new pilot.

During their time at Fort Rucker, Nam and his comrades were treated with great respect and hospitality by the American instructors. The flying itself was not too difficult. In fact, for Nam and his fellow pilots, this period felt more like a government-sponsored vacation to rest and recuperate than actual training.

The curriculum at Fort Rucker was entirely theoretical and not at all suited to the realities of the Vietnamese battlefield. This is evident from the fact that many U.S. aircraft were shot down in Vietnam because most American pilots, after graduating, were immediately sent to Vietnam without any real experience of the battlefield's psychological challenges. They relied strictly on the rules and standards they had learned in school. In contrast, while the Vietnamese pilots couldn't claim to be more skilled than their American counterparts, they had the crucial advantage of understanding their enemy's mindset.

The Vietnam War was largely a guerrilla war. The North Vietnamese communists could either concentrate large forces or disperse into small units, depending on the battlefield's needs. They knew that, given the overwhelming firepower of the U.S. Air Force, they couldn't win large-scale battles. Thus, they often organized small units at the battalion, company, or platoon level to block roads, disrupt transportation, and weaken the South's economy. In the mornings, when government forces cleared the roads, the guerrillas would retreat a bit further. In the afternoons, when government troops withdrew, the guerrillas would return to the highways to dig up roads or collect taxes from passing vehicles. So, if you flew at low altitude near the roads in the morning or evening, you were almost guaranteed to receive a hail of bullets from the guerrillas.

Additionally, in larger battles like Daksang, Ben Het, or Đức Cơ in the Second Military Region, the North Vietnamese would concentrate regimental or divisional forces to besiege outposts. They would stay very close to friendly forces to avoid artillery and B-52 bombers. In such cases, if you followed the textbook approach and made a standard approach to land, it would likely be your last mission. The Vietnamese battlefield had taught the VNAF pilots many lessons in "Blood and Tears." If you're still alive today (as the Buddhist saying goes), it's likely due to the blessings of your ancestors!

Time passed quickly. About four months later, the first CH-47 training course was successfully completed, and everyone was preparing to buy some gifts to bring home to their families in Saigon.

A few days later, Nam and his comrades reported to Air Force Headquarters in Biên Hòa to receive their orders. Since the CH-47 squadron hadn't been officially established yet, the division sent the group to Phú Lợi, Bình Dương, to fly combat missions with the U.S. 205th Aviation Battalion.

It's worth noting that during this time, the U.S. military had begun its rapid withdrawal from Vietnam. The Biên Hòa military airport and the areas designated for U.S. forces were vast and largely deserted. Many new VNAF helicopter, gunship, and observation squadrons were hastily formed. The Tactical Wing was upgraded to a Division. The VNAF expanded rapidly and was heavily reinforced during this period.

As the U.S. forces withdrew, the North Vietnamese regular army took full advantage of the situation, using the Ho Chi Minh Trail to transport heavy weapons, artillery, and tanks into the South to prepare for the decisive battles in the coming years.

A few months later, the 3rd Air Division received a number of CH-47 Chinook helicopters. Nam and his comrades were recalled to Biên Hòa to officially form the 237th Squadron. Major Hồ Bảo Đ. was a senior officer with extensive experience and a strong reputation in the helicopter community, chosen by Air Force Command as the first commander of the 237th Chinook Squadron. He was a firm believer in astrology and geomancy. He chose an auspicious day and time, set up a solemn altar under the open sky, and conducted the official departure ceremony, naming the squadron "Lôi Thanh" (Thunderbolt) in the presence of several high-ranking officers of the VNAF and the entire squadron.

Thus, the 237th Lôi Thanh Squadron officially became part of the VNAF family from that day forward. During the transition period, the squadron continued to conduct training flights to gain more experience and select personnel for the squadron's staff. The first deputy commander was Captain Nguyễn Phú C. The Operations Officer was Captain Nguyễn Văn H., the Safety Officer was Captain Nguyễn Văn M., and the Training Officer was Captain Nguyễn Thanh N., along with several assistant officers and flight leaders.

In record time, the pilots who had graduated from training courses in the United States were back, fully staffing the combat squadron. As the war escalated, with the U.S. military withdrawing rapidly, leaving significant gaps in the RVNAF, many outposts and district capitals were isolated, lacking the necessary support due to disrupted roadways. The North Vietnamese regular army had aggressively moved into strategic bases on Cambodian soil, near the northwest border of Tây Ninh province, preparing to launch a major offensive to capture the border provinces.

Recognizing the critical urgency of the situation, the General Staff of the RVNAF ordered the 237th Lôi Thanh Squadron to deploy earlier than planned. In early 1971, the squadron began regular operations in the hot battle zones of Tây Ninh, supporting friendly forces, infantry, and armored units advancing into the so-called inviolable strongholds of the communists on Cambodian territory to destroy their logistical and supply bases.

Following the Tây Ninh campaign were the battles of the "Easter Offensive of 1972," including An Lộc, Tống Lê Chân, and Chơn Thành, along the northwestern border of Saigon. The North Vietnamese communists sought to exploit the situation, sending in entire divisions equipped with heavy weapons and modern Soviet tanks to attack Phước Long and Bình Long provinces and their surrounding districts. Their goal was to capture the city of An Lộc and establish a provisional government for the "National Liberation Front of South Vietnam," creating a significant international impact.

The 237th Squadron once again demonstrated its versatility, supporting friendly forces through fierce and bloody battles that drew tears from the world's conscience! One example was a flight of three CH-47s heroically breaking through the enemy's anti-aircraft defenses to deliver ammunition to a heroic battalion of paratroopers besieged on "Windy Hill," northeast of An Lộc. Another example was the suicidal missions to land at the Tống Lê Chân (reference Chapter 2) outpost to reinforce and evacuate wounded soldiers of the elite Brown Beret Special Forces battalion defending the area. These positions became a testing ground for the VNAF's brave helicopter pilots from the UH-1 and CH-47 squadrons. These mighty eagles appeared from the sky, bravely soaring through the enemy's anti-aircraft fire, descending like angels to bring warmth and love to their fellow soldiers on the ground.

The 237th Lôi Thanh Squadron, though young in years, was seasoned and experienced. The pilots, honed by the fire of the Vietnamese battlefield, matured and demonstrated their superior capabilities through some of the most difficult missions in one of the most brutal wars of the twentieth century. To gain this bloody experience, the 237th Squadron had to endure and accept the painful and harsh losses that the Vietnamese nation was suffering.

Many eagle wings had broken mid-flight, entering the annals of history. These men truly flew away — flew into a realm of bright light in the afterlife, where there is no more pain, destruction, hatred, or bitterness. There, they will find eternal peace. Those of us left behind will always remember their names and mourn their loss. Their wives, children, and former sweethearts are still here. Though they bear great sorrow and longing, their eyes shine brightly when they hear the names of those who performed such heroic deeds. They are immensely proud and joyful to have once been the children, wives, or lovers of these men. We light incense in our hearts to silently pray and express our deep gratitude to those who sacrificed their lives so that we could live.

The 237th Squadron not only provided operational support in the Third Tactical Zone but also occasionally supported the Second and Fourth Tactical Zones when allied units faced difficult situations requiring urgent resolution.

Nam vividly remembers the year 1971. The battlefront at Đắk Tô was raging with fierce battles at places like Charlie, Tiền Đồn Năm, Plato.J, and Chu Prong, among others. Tiền Đồn Năm (Outpost Five) was held by a battalion of paratroopers stationed on a mountain peak. These brave red berets were isolated and surrounded by an entire regiment of North Vietnamese regular troops hiding in trenches at the base of the hill. They bombarded the outpost day and night, preventing helicopters from resupplying the battalion. The paratroopers called the forward command post at Tân Cảnh, urgently requesting air support. All the remaining American Chinook CH-47 pilots in Pleiku refused the emergency mission, fearing losses, before they withdrew from Vietnam. A message from the Second Tactical Zone Command was sent to the 3rd

Air Division, requesting the 237th Squadron to carry out the mission. Squadron Commander Hồ Bảo Đ. ordered a CH-47 to take off from Biên Hòa and head to Đắk Tô to execute the mission. Captain Nam, who was on operational duty that day, took responsibility for the mission. Taking off from Biên Hòa, they refueled at Buôn Ma Thuột and then proceeded directly to the forward command post at Tân Cảnh to receive orders. After landing, the helicopter hadn't even come to a complete stop when Nam noticed the battle commander driving a jeep out to meet him, taking him straight to the briefing room. With a voice full of emotion and seriousness, the commander said:

"I requested the Chinooks from the Americans stationed at Pleiku, but they refused, afraid of losses before their withdrawal. Now, I can only rely on you and your comrades in the Vietnamese Air Force to save the situation for the paratroopers who are holding out on the peak of Outpost Five..."

Nam was deeply moved and respectfully shook the commander's hand, promising to do his best to carry out the mission. After receiving the intelligence briefing, Nam knew he had to rely on the crew's technical skills and composure. The primary task was to sling load several tons of ammunition, food, and medical supplies urgently needed by the battalion and to retrieve a damaged 105mm howitzer. The mission was straightforward: complete it and return to Biên Hòa the same day.

Outpost Five was not far from the Tân Cảnh command post, only about a five-minute flight away. On clear days, the outpost could be faintly seen from the command post.

The crew was ready to take off. Nam ordered the rear ramp to be lowered to allow for easy escape if the helicopter was hit or caught in artillery fire. The large CH-47, with a massive load slung underneath, approached the hilltop from the northeast, escorted by two armed UH-1 helicopters flying close below, firing rockets and machine guns at suspected enemy positions. A minute later, they successfully delivered the supplies. Nam prepared to turn the helicopter left to pick up the 105mm howitzer when explosions suddenly erupted all around them, throwing up clouds of dust and smoke. The crew chief's panicked voice came over the intercom:

"Get out of here, Captain! They're shelling us!"

Nam, already prepared for this possibility, calmly pushed the control stick to the right and kicked the pedal hard, spinning the helicopter 180 degrees. Just as he did, a large artillery shell exploded near the rear of the helicopter, the blast wave pushing the aircraft's nose down towards the hillside. Nam struggled to keep the helicopter level, quickly gaining speed to escape the enemy's firing range. In those few tense seconds, the crew chief reported that everyone was safe and the instruments were still within safe limits. On the FM radio, Nam reported the incident to the forward command post. The battle commander personally thanked them for delivering the supplies that saved the besieged battalion.

Upon their return, the command staff at the headquarters came out to congratulate the crew on their safe return. The battle commander awarded the crew the Gallantry Cross in front of high-ranking Vietnamese and American officers. A staff officer later told Nam that they had been incredibly fortunate, as just the day before, a UH-1 crew had been shot down at the base of the hill. Nam reflected that life and death are determined by fate; a second's delay or haste could change everything in a person's life.

After inspecting the helicopter and finding only minor damage, Nam decided to return to Biên Hòa. Upon arrival, he filed a report of the damage caused by artillery fire, not considering it particularly noteworthy. In the past, when flying CH-34s and UH-1s with the 215th Squadron in battles at Đức Cơ, Đức Lập, Bồng Sơn, and Tam Quan, being shot at by the Viet Cong was a common occurrence. However, this time, the squadron commander insisted on nominating Nam to represent the 3rd Air Division in the delegation of outstanding soldiers from the Republic of Vietnam, led by Brigadier General Trần Bá D., to visit Taiwan.

In 1972, Lieutenant Colonel Đ. was promoted to Deputy Wing Commander, and Major C. was appointed Squadron Commander. Major Võ Châu P. became the Deputy Squadron Commander, and Nam replaced Major H. as the Operations Officer.

The 237th Lôi Thanh Squadron produced many outstanding officers who went on to hold important positions in newly established Chinook squadrons, such as Lieutenant Colonel Nguyễn Văn M., who became the Squadron Commander in Đà Nẵng, Lieutenant Colonel H., who commanded the squadron in Phù Cát, and Major Phạm Văn T., who led the squadron in Cần Thơ, among others. Many young pilots also rose to become exceptional aircraft commanders, holding key positions such as flight leaders and detachment commanders.

In the summer of 1972, during the fierce battles of An Lộc, North Vietnamese forces, equipped with advanced weaponry to counter the support of the RVNAF, such as heavy anti-aircraft guns and SA-7 heat-seeking missiles, turned the pilots of the 237th Squadron into knights of the sky. The crews, armored in bulletproof vests, sling-loaded massive supply crates, flying just above the treetops at high speeds. Groups of three or four helicopters would weave through enemy territory to deliver supplies to friendly forces operating in the forests south of An Lộc. Beyond pure combat support, they were also warriors of compassion. Women, children, the elderly, and the sick, having abandoned all their possessions and homes to flee the war, were often evacuated by the pilots who, despite the constant threat of artillery fire, would take the time to ensure they were safely on board before departing the battle zone.

As a child, Nam had only heard about, but never seen, Japan's "Kamikaze" pilots, yet he held them in high esteem as they embodied the ideal of manhood in times of war. Today, in the 237th Lôi Thanh Squadron, Nam saw with his own eyes those same Kamikaze warriors in the flesh. However, there was no warm sake or heartfelt farewell from superiors as there had been for the Japanese pilots. These men only had their unwavering loyalty and devotion to the words "Vietnam" as they volunteered to fly into the death zone of Tống Lê Chân.

After the grueling and drawn-out battles of the Vietnam War, the helicopter crews proved their versatility and heroism in all aspects. In the air, they were angels of death, raining down rockets and machine-gun fire that struck terror into the hearts of the enemy as their armed helicopters roared overhead. To friendly forces, they were fearless knights, braving fire to deliver ammunition, food, and medical supplies, earning their comrades' unshakable trust.

To the fleeing civilians, they were warriors of compassion, deeply appreciated by those they saved. The helicopter squadrons, alongside paratroopers, special forces, marines, and rangers, engaged the enemy from deep valleys to high mountain ridges and cliffside caves—places where no human had ever set foot before—in operations like Hawk, troop landings, and special reconnaissance across all four military zones.

Nam was immensely proud and thrilled to praise the men who had once written heroic chapters in the history of the Republic of Vietnam Armed Forces (RVNAF) and the Air Force in particular.

With such proud military branches and millions of determined citizens and soldiers of the Republic of Vietnam ready to defend the nation, it is tragic that the destiny of our country was not ours to decide. In the world of today, and for all time, the ultimate power rests with the strongest.

Vietnam, our small, impoverished homeland, proud of its thousand-year culture, must ask itself: what have we learned from the losses and suffering we have endured and continue to bear?

The 237th Lôi Thanh Squadron is one of the many witnesses to the tragic history of April 30, 1975. The Viet Cong began shelling Biên Hòa Air Base and the headquarters of the Third Tactical Zone on April 27, 1975. The squadron was ordered to relocate to Tân Sơn Nhất. During this time, Nam was the Deputy Squadron Commander, replacing Lieutenant Colonel Võ Châu P., who had perished in a training accident at Long Bình.

Nam was assigned to coordinate the flight crews at Tân Sơn Nhất. Lieutenant Colonel C., the Squadron Commander, stayed behind at Biên Hòa with Colonel T. and Brigadier General T., the commander of the 3rd Air Division, to oversee the evacuation of the airbase to Saigon.

On April 28, 1975, as the pilots were busy moving troops back and forth on the Tân Sơn Nhất airfield, Nam stood at the base of the control tower, monitoring the flights. Suddenly, he noticed two A37 jets flying at a very low altitude from the south. In an instant, two "Napalm" bombs dropped onto the runway, engulfing it in smoke and fire. Many soldiers working or standing nearby panicked and ran for cover behind aircraft shelters. After dropping their bombs, the two A37s climbed and made a 180-degree turn to attack again as if in a no-man's-land. From that moment, Nam realized that the command structure of the RVNAF and the Air Force was effectively non-existent. Soldiers deserted their units and fled home in a state of chaos.

In this critical moment, the pilots in Nam's unit were deeply confused. Nam called the "Paris" control tower to report and ask if there were any orders from the command center, but the officer on duty there seemed just as lost, responding with, "Just wait for orders..."

The hours of waiting dragged on, feeling like an eternity.

On the night of April 28 and into the early hours of April 29, 1975, Nam and his crew slept under the belly of the helicopter. Around 4 a.m., Nam was awakened by the deafening roar of 122mm rockets and 130mm artillery shells fired by the Viet Cong, hitting the fuel tanks and sending towering flames and thick black smoke into the sky. Through the radio, the airport security personnel reported that the enemy had breached the perimeter on the northwest side of the airfield. Nam realized they couldn't stay any longer and ordered the helicopters to take off. Four CH-47s flew to Vũng Tàu, as Tân Sơn Nhất was now practically abandoned. However, after refueling at Vũng Tàu, the Viet Cong began shelling the area again. Nam ordered the crew to restart the engines and prepared to fly to Cần Thơ. Once the aircraft reached 5,000 feet, Nam radioed "Paris" for new orders, but received no response. At the same time, Nam heard someone say, "There is no more authority now, Lôi Thanh has full decision-making power..."

That marked the end—the final day of the country and the forced disbandment of the 237th Lôi Thanh Squadron...!!!

As a warrior, Nam was immensely proud to have lived and shared the hardships, joys, and struggles of military life with all his comrades. Their holy battle and the sacrifices they made were not for personal gain or based on foreign ideologies. Their sacrifices were sincere and simple, with no hesitation or second thoughts, motivated only by their love for their ravaged homeland, Vietnam! We fought and learned to follow in the footsteps of our ancestors, who founded and defended this country a thousand years ago.

CHAPTER 7

Teacher Bảo

Original Title: "Thầy Giáo Bảo"

Bảo absent-mindedly opened the door and stepped onto the front porch, looking for an empty spot on the steps leading to the small garden in front of the house. He sat down, reached for a hot cup of coffee nearby, and quickly pulled out a pack of cigarettes, putting one between his lips and lighting it. He took a long drag, like a true smoker, and slowly exhaled white clouds of smoke, which floated in delicate waves before disappearing into some unseen space. Suddenly, he flicked the half-smoked cigarette forcefully into the garden in a gesture of frustration, something that rarely happened to him.

In fact, Bảo was upset about something trivial that had occurred a few days ago during a meeting of the local retired teachers' association. Someone had questioned him, "Why did you return to Vietnam while the communists were still there?"

###

Before 1975, Bảo was a Vietnamese literature teacher at a private high school in Saigon. Around 1963, he was drafted to attend the Thủ Đức Reserve Officer Training School, and upon graduation, he was assigned to a combat unit in the first military zone. About a year later, he was reassigned by the Ministry of National Defense to the Ministry of Education to continue his teaching profession.

For Bảo, the Vietnam War was a politically complex conflict, but ultimately, it was still a proxy war between two ideologies: "Capitalism" and "Communism."

Bảo didn't like to delve deeply into it. He was just a teacher who loved teaching, loved analyzing stories like The Tale of Kiều, and reading poems by Xuân Diệu, Lưu Trọng Lư, and Hàn Mặc Tử. Bảo had the demeanor of an artist more than a strict, exemplary teacher. He liked to smoke and always wore his hair long, down to his nape.

When teaching poetry, especially Hàn Mặc Tử's love poems, his voice would lower as he read each line slowly, carefully repeating each word as if they were musical notes rising and falling. The entire class would sit in silence, listening intently. His unique delivery created a strong allure, bringing characters to life and materializing the imagery in the poems he was reading. Because of these special qualities, students affectionately nicknamed him "Tú Uyên," after a poet from Bích Câu Kỳ Ngộ.

Bảo often told his students that art must serve art itself to express and fully understand the beauty and mysterious magic of literature. He compared a great poem to a beautiful rose created by nature or night-blooming jasmine with its intoxicating fragrance. The beauty of the rose and the scent of the jasmine existed naturally without any intention beyond themselves.

Bảo believed that to fully reveal the emotions of a poem, one must place oneself in the shoes of the poem's characters. Like a singer who must inhabit a song to truly convey its beauty, a reader of poetry must become part of the story to grasp its essence.

Then came April 30, 1975. In the initial period of societal upheaval, schools temporarily closed, and Bảo had to stay home, waiting. He began to see changes in local leadership at the grassroots level in his neighborhood and felt an intuition that something bad was about to happen.

His prediction wasn't wrong. Shortly after, the people in his ward were required to register their personal details. The ward chief was none other than Tư Thẹo, a thug who had dodged military service before 1975.

When Bảo went to register, Tư Thẹo sat behind the desk, glaring at him with cold eyes, his voice brusque.

"You're a teacher, right?" "Yes, I taught at a private high school in

Saigon." Tư Thẹo reached into a drawer, pulled out some papers, handed them to Bảo, and pointed to an empty table at the back of the room.

"Go over there and fill out all your details on these forms. A comrade from the education department will follow up with you later. Make sure you fill everything out truthfully. No hiding anything, got it?"

Bảo took the papers and, feeling like a lifeless shell, walked to the table. He never imagined that his life could change so drastically and so quickly.

Before that day, Bảo had never had any interaction with Tư Thẹo in the neighborhood. He only knew that Tư Thẹo had lived with the family of Mr. Hai, a cyclo driver, and had no real job. He'd wander around the neighborhood, picking up odd jobs when available. Occasionally, he would borrow Mr. Hai's cyclo to transport passengers at night. That was all Bảo knew about him, and no one else in the neighborhood seemed curious enough to inquire about his background either.

Because of the three-centimeter scar on the left side of his forehead, people called him Tư Thẹo ("Scarface"). He didn't bother anyone, so no one cared about the story behind his scar.

About half an hour later, Bảo finished filling out the forms and brought them back to Tư Thẹo. "I've filled everything out. Please review it." "Are you sure it's all complete?" Tư Thẹo asked. "I believe I included everything."

After a quick glance at the forms, Tư Thẹo smirked. "You call this a complete form? You're a literature teacher, and this is what you write?"

"I listed my profession and background as required. What else do you want me to include?" Bảo asked, confused.

Tư Thẹo crumpled up the form and threw it into the trash bin. He reached into his drawer for a new set of papers.

"You need to include every date, year, and detail of where your family lived, what they did for a living, everything. Now go back to the table and fill it out again."

Bảo felt his pride being crushed by this uneducated man, who now had the authority in this wretched ward. His cheeks flushed with heat as he swallowed his rage. He grabbed the new set of forms and silently returned to the table. This time, he wracked his brain, listing every family member, from his paternal and maternal grandparents down to his siblings, without missing a single person.

After more than an hour, he reviewed the forms one last time and believed they were as thorough as Tư Thẹo wanted, so he brought them back. Once again, Tư Thẹo looked at him with a contemptuous laugh.

"You seem to forget things quickly for a teacher. Didn't you once serve as an officer in the puppet army?"

Bảo, annoyed, replied directly, "Yes, I went through Thủ Đức Reserve Officer School from 1962 to 1964." "Then why didn't you include that in your form? What did you do during that time?" Tư Thẹo asked. "I was a low-ranking reserve officer just out of school. And it's been so long since then. I went back to teaching, so I didn't think it was that important."

Tư Thẹo's expression turned cold, and he spoke slowly, emphasizing each word.

"To you, it may not be important, but to us, you were an enemy of the people, a traitor who sold out the country. Do you understand?"

Bảo wanted to retort, "To you communists, everyone is an enemy! There are millions of enemies in the South alone..."

But he stopped himself in time, thinking, "There's no point arguing with this man. He's like a horse with blinders, only following the reins of his master."

Bảo glanced at his watch, then back at Tư Thẹo. "It's already past five in the afternoon. If it's not urgent, could I take these forms home to fill out and return them to you tomorrow morning?"

After a moment of hesitation, Tư Thẹo nodded. "Fine. I know where you live in the ward. I'll keep these forms here. You go home and fill them out correctly. Bring them back tomorrow."

Feeling thoroughly disheartened, instead of heading straight home, Bảo drove his motorbike to the Saigon Riverbank. He found an empty bench, parked his bike, and sat down, hoping the cool breeze from the river would ease the bitterness of his life's sudden upheaval.

Bảo remembered that just two weeks ago, the two-story house of Uncle Năm on the outskirts of town, where Bảo rented the upper floor while teaching, was his temporary refuge. It was a Sunday, just past five in the morning, when the family's brindle dog started barking furiously at something suspicious.

The dog's prolonged barking woke Bảo. He tossed off his blanket, stepped to the window, and peeked through the curtains. Five soldiers in green khaki uniforms, armed to the teeth, were standing in the front yard, blocked by the dog. Bảo quickly ran downstairs to inform the family. By then, Uncle and Aunt Năm had also woken up and stepped out of their room. Bảo whispered to them.

"Uncle Năm, it seems the police are here to search the house. There are five of them, and the dog is keeping them at bay." "Okay, Bảo, go back to your room and act as if nothing is happening. I'll take care of it," Uncle Năm replied.

He walked to the window, opened it, and looked out. The police officers stood huddled together, clearly afraid of the dog. When they saw the window open, one of them shouted, "We're the local police! We have an order to search the house. Control your dog, or we'll shoot it."

"Give me a moment to put on my shirt," Uncle Năm said, though the officer impatiently yelled, "Hurry up!"

Aunt Năm, trembling, asked, "What do we do now, dear?" "Go hide all the jewelry quickly," Uncle Năm instructed.

Hearing the familiar voice of the homeowner, the dog stopped barking and stood growling at the door. After donning his shirt, Uncle Năm turned on the house lights and opened the front door. The dog, seeing its owner, wagged its tail and ran into the living room. The officers followed closely behind. The leader spoke.

"We have orders to inventory all of your household's assets."

Several officers rushed to the back, closing the rear doors and checking for any other exits. Afterward, they gathered the family in the living room, and the leader asked Uncle Năm.

"Are these all the people in your house?" "Yes. I had a son, but he went missing after you guys 'liberated' the city. Bảo here is just a tenant, staying while he teaches at the local school."

"Your son went missing, or did he join the resistance? Or maybe he ran off with the Americans and puppets?" "My son went off for work and occasionally visited home. But we haven't heard from him for months. We don't know if he's alive or dead."

"We'll deal with that later. Right now, our task is to inventory and report all your assets to the state."

After this, the officer pulled out a stack of papers from his bag and handed them to Uncle Năm.

"Here, take these forms and list every asset, big or small, that you currently own. Remember to report everything honestly. We've been informed by the people that you are one of the wealthiest households in this ward. Make sure you don't leave anything out, understood?"

Aunt Năm, standing beside her husband, began to sob.

"What's left to report? Our shop in Chợ Lớn, where we sold fabric and clothes, was looted the day you 'liberated' the city. They smashed the doors, broke the walls, and took everything. Now you want us to report what? There's nothing left!"

The leader of the police squad grew more aggressive.

"We don't care about that. We only care that the people reported you as part of the capitalist bourgeoisie who exploited the labor of the people."

"But we didn't exploit anyone's labor! Everything we built came from our own hard work and intelligence over decades. The day you liberated us was the day we lost everything!" Aunt Năm cried.

The officer cut her off.

"Don't slander us like that—it's a crime. We came here to liberate the southern people, to give them freedom and happiness. Now, we don't have time to discuss this with you. We'll stay here to maintain security while you finish your report."

With that, the officer turned to Bảo.

"And you, you're the tenant, right?" "Yes, I rent a room from Uncle Năm while I teach." "So, you're a teacher. Do you have any assets to report to the state?" the officer asked, sarcastically.

"Yes," Bảo replied. "I have three or four sets of clothes for teaching, a blanket, a mosquito net, and a bed. Would you like me to report that?"

One of the officers who had searched the upstairs room reported back.

"We checked his room, but there's nothing of value."

Satisfied, the officer nodded at Bảo.

"You're a teacher, an intellectual, part of the proletariat class—very good. You can go back to your room, but don't leave the house until we're done."

The entire day, Uncle and Aunt Năm had to rewrite and adjust their asset report multiple times. Each time they finished, the officers would split up to search the house, tearing apart closets, beds, and even the kitchen. If they found anything remotely valuable, they would berate Uncle and Aunt Năm.

By noon, the police squad, having exhausted all possibilities, finally accepted the asset report. They gathered all the cash, jewelry, and valuables, stuffed them into a box, and locked it. They handed Uncle Năm a receipt, which was as worthless as trash.

"Here, keep this receipt. We'll take the items to the authorities for a decision."

Uncle Năm, looking defeated, slumped into a chair, unable to speak. Aunt Năm, devastated, wept uncontrollably.

"God, all the hard work of our entire lives, saved up for old age, gone! Now you take it all to get a 'decision'? What decision?"

The police, ignoring her grief, left with the box of valuables. Upstairs, Bảo had been confined, reading a book to pass the time, listening to the ruckus below. After the police left, the house was eerily silent, except for Aunt Năm's sobs of anguish. Bảo tossed his book aside and descended the stairs to survey the scene. Uncle Năm sat with his head buried in his arms at the table. Aunt Năm was still on the floor, her hair disheveled, crying. Bảo approached her.

"Have they left, Auntie?"

"Yes! Those damned scoundrels have taken everything from us!" she wailed.

Looking at the scattered furniture and clothes strewn across the floor, Bảo felt a lump in his throat, a silent rage against the inhuman regime. They used the pretense of rooting out landlords and capitalists, accusing anyone of exploiting the people, while targeting innocent, hardworking individuals.

Bảo had stayed with the Năm family long enough to know that they were honest, ethical businesspeople. Uncle Năm regularly donated money to support flood relief and orphanages. As a literature teacher with a degree in the arts, Bảo had read extensively about the philosophical debates between idealism and materialism, and he had a fair understanding of both capitalist and communist ideologies.

In the months following the fall of the South, Bảo witnessed firsthand the irrational and inhumane actions of the Northern communists. He often recalled the words of President Thiệu: "Don't listen to what the communists say; look at what they do." Now, those words felt more accurate than ever.

The sun was setting over the Saigon River. Apart from Bảo, only a few others sat on benches, each lost in their own thoughts. Some stared at the ground, resigned like enslaved workers, while others gazed at the flowing river, perhaps chasing distant dreams. These were people living in the present, yet seemingly unable to accept the reality of the society in which they now found themselves.

The noble image of the Vietnamese soldier, rising to overthrow colonial rule and win independence for the nation, was absent from the current regime. These officers were nothing more than mercenaries for a godless proletarian ideology under the dogma of Marx-Lenin. Wealthy people were lumped into the category of exploitative capitalists and were to be destroyed to pave the way for the establishment of a socialist state, the first step toward communism, in line with the spirit of the Third Communist International.

Bảo believed that sooner or later, he too would be classified as an intellectual with capitalist tendencies, just as intellectuals were during the land reform in the North or the purges under Mao Zedong in mainland China, where landlords were condemned and intellectuals were imprisoned.

Wherever communism exists, it follows the same pattern. Its goal is to proletarianize the people to make them easier to control.

In summary, Marx and Engels categorized capitalists, landlords, and feudal lords as exploiters of the working class, backing their arguments with charts and scientific formulas. They encouraged and led these oppressed classes to rise and overthrow the capitalists and landlords to claim ownership of the land and factories they worked on.

Most farmers and workers, with limited education and understanding, were easily swayed by these appealing promises. But after they overthrew the capitalists and landlords, the factories and land fell under state control, led by the party. The workers and farmers, supposedly now the owners, continued working for a new master—the state-owned enterprises and cooperatives. Their lives returned to the same oppressive conditions, if not worse.

Before, a worker or farmer could choose where to work. If treated unfairly, they could leave and find a better opportunity. But now, with only one employer in the state-controlled system, refusing to work meant starvation for the whole family. Worse, they could be labeled reactionary and accused of individualistic tendencies. The socialist system, with its centrally planned economy, was more corrupt and bureaucratic than the old capitalist or feudal systems. Farmers could no longer freely sell their produce. Instead, they were forced to sell at low prices to the government through cooperatives.

Southern workers and farmers had opened their eyes and realized that the new regime was even more exploitative than the old landlords. They resisted through passive non-cooperation, abandoning their fields and avoiding heavy labor. They decided that they would rather starve than be exploited further. This became evident in the years following the North's "liberation" of the South.

As Bảo sat lost in thought, a hand clapped him on the shoulder from behind.

"Hey, professor, what are you doing sitting here all alone?"

"Oh, Tâm! It's a bit hot today, so I came out here to enjoy the river breeze."

Tâm was a history and geography teacher at the same school as Bảo. He had recently gotten married to Loan, who used to work as an accountant for a large import-export company in Saigon. But after the fall of the South, the company's owner fled abroad, leaving Loan jobless, just like her husband.

Bảo asked, "Where's Loan? Why are you out here by yourself?" "She's busy selling goods at home." Bảo looked surprised. "Selling goods? What do you mean?"

Tâm glanced around to make sure no one was eavesdropping.

"Yeah, she's selling cigarettes and coffee under the table to people in the apartment complex."

Bảo scratched his head and sighed. "This society is so frustrating! Knowledgeable, capable people like us are ignored, while the ignorant are promoted just because they're tied to the regime!"

Sensing Bảo's frustration, Tâm interrupted.

"Let's not dwell on that. We do what we can to survive. Complaining will only get us in trouble. Last week, I took Loan back to her hometown in Bình Tuy. Officially, it was to visit her family, but really, it was to borrow some money from her mother to start a small business. After we got married, all our savings went into buying an apartment. Then, after a month, Loan lost her job, and I did too, at the same time as you. We had to sell all our furniture piece by piece to survive. Now we have nothing left to sell, so Loan decided to take a risk and asked me to go to her family for a loan to buy cigarettes and coffee to sell."

Bảo looked at his friend with sympathy.

"You and Loan were the most well-off among us back when we were teaching. Now, look at this."

"Who could've predicted this?" Tâm said. "But let me tell you a funny story, or maybe it's a sad one for our nation. When we arrived in Bình Tuy, we immediately registered with the local police. After they processed our papers, they invited us to a neighborhood meeting that evening to hear a report on the activities of the community. Since Loan was born and raised there, she wanted to hear what was happening in her old neighborhood. So, after dinner, we dressed up a bit and went to the meeting with her parents. The local office, which used to be a village hall, had been slightly renovated for meetings. When we walked in, I noticed two bright red banners with bold black letters on either side of the hall's entrance. Curious, I stopped to read them: 'Down with the reliance on machinery! Long live the spirit of manual labor!' I was confused and pointed it out to Loan."

"Look at this, Loan. In Saigon, the regime's newspapers are promoting industrialization, but here, they seem to be completely against it! Is the king's law weaker than the village rules?"

Loan just smiled and said, "You don't understand. The communists here are pushing for self-reliance."

"What do you mean?"

"Remember the dinner we just had at my mom's house?"

Tâm grinned, "Oh, you mean the fish stew we had? That was a treat because we were visiting."

Loan laughed. "No, that's not it. If we stay longer, you'll see that the next meal will just be rice with fish sauce."

Tâm looked puzzled. "What do you mean?"

"You haven't heard the slogan about local self-reliance and self-sufficiency? Whatever a place produces, that's what people have to live on. In rural areas, it's rice instead of fish. In coastal areas, it's fish instead of rice. Trying to transport goods between regions is nearly impossible because of all the checkpoints and penalties."

Before Tâm could respond, the neighborhood meeting began. Everyone found a seat, with the elderly sitting up front and the younger people standing in the back.

After introductions, a political instructor started lecturing about the new policies of the socialist regime. It was the usual rhetoric condemning the old capitalist regime, the imperialist American puppets, and their exploitation of the people.

After more than an hour of this, the floor was opened for public comments. A local political officer encouraged the neighbors to share their thoughts.

An elderly woman, Aunt Tám, raised her hand. She was respected in the community, not only because she was a poor peasant but also because she had lost a son fighting for the communists. After the war, she was honored as a "Mother of a Soldier with Meritorious Service to the Revolution." Her comments were often praised, and she usually took the opportunity to praise the government's accomplishments. Today, she stood up and spoke loudly.

"We are very grateful to the revolutionaries for liberating us early. If it weren't for you, the Việt Cộng would have bombed us to death."

The crowd burst into applause.

Hearing this, Bảo couldn't help but laugh.

"What? That doesn't make any sense!"

"She's a countrywoman. She says what she thinks, even if it doesn't make sense," Tâm said with a grin.

About a month after Bảo submitted his personal details to Tư Thẹo, he received an invitation to attend a political reeducation session, organized by the communist government for teachers and artists. Unlike the cramped and uncomfortable meetings at the local ward office, this session was held in a relatively decent venue with chairs and tables.

When Bảo entered the room, he immediately saw Vân and Tâm, already seated. They waved him over excitedly. Vân, a beautiful young English teacher from the same school, smiled warmly as she moved over to make space for him.

"Come sit with us, Bảo!" Vân called out, waving her hand. Bảo smiled and shook hands with Tâm as he sat down next to them.

"When did you and Tâm arrive?"

Vân quickly responded,

"Anh Tâm and I just got here about five minutes ago. We saw that you weren't here yet, and we were starting to get worried!"

"I passed by your house, but the kids playing out front said you'd already left."

Tâm interjected into the conversation,

"My wife also wanted to come early to see if there were any familiar faces, and we ran into Vân here. Now that you're here, it's a bit less boring."

The training class had about thirty people, most of them local teachers and artists. Bảo scanned the room, trying to see if he could spot any other friends or acquaintances. Sensing what he was thinking, Tâm whispered,

"They all bolted!"

Bảo nodded slightly and sighed.

After the initial greetings, one of the two lecturers began speaking.

"You all probably already have a general idea of the purpose of this political training class today, aimed specifically at you. The Party and the State have classified you all as intellectuals of Saigon. Now that Vietnam is unified, independent, and free, the Party wants all southern artists and teachers to study and follow the guidelines and principles of our socialist society. The length of your study will depend on how quickly or slowly you comprehend the new guidelines. After these lessons, we will have a review and critique session, and you will have the right to express your individual opinions freely, and any concerns will be addressed satisfactorily."

During the first lesson, the entire class was completely silent—no one had any questions, and no one offered any opinions. The lecturer kept droning on like a machine, reciting lines from the Party's prepared materials. Every word came from "Comrade Marx," "Comrade Lenin," and "Comrade Hồ," and thus, we were expected to listen and follow precisely.

Throughout the lecture, Bảo felt deeply bored. They kept repeating the same terms about how southern culture was degenerate, unable to serve the needs of the people, that capitalist traders exploited labor, and how all this went against "Marx and Engels'" ideology, following in the footsteps of American imperialism, and so on, and so on.

Finally, at five o'clock in the afternoon, they ended the first day's lesson. As soon as they left the class, Tâm quickly said his goodbyes, explaining that he had to hurry home because his wife was waiting. Bảo turned to Vân and asked,

"How did you get here, Vân?"

"I took a cyclo (pedicab)."

"Then hop on my bike. I'll give you a ride home," Bảo offered.

Vân smiled and nodded in agreement. Bảo checked his watch and then suggested,

"It's still early. How about we take a quick ride down to the Saigon docks, enjoy the cool air for a bit, and then head home? What do you think?"

"Sounds good! It's still too early for dinner at my house anyway," Vân replied.

Bảo and Vân had developed a special bond due to a chance occurrence. Vân's brother, Sơn, had been Bảo's classmate back in 1962. After graduating from high school, Sơn joined the Air Force, while Bảo pursued his passion for literature and continued his studies at the Saigon College of Letters, eventually earning his bachelor's degree and beginning his teaching career.

By chance, during a flood relief mission organized by the private teachers' association to aid victims in the southern provinces, Bảo was leading the relief team. The Republic of Vietnam's military transport command assigned two helicopters to the mission. Sơn, who at the time was the deputy commander, was tasked with assisting the relief team and guiding the helicopter crew on the mission.

As Sơn walked out of the flight command office with his flight helmet in hand, he stopped in surprise when he saw Bảo stepping out of the relief team.

"Hey! Is that Bảo?"

"Yes, it's me! Is that you, Sơn?"

"Yes, it's me!"

The two old friends, who hadn't seen each other in over a decade, greeted each other warmly. Sơn gripped Bảo's hand tightly,

"It's been more than ten years since I last saw you!"

"At first, I didn't recognize you at all, but it's really you."

"Yeah, I was stationed in Nha Trang before they transferred me here. What about you? Are you married yet?"

"No woman wants me! I'm still single. How about you, Sơn?"

"You're a teacher and still haven't found anyone? As for me, I'm in the military, so no woman dares to marry me. They only like us for fun but not for the long haul!"

Both laughed, agreeing with the joke. Later that day, after the relief mission, Sơn invited Bảo to his house to catch up, as he was living with his mother and younger sister.

The two rode a Honda together, and as they arrived at the house, Vân had just returned from school. She greeted them cheerfully,

"Oh! You're home early today, brother!"

Sơn turned to his sister,

"It's almost five in the afternoon. How is that early, missy?"

"How would I know? You usually don't come home until eight or nine at night."

"Well, I had a duty to attend to," Sơn explained.

Vân wasn't convinced and playfully narrowed her eyes at her brother,

"Yeah, right! More like you were 'on duty' with your girlfriends."

"Enough with that! You're always meddling in my personal life!"

Sơn turned to Bảo and introduced him to his sister,

"Do you know who this is?"

Now realizing there was a guest, Vân looked a bit shy and quietly replied,

"He's your friend, right?"

"Yeah! This is Bảo, my classmate from over ten years ago."

Feeling more comfortable now, Bảo smiled at Vân and nodded in greeting. Vân nodded back before heading inside. Once she was out of earshot, Sơn spoke up,

"That's my sister, Vân. She's studying English at Saigon University's College of Education. She'll be graduating this year and will start teaching, just like you."

Bảo grinned,

"So, I'll have a new colleague soon."

Sơn's mother, who had just come out carrying a vase of marigolds, saw Sơn and called out,

"Sơn, you're back?"

"Yes, mom. This is Bảo, my old school friend. He's visiting."

"Alright, come inside, both of you. I prepared an offering earlier today for your grandfather, so I'll have Vân set the table, and we can eat."

Bảo bowed his head in greeting to Sơn's mother, then followed Sơn inside. Vân had already changed into a new outfit—a stylish pink blouse with white trousers. She stepped out into the living room and asked,

"Do you want me to set the table now, or should we wait a little longer?"

Sơn's mother chimed in,

"It's getting late. Go ahead and set the table so we can enjoy the meal."

Feeling a bit awkward, Bảo leaned over to Sơn and whispered,

"I just stopped by to see your house, but now I should probably head home. We can meet up another time."

Sơn laughed,

"If you leave now, I'll have Vân pack some food for you to take with you. Got it?"

Sơn's mother, hearing Bảo's polite refusal to stay for dinner, came closer and said,

"Don't worry about it. Any friend of Sơn's is like family here. Don't feel like you're inconveniencing us."

Sơn added,

"My mom always says that guests who leave the house full are a sign of good fortune. You're new here, so you're not used to it, but my other friends come over all the time. Now, let's have a beer together before Vân sets the table."

Sơn went to the fridge and grabbed two bottles of 33 Beer, handing one to Bảo.

"Let's sit outside for a bit and enjoy the cool air. It's been so long since I last saw you, Bảo. So many of our old classmates have disappeared."

"Yeah, most of them went into the military, so it's rare to meet again. I had no idea I'd run into you today," Bảo agreed.

Sơn pulled a pack of Captain cigarettes from his flight jacket, offering one to Bảo before lighting one himself.

"I remember back in school, your family lived in the countryside, right?"

"Yeah, your memory is still sharp. But I don't visit there anymore because the Viet Cong have started showing up regularly."

After a long drag from his cigarette, Sơn blew the smoke out slowly and stared at the ground in silence for a moment, as if deep in thought. Then he turned to Bảo,

"What do you think about the war these days, Bảo?"

Bảo didn't answer right away. He brushed his hair back from his forehead and finally replied,

"A few years ago, I served in the military, just like you. I was stationed in Da Nang for over a year before I was reassigned to teaching. During that time, I asked myself the same question you're asking me now. In my opinion, this war we're fighting is a proxy war between two ideologies: 'materialism' and 'idealism.'"

After taking another puff from his cigarette, Bảo flicked the remaining stub into the garden and continued,

"Let me explain it a bit more. These two ideologies, when it comes to the relationship between matter and consciousness, between 'existence' and 'thought,' form the basic issues of philosophy. Marx and Engels are the two key figures who founded 'Dialectical Materialism' in the mid-19th century, and Lenin further developed it in the early 20th century. To put it simply, the two ideologies we're dealing with are 'capitalism' and 'communism.' The superpowers leading these two sides are the U.S. and the Soviet Union. Small, underdeveloped countries like ours can't stand alone; we have to choose one of these two systems."

"In my opinion, the key for national leaders is to wisely choose the right path for their people if they want their country to survive in the long run. Take modern Chinese history, for example. During the Qing Dynasty, China was fragmented, with different regions following different political systems. The north was under the Manchu monarchy, while the south was influenced by Sun Yat-sen's revolution, which aimed to overthrow the monarchy and establish a democratic system under the 'Three Principles of the People.' At the same time, Stalin was supporting and training the Chinese communist revolution. Western powers like Britain, France, Germany, and Japan were carving up China into different spheres of influence to expand their imperialist empires."

"The Chinese people were already poor, and this only made them poorer. They had to plow fields like beasts of burden just to pay taxes to their landlords. Local officials were corrupt, and the people's poverty, combined with national humiliation under foreign domination, left them with no way out. The Chinese Communist Party seized this opportunity and eventually succeeded in pushing Chiang Kai-shek's Nationalist Party to Taiwan."

"I'm sure you've heard of the Chinese Communist leader who recently said, 'It doesn't matter if the cat is black or white; as long as it catches mice, it's a good cat.' This shows me that communism wasn't their ultimate choice; they followed it because there wasn't a better option at the time."

Sensing that he had been talking for a while, Bảo paused to gauge Sơn's reaction.

"Am I rambling, Sơn?"

Sơn shook his head,

"No, you're not rambling. You made some very valid points. I respect that. You were meant to be a literature teacher. In Vietnam, everyone knows that Hồ Chí Minh was a member of the Communist International, and his ultimate goal was to spread communism throughout Indochina, not just South Vietnam. Whether or not the 1956 Geneva Agreement had led to elections to reunify the country, he still would have sought to conquer all of Vietnam by military force. There's no way he would have adopted any political system other than communism."

As the two friends continued their conversation, Vân came out and called them in,

"Brother, Mr. Bảo, dinner is ready."

And so, their discussion about current affairs was put on hold. Sơn stood up with his beer in hand and nudged Bảo,

"Come on, let's eat. Don't keep my mom waiting."

The dinner that evening was a cheerful affair. Sơn introduced Bảo to Vân, explaining that Bảo was teaching Vietnamese literature at a private high school in Saigon. Upon hearing this, Vân, who would soon be entering the same profession, commented,

"So, you're a teacher, Mr. Bảo? All this time, I thought you were in the military with my brother!"

Bảo smiled,

"I did serve for about a year in the First Military Zone, but then I was reassigned to teaching."

Sơn chimed in,

"Yeah, Bảo was drafted into the Thủ Đức Reserve Officer School. He didn't even finish his military rations before they sent him back to school."

Turning to his sister, Sơn added,

"Vân, once you graduate later this year, if you have any trouble finding a job, you can ask Bảo to recommend you to some private schools in Saigon."

Vân didn't seem pleased and interrupted,

"Brother, that sounds so inappropriate! I haven't even graduated yet, and you're already asking for favors."

Sensing Vân's discomfort, Bảo joined in,

"Sơn is right, Vân. I have a few friends who were assigned to remote areas after graduation, and they wrote letters home saying they were miserable and wanted to quit their jobs and return to Saigon. I can't promise anything, but I've been teaching for a while now, so I know a few people. If I can help, I will."

Sơn's mother, who had been listening quietly, now spoke up,

"Your brother and Mr. Bảo are right, Vân. When you graduate, if you don't know anyone, you could end up in a difficult situation, especially as a young woman. Being assigned to a faraway place is much more challenging for a girl than for a man."

In a slightly playful tone, Vân responded to her mother,

"Well, I didn't mean it that way, but still..."

In the "Paris Agreement," Dr. Kissinger and President Nixon of the United States pressured President Nguyễn Văn Thiệu and Foreign Minister Trần Văn Lắm of the Republic of Vietnam to sign a peace accord with Lê Đức Thọ and Xuân Thủy representing the North Vietnamese Communists, and Nguyễn Thị Bình representing the National Liberation Front of South Vietnam.

From a U.S. political perspective, the goal was for the South Vietnamese military to take over the burden, allowing American forces to withdraw.

But as soon as the ink was dry, violence erupted across the entire southern region. The "Red Summer" saw brutal battles at An Lộc, Phước Long, Ban Mê Thuột, Pleiku, and many other places.

Southern Vietnam, once hailed by the free world as the frontline against the spread of communism, was beginning to fall.

At this critical juncture, Vân graduated from university. The smile had barely settled on the lips of the young English teacher before she received an assignment from the Ministry of Education to teach at a high school in the First Military Zone. When the official letter arrived at her home, Vân was puzzled by the unfamiliar name of the district and the school. She read it over and over, even consulted maps, but couldn't pinpoint where it was located on this strange, distant land.

Seeing her daughter's anxious expression as she turned the letter over repeatedly, Vân's mother, who was chewing betel nut while lying in a hammock, asked,

"What's that paper you keep looking at, Vân?"

Vân replied in a tired voice,

"This..."

"What's this? Let me see."

Vân stood up and handed the letter to her mother,

"I just got this letter. It's my assignment from the Ministry to go teach, but I don't know where it is!"

After glancing at the letter, Vân's mother shook her head,

"I've lived for almost eighty years, and I've never heard of that district! Let's wait for Sơn to come home. Maybe he'll know."

Vân's face brightened as if she had just found the answer to a difficult puzzle,

"Yes, you're right, mom. Brother has been stationed at Da Nang Air Base before. He probably knows these places well."

"He flies all over, so he must know."

When Sơn came home, Vân rushed to him before he even stepped inside, excitedly sharing her news,

"Brother, I've been assigned to teach!"

Sơn turned to his sister,

"Wow, that was fast! I thought you'd have a few weeks off. Where are you assigned?"

"I don't even know where this district is! Here, take a look at the assignment letter and tell me where it is!"

Vân handed the letter to her brother. After a moment, Sơn shook his head and gave it back,

"I think the Ministry made a mistake. They shouldn't be sending you out to a place like this. Normally, people are assigned to work in their local area for convenience, not faraway places like this!"

Growing impatient, Vân pressed,

"So do you know where this is or not?"

"Of course, I know! I landed there a few times. It's a fairly large district in Quảng Trị Province, but it's near the DMZ (Demilitarized Zone). I don't think it's a good fit for you."

Hearing that her assignment was in such a remote, desolate place made Vân anxious,

"So what should I do? I can't just refuse the job, can I?"

"Why don't you apply for a transfer to the Third Military Zone, citing family reasons? You have a brother in the military, and you need to take care of our elderly mother. I think they'll understand."

"I talked to Uncle Tư, Hồng's father, who works in the Ministry's personnel department, and he said that new graduates have to accept remote assignments for about a year before they can apply to transfer back home. It's the same for everyone, not just me!"

Sơn shook his head,

"I doubt you'll last more than a month, let alone a year. Besides, I've heard things are pretty chaotic out there right now, with Viet Cong shelling the district regularly. It's dangerous."

"So, are you saying I shouldn't accept the assignment?"

"Yeah, I don't want to say it outright, but think about it—our family only has the two of us. I'm always off on military missions, and you're the only one at home with Mom. If you go, who will be here for her?"

"So, does that mean I'm going to be unemployed after all that schooling?"

"Why would you be unemployed? If you can't teach at a public school, just find a job at a private school. No one's going to starve."

Suddenly, Sơn seemed to remember something,

"Oh yeah! I'll talk to Bảo and see if he can help you out."

Vân didn't seem too happy with this idea and gave her brother a side-eye,

"You always rely on other people too much."

Sơn quickly defended himself,

"He's my close friend, not some stranger. We're just asking for a recommendation, not taking advantage of him. Why are you so worried?"

Vân muttered under her breath,

"Asking for a favor is just a polite way of taking advantage."

Sơn burst into laughter,

"Do you know what? When guys like me help out pretty ladies, people call it being 'gallant.' Especially when the lady is smart and single, like you. Don't call it 'taking advantage'—that sounds so crass!"

Vân retorted,

"With soldiers like you, you're always 'gallant,' so wherever you go, women fall for you, and then you end up breaking their hearts. Mr. Bảo, on the other hand, is a teacher, a man of education. I'm sure he's different from military men like you."

"I beg to differ! I can guarantee you that out of a hundred men, ninety-nine of them act like me, and there's only one who swims against the current!"

"If that's your logic, then one day you'll pay the price for your behavior," Vân shot back.

Sơn laughed heartily,

"Alright, let's drop this. But I'm telling you, Bảo would be a great match for you. Who knows, maybe one day he'll be calling me 'brother-in-law!'"

Understanding the implication of her brother's teasing, Vân playfully slapped him on the back,

"Stop it, brother! I'm not talking to you anymore."

In the end, Vân listened to her mother and Sơn's advice and applied to the Ministry of Education for a transfer closer to home, citing family reasons. The personnel department approved her request, but she had to wait until a position became available.

During that time, Bảo reached out to the administration at his school and arranged for Vân to teach English there.

Eventually, everything fell into place, and the young teacher, Vân, secured a relatively stable job. Each morning, Bảo would arrive at school early, stopping by the break room for coffee, hoping to catch a glimpse of Vân before class. Occasionally, if they finished teaching at the same time in the afternoon, Bảo would give Vân a ride home on his Vespa, saving her from waiting for a taxi.

Bảo and Vân had known each other for over a year, and they got along well, but neither of them dared to step beyond the boundaries of friendship and confess their true feelings. Bảo had noticed Vân's special gestures toward him many times, and there were moments when he wanted to tell her that he had feelings for her. But he always held back, fearing that Sơn would think he was taking advantage of their friendship to pursue his sister.

In contrast, both Sơn and their mother were just waiting for Bảo to make a move. After all, they didn't want to be too forward in offering their daughter to him.

The Vietnam War was intensifying. Sơn was constantly on duty or sent on missions to distant provinces, leaving him rarely at home. Every evening, after finishing his classes, Bảo would make time to visit Sơn's house to check on the family. Southern Vietnam's situation grew more dire as the United States, along with its allied countries, withdrew support, leaving the Army of the Republic of Vietnam (ARVN) to fend for itself, deprived of the resources it needed to survive. It became easy prey for the ravenous communist forces.

Then, on April 30, 1975, President Dương Văn Minh officially surrendered to the North Vietnamese communists. And with that, it was over—a new chapter in Vietnam's history had begun. It was a rare, costly lesson for those who had placed their faith in the strength and support of allies only to offer up their nation as the world's frontline in the Cold War.

Hồ Chí Minh's dream had come true. All of Vietnam was now painted red. Millions fled the country in search of freedom, while tens of millions remained behind, clinging to fragile hopes.

That evening, as on every other evening before, the Saigon docks were eerily quiet, with only a handful of people wandering aimlessly or sitting in a daze, staring blankly at the flowing river like lost philosophers. Who were they? They were the remnants of old Saigon, the strange products of a changed world. On the outside, they had been painted red, but their minds remained untainted. Were they once capitalists exploiting the people? Or were they once peasants dressed in rags? Now, they all looked alike, and no one could tell what their pasts had been.

Bảo didn't have the time to figure out the answer. These days, the police were everywhere, from the most remote corners to the narrowest alleyways. Everyone had to mind their own business if they wanted to survive.

After parking his bike by a tree, Bảo invited Vân to sit with him on an empty stone bench by the riverbank.

"So, Vân, what did you think of today's first political training session?"

Vân raised an eyebrow at him,

"You're a literature professor, and you're asking me?"

"They told us that from now on, we have to change the way we teach to align with the Party's guidelines. That means you're no longer allowed to praise The Tale of Kiều as a cultural treasure of Vietnam. You also can't say that Hàn Mặc Tử's romantic poems are brilliant. Now, you have to extol the virtues of Stalin, Marx, Lenin, and Hồ Chí Minh's revolutionary ideals as the pinnacle of practical scientific thought."

Bảo stayed silent, gazing far into the distance across the river as he listened to Vân. He was still grappling with the thoughts the lecturers had drilled into them earlier—that all forms of art must serve the people and politics according to the principles of socialism.

When Vân finished speaking, Bảo pointed to a clump of water hyacinths floating by,

"Vân, do you see those water hyacinth flowers drifting by? Aren't they pretty?"

Vân turned to look at him,

"Yes, they're beautiful, but what does that have to do with your question?"

"It's related to what the lecturers were saying earlier. They claim that everything we think or do must serve the interests of the people and benefit the working class. In their view, all art created by humans should serve humanity and not just be art for art's sake. In other words, art must fully serve politics and society. But look at those water hyacinths. Their beauty is a natural creation by God, with no specific purpose to serve humanity. Yet, the vibrant colors of the flowers give us both a moment of relaxation, soothing our souls. In that sense, they inadvertently have an impact on human well-being. When we feel mentally refreshed, we're more productive in our work, aren't we?"

"So, you're saying that all artistic creations—whether made by humans or found in nature—carry both functions?"

"Let's broaden the discussion a bit, Vân. Water hyacinths, or any flower for that matter, are natural creations born from the harmonious balance of the heavens and the earth. Some are beautiful, some are not; some have a pleasant fragrance, while others smell foul. The same goes for humans—there are beautiful people and not-so-beautiful ones, smart people and slow learners. Over time, human minds absorb knowledge through research, understanding various aspects of society and the sciences passed down by previous generations. This learning process must be free and unrestricted by rigid rules. Each of us, whether rich or poor, educated or not, carries some degree of artistic potential within us. A true writer or poet, when creating a story or crafting a beautiful poem, can't just do so at will. It requires a blend of mental inspiration and the right environment, much like how a fresh flower needs the right conditions to bloom perfectly. Creative expression must be free and open. If it's restricted, the result will be nothing more than cheap, artificial flowers created to serve specific agendas. That cannot be called true art."

After listening to Bảo, Vân offered her own opinion,

"Your words remind me of something. Not long ago, I read a poem by Tố Hữu that left me shocked. I couldn't tell if he was being serious or just being sarcastic. In it, he praised Stalin with lines like, 'I love my father one-fold, but I love you tenfold...' And when Jiang Qing, Chairman Mao Zedong's wife, visited Moscow, she praised Stalin by saying, 'Your health is the happiness of the entire Chinese people.' Don't you think they were glorifying these communist leaders like emperors of old?"

Mr. Tố Hữu is a renowned figure in the country, yet he loves his own father less than he loves the white, big-nosed Stalin by a factor of ten. So, does that mean the teachings of our ancestors are no longer correct?

"Father's labor is like Mount Tai,

Mother's love is like the water flowing from the source..."

Is all of that invalid now? Because today, love and loyalty must be reserved for the Party and its leaders. Family and morality—those must be discarded, leaving only the Party and its leaders as supreme.

As she spoke, Vân's voice seemed to rise in irritation. Bảo gently placed a hand on her shoulder,

"Calm down a bit, Vân. If they hear us, we'll both end up in jail counting the days. Don't you know, no matter where they are, communists are all pretty much the same. They follow the teachings of Marx and Lenin, which they believe to be the one and only path to liberating humanity—a philosophy rooted in dictatorship and violence. As Marx said, 'The weapon of criticism cannot replace criticism by weaponry. Material force must be overthrown by material force.' Do you see that now?"

After regaining her composure, Vân continued,

THE LAST FLIGHT OUT

"But does that mean one day the whole world will be painted red?"

"I don't think so, Vân. History has clearly shown that no single ideology remains absolutely true across generations. As you know, society and people create history; history merely records the phenomena that occurred in society at a particular point in time. In my opinion, Marx and Engels came up with a new ideology because, in the mid-19th century, capitalism had become the dominant economic system in Western Europe. The growing injustice between capitalists and workers sowed the seeds for an increasing class struggle between labor and capital and between the proletariat and the bourgeoisie. This is what led Marx and Engels to shift from idealism to materialism, from democratic revolution to communism, influenced by the social class struggles of that era. Following that, in the early 20th century, Lenin developed Marx's theories into strict doctrines that became the guiding principles for later communist states."

"So, according to you, communism only succeeds in certain periods or at particular times?"

Bảo glanced at Vân, as if probing to see whether she was genuinely interested in understanding communism.

"You're asking good questions, but I'm afraid I'm talking too much, and it might bore you. These topics are a bit dry."

Vân laughed and gave Bảo a playful shove,

"You're a Vietnamese literature teacher! If it's too dry to swallow, just add a little water, no big deal!"

"Alright, I'll try to make it easier to digest. Here's a challenge for you, Vân."

She widened her eyes,

"I haven't read much about communist theory, so how can I answer your questions?"

Bảo smiled,

"It's a simple one. Do you know the difference between religion and communism?"

"Umm...well, religion deals with theology, idealism. It generally teaches us to live righteously, build merit, and when we die, we'll go to heaven or some other paradise. Communism, on the other hand, follows materialism. But honestly, I don't really know what they want. So I'll pass the question back to you."

Bảo clapped his hands softly,

"That's close! Let me fill in the rest. Religion and communism are somewhat similar in their goals, with one key difference. Religion says that after death, the souls of the righteous ascend to heaven or paradise. Communism, on the other hand, seeks to establish paradise on Earth while we are still alive. So, the difference lies in what happens before and after death. That's why religion has persisted and evolved for so long, from ancient times to the present."

"But communism will never truly establish a paradise on Earth. Communism aims for equality, justice, and the end of exploitation between humans. On the surface, it sounds ideal, but if you think deeply, you realize people are not that simple. Humans are complex, more so than any other living creature, because we have intelligence, free will, and reason."

"According to Marx, 'Theory also becomes a material force once it has gripped the masses.' In my view, that means if you repeat something often enough, sooner or later, people will internalize it, and it will become second nature."

"According to Marx-Lenin theory, the first step is to seize the property of the bourgeoisie so everyone becomes equally proletarian. This is called the 'socialist stage,' leading up to communism, which is supposedly their final goal. As you've seen, one way they achieve this is through actions like currency reforms and property confiscation, which happened not too long ago."

When Bảo mentioned this, Vân could barely contain her anger and responded,

"They claim to be punishing exploitative capitalists, but are they really just robbing the people? If they were confiscating the wealth of corrupt officials who engage in dirty dealings, I'd applaud them! But I've seen so many hardworking, honest people who have become successful through their own thrift and labor, only to be labeled 'capitalist' so the government has an excuse to rob them. It's utterly unjust. It's like something out of prehistoric times when people weren't yet civilized."

Bảo chimed in...

"Look at Communist China. When Mao Zedong took power, he launched the 'Land Reform' campaign, and countless people were brutally persecuted and unjustly killed in struggle sessions. Then came Mao's 'Hundred Flowers' campaign, which was a trap to weed out intellectuals who opposed him."

"Finally, there was the 'Cultural Revolution,' led by Jiang Qing and the Red Guards, which wreaked havoc on China's intellectual and artistic communities. Even within the Communist Party, internal power struggles were common. So tell me, how can we say communists are good when they use such devious methods?"

At this point, Bảo realized they had been discussing the new regime for quite some time. Glancing at his watch, he turned to Vân,

"We've been so wrapped up in talking about society that we've lost track of time. Let's not keep your mother waiting and worrying about dinner."

Vân smiled,

"Why don't you come over and join us for dinner?"

"Thanks, Vân, but I'll pass this time. I didn't let Aunt Năm know I wouldn't be home, and I don't want her to wait for me."

Time dragged on, filled with the dull, fearful days of a Saigon under new rule. The once bustling streets with their spacious shops, where buyers and sellers used to come and go in droves, were now unrecognizable. At night, dim yellow lights barely illuminated the gaunt faces of the hungry, scavenging through piles of trash to collect plastic bags or wandering in search of a place to sleep until morning. These people were now the true embodiment of the proletariat, the very class the Party had claimed to liberate.

Uncle Năm, who used to run a clothing store in Chợ Lớn, had been 'elevated' to the role of a bicycle tire repairman, working under a tamarind tree in front of his house. Aunt Năm had fared slightly better, growing a few rows of sweet potatoes in the backyard, which she planned to secretly dig up to sell in the neighborhood for some extra cash to buy rice, all while evading the local police.

Bảo still rented a room at Uncle Năm's house, but with his meager teacher's salary, he handed most of it to them for food and rent, keeping only a little for personal necessities.

Since Uncle Năm's eldest son had gone missing, the couple treated Bảo like their own son and never fretted about the money he contributed each month.

The bureaucratic state control and the family registry system were like shackles, tightly binding the people to the oppressive communist regime. Even in Russia, the birthplace of communism, where it had ruled for over sixty years since 1917, people were still standing in line for bread, and the secret police continued to monitor them in various ways. So, one might ask, is this the 'paradise' of communism, or is it hell on Earth?

After two years of living under communism, all Bảo saw was lies and manipulation. During his lectures, he sometimes felt like he was losing his mind. His body was nothing more than a machine, spouting lifeless words to glorify a godless regime. He was no longer the 'Tú Uyên' of old, whose literary critiques once brought the classroom to applause.

Today, after a particularly dull lesson, Bảo stopped by the tamarind tree to chat with Uncle Năm, who was busy repairing a bicycle tire. As Bảo approached, Uncle Năm called out,

"Your bike needs its brakes adjusted. Bring it here, and I'll fix it for you."

"I just tightened the bolts yesterday. Maybe they've loosened again."

"Well, you have to tighten them really well; otherwise, they'll slip when you brake. Oh, by the way, Vân stopped by earlier to ask about you. She said she hadn't seen you in days and was worried. If you have time later, you should go visit her."

"I've been busy preparing lessons for the past few days, so I haven't been able to visit."

As soon as Bảo finished speaking, Uncle Năm stepped closer and asked quietly,

"Any news about her brother? That Air Force major—have you heard anything?"

"No, I haven't, Uncle. Vân was let go from the school because they said she had connections to an American-backed family."

Uncle Năm shook his head and sighed,

"What a strange system. They won't use talented people; instead, they employ fools. How can society progress like that? Well, your brakes are fixed. Try it out, and then go have dinner."

Aunt Năm was outside, picking amaranth and vine spinach leaves near the front fence. When she saw Bảo, she immediately set down her basket and opened the gate,

"Where have you been these past few days? You're not sick, are you?"

"No, Aunt. The school's administration gave me a stack of materials to study for my lessons, so I haven't had time to visit."

"Well, bring your bike in, dear. Vân's in the kitchen cooking dinner.

She's been asking about you every day."

Vân was stoking the fire on the stove, preparing dinner. She heard Bảo's voice outside and was about to step into the living room just as he arrived. She greeted him with a mix of happiness and mild reproach,

"Bảo, where have you been these days? I thought you might have fallen ill. I even stopped by to ask Uncle Năm this afternoon, and he told me you were still teaching as usual."

"I'm fine. I've just been swamped with work. If I don't keep up with it, they'll criticize me, and the review meetings are a real headache."

Vân smiled sympathetically,

"More of that Marx-Lenin or Uncle Hồ material, right?"

"You already know, Vân. It's the same thing over and over again. The people are starving, with barely enough clothes to wear, looking like skeletons, yet the Party only talks about how glorious the path of Uncle Hồ is. They make it sound like our nation is brimming with heroes, as if every corner you turn, you run into one."

Vân chuckled mischievously,

"Haven't you heard the kids joke about it? They say, 'Uncle Hồ's path leads to tragedy, and the moment you step outside, you run into a lunatic!'"

Bảo waved dismissively,

"Let's drop that. Is there anything new with you these days? How's business?"

Vân suddenly remembered something important. She moved closer to Bảo and whispered,

"I just got a letter from my brother. It came through France, and my mother and I were so relieved."

Bảo's face lit up, and before Vân could finish, he asked eagerly,

"Your brother is in France now?"

"No, he's actually in the United States. He had someone in France forward the letter because he couldn't send it directly from the U.S."

"Ah, I see. So he's doing well, I hope? Has he found a job?"

"In the letter, he says he's doing fine. He's working part-time and studying electronics in the evenings. It's been nearly two years since we last heard from him. Now we finally know he's alive. Before this, my mother would cry and pray for him every day, hoping he'd survive."

Bảo stroked his forehead,

"After you lost your job because of your family connections, and with so many friends who couldn't make a living under this regime leaving by boat, it's hard to find anyone from the old days at school. It's so disheartening, Vân. Your brother was lucky to escape. If he had stayed behind, he would have suffered greatly."

Vân nodded and added,

"You still haven't answered my question about how business is going."

Vân tilted her head and smiled,

"Are you asking about my new line of work?"

"Who else would I be asking about?"

"Well, on good days, when I don't run into the police, I manage to scrape together a little. But if I'm unlucky and get caught in one of their raids, I sometimes lose everything. You know the second-hand clothes I sell? They call them 'Sida' clothes. I buy them, wash them, iron them, and resell them for a profit. It's kind of like how the old Chinese used to collect and resell empty bottles and duck feathers."

Bảo shook his head,

"At least these days, if you can earn a little to keep going, that's a success."

Today is a major communist holiday, commemorating the victory that painted all of Vietnam red. Schools are closed, and flags hang from government offices and along the main streets. Large banners proudly celebrate the Party's revolutionary achievements.

Bảo pedaled to a nearby state-run store to buy a newspaper to pass the time. The sun was now high in the sky, its rays piercing through the tamarind trees and casting small patches of light on the black asphalt, like a torn shirt patched with white fabric. As he headed back to his lodging, Bảo suddenly remembered Tâm. He thought to himself, 'Maybe I should stop by and visit him and his wife before heading home.'

Tâm and Loan's apartment wasn't far, only about fifteen minutes away by bike. Without rushing, Bảo turned around and slowly made his way there. The streets were still mostly empty, with only a few sluggish cyclo drivers pedaling in the opposite direction. Seeing the thin, worn-out appearance of one of the drivers reminded Bảo of something a friend had told him recently,

"These days, rickshaw drivers are better off and making more money than we are as teachers."

Reflecting on his friend's words, Bảo smiled wryly and thought, 'Maybe this is the golden age for laborers and the working class, just as the Party intended. The government controls everything and owns everything. Soon, perhaps every heroic citizen of Vietnam will be a cyclo driver or manual laborer!'

Tâm was sitting with his back to the door, carefully polishing his bicycle, when Bảo arrived and called out,

"Hey, professor! If you keep polishing it like that, you'll wear off all the paint!"

Tâm was so focused that he jumped in surprise before turning around,

"Oh! Bảo! I haven't seen you in a month. Where've you been?"

"The school's administration changed the schedule, so I couldn't see you as often. Today's a holiday, so I finally had some free time and thought I'd stop by to visit you two."

Tâm leaned his bike against the wall,

"Come on in and have a drink."

Hearing the commotion, Loan stepped out and smiled warmly when she saw Bảo,

"We were just talking about you and Vân yesterday. Anything new with you, Bảo?"

Placing the newspaper on the table, Bảo quickly replied,

"Nothing new. It's the same old stuff."

Loan teased him,

"Don't keep any secrets from us! When's the wedding? Make sure you invite us to crash the party!"

"Go chat with Tâm while I make some coffee," she added as she headed to the kitchen.

Tâm used a feather duster to wipe the dust off a chair for Bảo,

"We thought you'd be out in Saigon today watching the parade."

Bảo shook his head,

"I'd rather not listen to them boast and lie. I picked up a newspaper to catch up on the news later."

Tâm laughed and pointed at the newspaper,

"You believe what those reporters write? They're the kings of propaganda. Everything the Party and Uncle Hồ do is the best."

"Well, if you don't praise those things, you'll end up in prison counting the days. Last week, I bought a book on Vietnamese proverbs, folk songs, and folk literature. But as I read through it, I grew increasingly frustrated. I got the impression the author was a communist propagandist, not a genuine scholar of folklore. In the proverbs and folk songs, he tried to tie everything to the class struggle between the proletariat and capitalists. It was so forced, even to the point where he claimed that thanks to those proverbs, communist guerrillas were able to defeat the American-backed regime!"

"Didn't you say before that all writers, poets, and musicians nowadays have the same mindset? Dialectical materialism and 'art for the people'? For the communists, whatever comes from Marx-Lenin, Stalin, Mao Zedong, or Uncle Hồ is absolutely right. Anything opposed to it is wrong, counter-revolutionary, reactionary, and labeled as foreign espionage. The lightest punishment is prison, the worst—death. Even high-ranking, long-standing members of the communist party fear each other, let alone the rest of us. Look at the Chinese Communist Party under Mao Zedong. Lưu Thiếu Kỳ, Lâm Bưu, Chu Ân Lai, and Đặng Tiểu Bình—these were powerful figures in the Central Committee, ranked just below Mao. Yet in the end, they were accused of revisionism and counter-revolutionary crimes during the Cultural Revolution led by Jiang Qing."

As the two friends chatted, Loan returned with the coffee and invited Bảo to try it,

"Have a sip, Bảo. It's really good. I bought it from someone who brought it down from Ban Mê Thuột. They sold it to me to raise money for bus fare to visit their son serving in Tây Ninh."

Bảo took the cup from Loan and chuckled,

"You don't need to advertise it. I've been smelling the sweet aroma of the coffee from the kitchen since I got here. We'd better hope none of the neighbors pass by and mistake it for capitalist coffee!"

"Don't worry," Loan reassured him. "We know everyone around here. By the way, how's Vân doing these days, Bảo?"

"Since she lost her job, she's been buying and selling second-hand goods to make ends meet. She's doing alright."

"And have you heard any news about her brother?"

"Last week, Vân told me she received a letter from Sơn. He's living in the U.S. now."

Tâm chimed in,

"That's great news. Anyone who can escape should take the opportunity. Sometimes, I think we need to be realistic. Who doesn't love their family and homeland? But these leaders manipulate concepts like patriotism and love for one's heritage to demonize those who leave. They try to hide and deny their own mistakes and wrongdoings."

After her husband's remark, Loan glanced at Bảo and smiled,

"My husband is really bitter about the communists!"

Bảo nodded in agreement,

"How can you not be bitter, sister? The sad part is that we hate it, but we still have to work with them to survive. When they first arrived, people had high hopes because they thought this would finally bring peace and reunify the country. We all knew the North was communist, but with the spirit of unity and our close ties to the land and our nature as peaceful people, we believed that since we were all one nation, they wouldn't mistreat us."

Loan laughed and added,

"Like they say, 'You don't know until you live with them and see the lice!' Right, Bảo?"

"They've bitten me enough that I know now, sister! Sometimes, I think the problem is just that our bodies keep causing us grief. All day long, we're consumed with finding food, clothes, and shelter. It's like being stuck in an endless cycle."

Tâm sipped his coffee and stroked his chin,

"Well, we're no different, are we? If we weren't poor and hungry, we wouldn't need them. If people had full bellies and time to think, the Party wouldn't be able to rule. Communism is the society of the impoverished and the wretched. Only the top leaders live in luxury."

Bảo nodded and sighed,

"It's frustrating. But if you lose your household registration or your job, your life is over."

After multiple invitations from the deputy head of the school, a communist official, to join the Party, Bảo had always found a way to politely decline. Perhaps because of this, despite over a decade of teaching, Bảo remained an ordinary teacher, while those who joined the Party quickly rose to higher positions in the education system.

Today was the end of the month. After receiving his paycheck, Bảo sent money for rent and food to Uncle Năm. With a little left over, he decided to visit Aunt Hai and invite Vân out for an evening stroll and bánh xèo.

It was now half past five, and the sun was still casting a warm glow. Bảo brought his bike to the front yard, where Uncle Năm added some oil to the chain and pumped up the tires. At this time of day, laborers returning home from work walked in small groups of three or four, chatting about their day's work. Most of the old private businesses remained shuttered, adding to the sense of isolation Bảo felt as he pedaled down these once-familiar streets. The tiny yellow tamarind leaves fluttered down to the road, reminding him of something his mother had said last Sunday when he visited her,

"Bảo, the situation has stabilized. You're 36 years old now, not exactly young. You should find someone so I can finally have a daughter-in-law."

Bảo had stammered,

"Well…um…in time, Mother, in time."

Her words echoed in his mind like an alarm clock going off. He thought to himself, "Time flies so quickly. I'm already 36." All his old friends, like Tâm, Tiến, and Mỹ, were married, leaving only him to face the loneliness after each school day. Sometimes, sitting alone in his empty room, he would feel a twinge of sadness, but it was easily forgotten amid the everyday chaos.

He had long harbored feelings for Vân, but he kept his behavior cautious, treating her like a younger sister rather than a love interest. This often irritated Vân, and she didn't hesitate to criticize him directly,

"You're a Vietnamese literature teacher, but sometimes you act like an old man!"

Instead of being offended, Bảo would smile and respond lightly,

"There's still plenty of time. Who knows if your predictions will come true?"

And she would lower her voice, feeling a bit regretful for speaking too harshly,

"I meant to say you're like Tú Uyên, staring at a painting while waiting for your Giáng Kiều to come to life!"

Bảo would laugh,

"I already have a real Giáng Kiều; no need to dream about a painted one!"

This would make Vân blush,

"Then introduce me to her!"

"You already know her."

Hearing this, Vân's heart would race, and she would playfully push him away,

"You're talking nonsense again!"

When Bảo arrived at the gate, Vân had just finished watering the rows of okra by the side of the house. Spotting him, she called out excitedly,

"I knew you'd come by this evening! My mother went to Rạch Kiến to visit Uncle Tư this morning. She won't be back until tomorrow."

"So you're home alone? Aren't you scared?"

"I'll just lock the doors and go to sleep. Nothing to be afraid of. Come in, Bảo. I'll put on some hot water and brew some tea."

Bảo waved her off,

"No need to go to the trouble. I actually came to invite you for a walk by the Bạch Đằng Wharf to enjoy the breeze."

Vân looked at him,

"But we only have one bike."

Bảo laughed and pointed to the back of his bike,

"I'll give you a ride!"

"Really? Let me change into something more comfortable, and we'll go."

The sunlight was fading over the other side of the river, and the cotton-like clouds on the horizon had turned into castles and palaces, their colors shifting in the light's refraction. A gentle breeze occasionally drifted across the river, bringing a refreshing coolness that was hard to resist.

Vân sat in contemplative silence, her eyes following the clumps of water hyacinth lazily floating downstream. Bảo lit a cigarette, and the fragrant blue smoke swirled into the air. The scent brought Vân back to the moment, and she turned to Bảo, pointing at the drifting plants,

"I bet you can't tell me where those water hyacinths are headed."

Bảo didn't answer right away. He looked tenderly at Vân as if trying to understand the deeper meaning behind her question,

"Are you asking seriously or just teasing?"

She gave him a playful look,

"You always think I'm teasing. Don't you realize we're both adults now?"

A slight sadness crept into Bảo's expression,

"Yeah...sometimes we lose track of time in this society. We spend our days just trying to scrape together enough to survive. Who has time to think about their personal life?"

Sensing she had unintentionally touched on Bảo's insecurities, Vân quickly shifted the topic,

"Hey, you still haven't answered my question."

"You asked where the water hyacinths are going, right? When the tide rises, they'll drift inland, and when it ebbs, they'll float out to sea. That's where they're headed."

Vân laughed,

"So, according to you, they'll just keep circling back and forth in this section of the river forever?"

She paused, then lowered her voice, speaking softly enough that only Bảo could hear,

"Kind of like the lives of the two of us, don't you think?"

Bảo remained silent. Gently, he reached out and stroked her shoulder-length hair, nodding in quiet agreement.

"You know what I've been thinking? We are people digging our own graves, burying time, burying the lives of our youth. Many times I feel too weak, unable to face what I truly dream of, so I keep hiding it, deceiving myself."

As soon as Bảo finished his sentence, Vân turned and softly asked:

"So what you mean to say is..."

"Yes, I mean to say that I love you. I miss you so much when you're no longer teaching at the same school as I do. Every time I come to visit you, I feel nervous. Being teachers like us, people often scrutinize and gossip about us. That's why sometimes you told me I talk like an old man, remember?"

Vân joyfully leaned her head against Bảo's arm.

"I feel the same as you. Each afternoon when you don't visit, I feel so lonely. Sometimes, my mother notices I'm acting strangely and asks me about it, but I don't know how to respond, so I just brush it off, saying I have a headache to avoid the subject."

Bảo spread his arms wide and hugged Vân tightly.

"We don't need to keep our distance like before, do we?"

Vân looked up at Bảo, overjoyed, gazing at the love she had just claimed for herself. Bảo gently leaned down and placed their first kiss on her lips.

Outside, the sky had started to dim. A pair of white storks were flying leisurely side by side towards some distant horizon.

Three months after Bảo and Vân's modest wedding, they found it increasingly difficult to adapt to life in this new socialist regime. Bảo continued teaching, but his salary barely covered his own expenses, and Vân had to sell goods at street markets to help support their household.

In the communist regime, intellectuals were often considered to harbor capitalist ideals. Moreover, Bảo had married someone from a family connected to the previous government, meaning that the couple's future was clouded in uncertainty. Vân's mother, aware of this situation, had witnessed similar events after the communists took control of their village. She remembered the time when Uncle Năm Can's house, in the same neighborhood as theirs, was burned down by the Viet Cong because his son had been taken by the French army. All the wealth and savings of the old couple went up in flames. The devastated couple cried and had to rely on neighbors to help them build a small hut to live out the rest of their days.

That day, Vân came home earlier than usual. Her mother noticed and asked:

"Hmm, did you get a good deal today and sell all your goods? Why are you home so early?"

Vân shook her head.

"No, the police were cracking down hard today, so I couldn't sell anything. I decided to come home early instead."

"Just put your goods away and try again tomorrow."

After a brief pause, as Vân went to store her unsold items, her mother approached her and spoke quietly:

"I've been thinking, my dear. You and Bảo will always struggle if you stay here. I'm thinking of selling this house so you two can buy some gold and escape by boat."

Surprised by her mother's suggestion, Vân turned quickly.

"Sell the house so we can escape by boat?"

"Yes, I've been thinking about it for a while now. I didn't say anything before because you weren't married yet, but now that you have a husband to take care of you, I feel I should mention it."

"But if you sell the house, where will you live? Who will take care of you?"

"I'll go live with your Uncle Tư in the countryside. He and your aunt told me I should move in with them to keep them company. Their house, which was left to them by your grandparents, is quite large. Their children have grown up and moved out, so they're living alone and feeling lonely. You should talk to Bảo and see what he thinks."

That whole day, Vân couldn't stop thinking about her mother's suggestion that they flee the country. It was something she had never seriously considered before. She wasn't sure how Bảo would react. He had once told her that he looked down on people who fled the country with the French army when they left, calling them cowards. He had also refused several offers from friends to escape during the final days of April 1975.

Lost in thought, Vân didn't notice Bảo's arrival until he spoke.

"When did you get home? I didn't even hear you come in."

"I just got back. How did business go today?"

Vân pouted.

"It was awful. The police were all over the place, so I couldn't sell anything."

Bảo chuckled.

"That's just part of selling at the street market. Let me make it up to you."

He walked over and embraced Vân, planting a soft kiss on her cheek. Vân, slightly embarrassed, pushed him away.

"Stop! Mother might see us and laugh. You should wash up and have dinner. I have something important to discuss with you later."

After dinner, while Vân tidied up in the back, Bảo went outside, lay down on a hammock, and watched the sparrows flutter around the spinach vines. After marrying Vân, Bảo had given up his rented room at Uncle Năm's house to live with Vân and her mother, saving on expenses. He loved spending his evenings relaxing in the hammock, watching the little birds happily peck at the ripe purple spinach berries. Sometimes, in his daydreams, he wished he could be as carefree as those birds flying freely across the skies.

Vân approached Bảo, breaking his reverie.

"Did you see any new birds today?"

"No, just the usual sparrows."

After a pause, Bảo added playfully:

"But wait, I do see one new bird coming my way!"

Vân, pretending not to understand, looked out at the fence.

"Where? I don't see anything."

Bảo pointed to Vân.

"Here!"

Realizing his teasing, Vân laughed, a bit embarrassed.

"You're so silly! I'm not asking you anything anymore."

Bảo stood up and invited Vân to join him in the hammock.

"This special bird is just for me!"

Vân playfully covered his mouth.

"You're such a joker! You'll make Mother laugh! Anyway, I have to tell you what she mentioned earlier today."

"Go on, what's the matter?"

"Mother is thinking of selling the house."

Bảo, shocked, sat up straight.

"She's selling the house?"

Vân grabbed his arm and held him close.

"Yes, she's thinking of selling the house to get enough gold for the two of us to escape by boat. But she wanted me to ask you what you thought first."

Bảo remained silent, glancing around. After a long drag of his cigarette, he asked:

"What do you think?"

Vân shook her head.

"I don't know."

"If we go, will mother come with us?"

"No, she said she'd go live with Uncle Tư. My brother Sơn wrote in his letter that we should find a way to escape, and he would try to help us once we're gone."

Bảo flicked the cigarette away.

"We should think this through carefully before telling mother."

"We can talk to her tomorrow."

Over a year later, Sơn had sponsored Bảo and Vân, allowing them to leave the refugee camp in Malaysia and move to the United States.

Thanks to her strong English skills, Vân had found work as a teacher's aide at a local elementary school for the past two months. Bảo worked part-time during the day and attended English classes in the evening. He planned to enroll in university in the fall to change careers.

It was a Saturday evening during the beautiful Pacific Northwest summer. The sun was still shining brightly through the pine trees outside their home, and squirrels with fluffy tails were chasing each other, competing for peanuts Vân had thrown out.

Bảo stepped out of the house, intending to invite his wife for a walk in the nearby forest. Just then, Sơn and his wife, Hằng, arrived by car. Vân excitedly called out:

"Brother and Sister Hằng are here!"

Sơn got out, waved to everyone, and said:

"You two look like you're doing great this evening!"

Bảo walked out to greet them.

"Weekends here are so relaxing. I was just about to invite Vân for a walk in the woods to pick some blackberries when you arrived."

"I just finished work and came home when Hằng suggested we stop by and invite Vân to go shopping for fun."

Hằng, wearing a stylish light green summer outfit, turned to Vân and said:

"I came to take you shopping!"

Vân smiled in delight.

"I'll never turn down a chance to shop with you, Sister Hằng. You dress so beautifully and look so elegant. It's no wonder my brother is head over heels for you!"

Hằng glanced teasingly at Sơn.

"And yet your brother still says I'm clumsy and country!"

Vân protested:

"He's the one who's country and clumsy! I love your style, and I'm sure other women do too!"

Hằng turned to the men and asked loudly:

"Do you two want to come shopping with us?"

Sơn waved his hand dismissively.

"No, sitting around while you shop is too tiring. Just be sure to buy some beef for stir-frying; we'll drink and snack later."

As the car disappeared from view, Bảo went back inside.

"I'll make some coffee while we wait for them, Sơn. I bought it at Safeway yesterday. My coworkers say it's good, so try it and see."

"Coffee here is good, but it doesn't seem as fragrant as the coffee back home."

"You're right. There are so many brands in the stores here that it's hard to know which one to choose. Back home, you had to stand in line at state-owned stores, and even if the products were terrible, you had to take them or go without them. It was so unfair."

"After years of teaching and living under the communists, do you think their propaganda was true?"

After a deep sigh, Bảo shook his head.

"Living with them showed me the truth. As a teacher, they forced me to become a propaganda tool for the Party, training students to know only communist literature and history. Everything else was labeled degenerate, the remnants of imperialist capitalism. Their policy was to completely replace the cultural heritage of the free South with the socialist culture of the North."

Sơn sipped his coffee and added:

"It reminds me of Mao's Cultural Revolution and the Red Guards in China during the 1960s and 70s, which was like Emperor Qin burning books and burying scholars."

"Exactly. History is past, but we can learn from it. The communists' political strategies were ruthless."

Sơn poured himself more coffee.

"At least we escaped. If we'd stayed, we'd probably have been sent to labor camps in the North."

"No doubt. Especially for Air Force officers like you, they would have given you an even harsher sentence. When they first took over, people in Saigon were excited and welcomed them like heroes. At first, I felt a bit optimistic and happy that the country was finally unified. When my friends invited me to escape, I refused, wanting to help rebuild the country for the next generation. But soon, their true colors showed, and it was too late. People in Saigon could only look at each other in dismay. The communists' speeches were inspiring at first, making people feel like wealth was just around the corner. But after being promised the world too many times, people just hoped for a full bowl of rice each day."

Sơn shook his head sadly.

"Karl Marx and Lenin dreamed of a utopia where everyone would be equal, and one day, the state and military would no longer be necessary. Wouldn't that be a paradise on earth, Bảo?"

Bảo laughed and set his coffee down.

"That's the communists' utopian dream. They believe Marx and Lenin were infallible, and if we follow their path, the world will eventually unite in harmony. If that were true, humans would become senseless creatures. Religion has existed for thousands of years before communism, yet has it made us all saints or gods? The truth is that human nature is inherently selfish. Even a baby, when given one breast to feed, will clutch the other with its hand, instinctively protecting what it considers its own. Even animals mark their territory and will fight to protect it. The idea that all people can live in harmony is a delusion. Even in communist countries, comrades compete and clash with each other. The so-called 'world commune' of the communists is nothing but a fantasy of fanatics."

At that moment, Vân and Hằng returned from shopping. Vân called out:

"Bảo, help me carry the bag of rice inside."

"Coming! Did you buy anything for us to snack on?"

"Of course! Sister Hằng picked out some prime beef for you two to stir-fry and enjoy with drinks. And tonight, we'll have sour fish soup and crispy fried tilapia with chili sauce."

Hearing the menu, Sơn clapped in delight.

"That sounds perfect! But Hằng has to make the sour soup her way. It's the best!"

Vân teased Hằng:

"See how much my brother adores you? He's always saying how well you cook."

Hằng laughed.

"Vân, he only says that because I had to learn how to cook his favorite dishes from Rạch Giá after he raved about them so much!"

After a delicious dinner filled with the flavors of their homeland—sour shrimp soup and fried fish with tomato sauce prepared by Vân and Hằng—Bảo and Sơn couldn't stop complimenting the meal.

Time passed peacefully. Nearly twenty years had gone by since the communists had taken over the South. Bảo was no longer the high school literature teacher or the romantic 'Tú Uyên' admired by his students. Now, he was a mechanical engineer working for a large company in the city. Vân had become a full-time employee of the state's Department of Education. After arriving in the U.S., the couple had a son whom they named Lam Sơn. He was now sixteen and attending high school. Their life was comfortable and happy.

That morning, Bảo was dusting the liquor cabinet near the dining table when Vân, having just finished dressing, called to him:

"Hurry up, we're going to be late!"

"Just tell Lam Sơn to get ready. I'll finish dusting this and be right there."

"He's all ready and waiting. It's just you we're waiting on."

Bảo hurriedly put away the feather duster.

"Okay, okay, I'm getting my coat."

Vân was anxious because they had promised to arrive at Hằng's house by nine to go to the temple together for the Lunar New Year.

Bảo locked the door as Vân and their son sat waiting in the car. Glancing at his watch, he said:

"We still have ten minutes."

"I don't want to get stuck in traffic. Let's get to Brother's house in time."

Bảo nearly reminded her that they were in a car now, not on foot or bicycles as they were back in Vietnam, but he stopped himself and smiled.

Bảo often reflected on how different life in the U.S. was compared to Vietnam. No matter the circumstances, he remained proud of his Vietnamese identity. He had never considered changing his name, Nguyễn Trọng Bảo, which his parents had given him. He believed that Vietnam's poverty and underdevelopment were not due to laziness or ignorance but poor governance, stemming from the remnants of feudalism, colonialism, and now, communist dictatorship. The communists had gained power by promising a better life for the poor, using them as tools for the Party's benefit. Once successful, they divided the spoils among themselves while continuing to oppress the very people who had helped them rise to power.

One day at work, a colleague asked him, "I've heard that Vietnam has an ancient civilization. It must be very impressive, right?" Bảo had simply smiled, nodded, and found an excuse to leave. Later, he recounted the story to Vân, and she laughed, teasing him.

"That was a question tailor-made for a literature scholar like you. You must have given them a lengthy answer!"

Bảo shook his head seriously.

"No, I just nodded, said I was busy, and walked away."

"Why? You should have told them about it!"

Bảo chuckled.

"People here in the West are very practical. They care about the present, not the past. What matters to them is what works now, not what happened before. I couldn't lie like the communists. You remember how, after they took the South, they constantly bragged about defeating the

French and the Americans, calling themselves the pinnacle of human wisdom. Yet not long after, those same 'heroes' were begging for help."

"Now that you mention it, I remember when they first arrived, they told so many lies that even I believed them. When Brother Sơn escaped, Mother and I were heartbroken, thinking he'd be fine if he stayed. Soon, we realized the truth when all of his friends who stayed were sent to labor camps in the North, and the markets were closed. That's when we saw their true face."

"Yes, I was no better than you. I thought that once the war ended, we'd have the chance to rebuild the country. But we were wrong."

As Bảo was lost in thought, their son suddenly pointed ahead.

"Dad, there's Uncle and Auntie waiting for us."

"Yeah, I see them."

Vân checked her watch.

"We're only a few minutes late, and they're already out waiting."

The Vietnamese Buddhist temple was beautifully decorated for the New Year, with yellow apricot blossoms, chrysanthemums, and marigolds brought from warm states like California and Florida to welcome the spring. The sound of bells and chanting echoed from the main hall, drawing people into a peaceful spiritual atmosphere. Young women in traditional áo dài strolled gracefully in the morning sun, their faces glowing with joy as they greeted the New Year with the blessings of Buddha.

After a tour of the temple grounds, Hằng and Vân wanted to enter the main hall to light incense and offer prayers. Hằng turned to her husband.

"Do you and Bảo want to come in?"

Seeing the long line of people waiting outside, Sơn shook his head.

"No, you two go ahead. We'll wait outside and give others more space. Bảo and I will take Lam Sơn to the front and wait for you."

As they turned to leave, Lam Sơn grabbed his uncle's hand.

"Uncle, is New Year's here as fun as in Vietnam?"

"Oh, when I was your age, New Year's in Vietnam was so much fun! On the first day, your mother and I would receive lucky money from our grandparents and elders. We'd wear new clothes, visit relatives, set off firecrackers, and play traditional games like bầu cua cá cọp with the neighbors. But after I fled the country, I don't know if it's still the same.

Your father can probably tell you more about what it was like after the communists took over."

After answering Lam Sơn, Sơn turned to Bảo.

"What were New Year's celebrations like in Vietnam after that? Were they different from before?"

Bảo nodded.

"Yes, there were more beggars than before. As for people visiting the temples, there were only a handful because survival was so hard. Many temples became desolate because the monks and nuns had to work to survive. Marx called religion "the heart of a heartless world," and that's exactly how they treated it. Vân and I saw it before we escaped. Now, I only know what I read in the news."

"The way they treated people's religious beliefs was terrible. They only believe in materialism. After Lenin's revolution in 1917, communists around the world followed his example. They believed they were liberating oppressed nations. But if Marx and Lenin's theory were correct, then after more than seventy years, wouldn't those countries have flourished by now? Instead, the Soviet Union—the birthplace of communism—was one of the first to collapse. What does that tell you?

After Sơn finished speaking, Bảo glanced at his son, who stood nearby, scratching his head in confusion. Not understanding the conversation, the boy looked bored, so Bảo told him:

"Lam Sơn, why don't you go check on your mom and aunt to see if they're done? We need to go out for breakfast afterward."

Relieved to escape, the boy eagerly ran toward the temple. Bảo pointed to a nearby stone bench.

"Let's sit down and wait. No need to stand around."

He handed a cigarette to Sơn and lit one for himself.

"Your point about communism was spot on. It's been over seventy years, and if they were right, how could the Soviet Union have failed? Westerners are pragmatic. They evaluate ideas based on results, unlike some Eastern countries that try to cover up their mistakes."

"Exactly. Look at China under Deng Xiaoping. They slowly shifted from a state-controlled economy to a market economy, and now Jiang Zemin is continuing that approach, attracting foreign investment and trying to join the global market. After the fall of the Berlin Wall, Russia and its allies had to reevaluate their political systems, and that made the Vietnamese communists nervous. They saw the failures of a centrally controlled economy, where corruption and incompetence were rampant. With the pressure from both inside and outside the country, they had no choice but to loosen their grip and allow a more market-oriented economy, which they now call a "socialist-oriented market economy."

Sighing, Sơn shook his head.

"We're lucky that the Soviet Union and the Eastern Bloc collapsed like dominoes. If not, the communists in Vietnam might never have loosened their control."

Bảo laughed and patted Sơn on the shoulder.

"Don't think that today's communist Party members are paragons of virtue. They still preach against capitalist exploitation, but their pockets are full of dollars. Their children study abroad in the U.S. and Europe. People now call them the "red capitalists." Meanwhile, the rank-and-file Party members are genuinely poor because they lack the connections to get rich. But as in any society, there are good and bad people. We can't ignore those who truly care about the country, but sadly, they are few and far between."

Turning to a more practical matter, Sơn asked:

"What's your take on capitalism here in the U.S.? Do they exploit workers the way communists claim?"

Bảo didn't answer right away. Glancing at his watch, he said:

"I'm sure the women are still inside, so we have time for a bit more chatting. To answer your question, I have two conflicting answers—one from the communists and one from myself. According to Marx and Engels, the answer is yes. They argued that the means of production—machines, equipment, and tools—are controlled by the capitalists, who exploit workers by extracting surplus value from their labor. They believed that since workers create the value, they should be the true owners, while capitalists merely steal their labor."

"But from my personal perspective, the answer is no. As a worker myself, I view fair competition in a free market economy as the key to progress. I don't know how workers were treated in the past under capitalism, but..."

"I don't know how capitalist societies treated laborers in the past, but at least from what I have witnessed since moving to America, the workers here are fully protected. They have unions and legal representation to defend their rights. Laborers are able to collectively bargain for wages, working hours, and benefits. Furthermore, the government has laws that regulate workplace safety and ensure fair treatment. I feel that I am compensated fairly for my work, and I have the freedom to seek better employment elsewhere if I choose to."

"In contrast, back in Vietnam under the communist system, there was no real protection for the common workers. Government-owned companies or state-operated collectives employed people, but they did not offer a fair or competitive environment. Instead, everything was controlled by a handful of officials who used their power to exploit the system. The people were told that they were all equal, but in reality, it was the party officials who lived comfortably while the rest of us struggled to survive on the meager salaries they provided."

Sơn sighed deeply and nodded in agreement.

"You're right. Even though communism claims to be for the people, it's often the people who suffer the most under that system. I remember the early days after the war when everything was rationed. We had to queue for hours just to buy basic necessities, and most of the time, they weren't even available. Meanwhile, the high-ranking officials never had to worry about food or supplies; they had everything they needed. It was a complete farce."

Bảo smiled faintly and placed his hand on Sơn's shoulder.

"Well, at least we are fortunate to be here now, where we have the freedom to live and work as we choose. It's not perfect, but it's far better than the life we left behind."

At that moment, Lam Sơn came running back from the temple.

"Mom and Aunt Hằng are done now! They're coming out!"

Bảo and Sơn both stood up, smiling at the boy's enthusiasm.

"Alright, let's go," Bảo said, adjusting his jacket. "I'm starting to get hungry anyway."

As they walked toward the entrance of the temple, the sun began to set, casting a warm glow over the quiet courtyard. Despite everything they had been through, Bảo felt a sense of peace wash over him. It wasn't an easy road that had brought them to this point, but for the first time in a long while, he felt hopeful about the future.

Now all the big companies have unions to protect the legitimate rights of workers. Moreover, workers can buy shares in the very company they work for and share in its profits. And if an employee is capable and seizes the opportunity, one day, they might even rise to the highest position to manage the survival of the company. In other words, in daily life, if for some reason one doesn't like their current job, they can look for a better one. If someone falls ill and is unable to work, the social welfare department will provide assistance for a time, such as food stamps, medical vouchers, etc. Think about it: living in a society that respects human dignity, where people enjoy personal freedoms to pursue happiness and manage their own lives without fearing government oppression. What more could ordinary people like us want!?

"Bảo's assessment is very accurate. The communists developed a strategic, scientific philosophy to underpin their arguments. First, they pointed out contradictions in capitalist society, accusing it of exploiting the working class, and then they enticed people with the idea of a utopian paradise under communism. So, how could poor, destitute people not be tempted? But if we look closely and think carefully, we see that things aren't exactly as they claim."

"Contradictions in human nature and society always exist; when one is resolved, another arises immediately afterward. Only in death can contradictions end. As for the communist paradise, it's a fantasy, like an exquisite cake that you can admire but never actually eat. That's why the Soviet Union and Eastern European countries had to abandon it after waiting more than seventy years for this fantasy to become reality."

Sơn had just finished speaking when Lam Sơn came running out of the main hall, calling out loudly:

"Mom and Auntie are out, Dad! Let's go eat pho, okay?"

Sơn ruffled his nephew's hair:

"Today, Uncle's treating you to whatever you want!"

"I want to eat pho!"

"Alright, let's go have pho."

Just then, Hằng and Vân, wearing traditional Vietnamese áo dài in a lovely sky-blue color, approached. Hằng asked her husband:

"Have you two been waiting long?"

Sơn squinted playfully at his wife:

"Of course, we've been waiting forever. The boy here is starving!"

Vân laughed and teased:

"Don't believe him, sis. He's the one who's hungry and blaming it on the boy."

Sơn gave his sister a mock scolding look:

"With you around, I can't keep anything secret!"

Hằng playfully scolded her husband:

"Who told you to be hungry and then blame the child?"

"I can't very well admit that I'm hungry, can I?"

Hằng mimicked a northern accent to tease:

"Oh, so that's how it is, sir!"

Everyone burst into laughter at her playful teasing.

Today marks the beginning of the 21st century (Y2K). People are worried about possible disruptions in computer systems. Young workers are anxious about the U.S. economy slowing down, which could affect their jobs. Retirees are concerned about banks and investment firms potentially messing with their savings accounts, etc.

Life in a capitalist country is different; people are always busy and competing daily. Perhaps that's why it seems like time passes faster here. Suddenly, Bảo found himself approaching 60 years old — the age of retirement. People often reserve this time for walking, which is the final stretch of life before returning to dust.

The situation in Vietnam has eased somewhat, so Bảo and Vân decided to return for a visit after nearly 20 years away.

Saigon before him now was quite different from the years immediately after the communists took over. He still remembered how, back then, Saigon was so sad and desolate. The streets were empty, with shuttered shops lining both sides. At night, the faint yellow streetlights cast eerie shadows through the trembling tamarind trees, like ghosts leading each other into hell.

Today, Saigon has been renamed, but nobody bothers calling it by its new name. Nearly 25 years have passed since the communists took control. After World War II, Japan was almost bankrupt, yet within 20 years, it had rebuilt itself into a global economic power. But Saigon's people remain poor, far from the promises Hồ Chí Minh made decades ago. Meanwhile, the Communist Party is stuck in a catch-22, unable to swallow or spit out the "bitter pill." They blame one another, accusing each other of revisionism, sabotage of the proletarian revolution, and remnants of capitalist American puppets, etc.

They hold party meetings to discuss reforms. If they don't get it right this time, they'll meet again to discuss more. The country's economy is like a patient suffering from a chronic illness — if it doesn't go lame, it will end up blind or maimed. This shows that after years of consecutive failures, the Vietnamese communists had to open up the economy and allow some private enterprise while maintaining control over the most profitable sectors through state-run enterprises. These are the economic engines used to support and protect the Party.

Today, thanks to this limited market economy, Saigon and Chợ Lớn have regained some of their former hustle and bustle, looking a bit brighter than in previous years.

The people of Saigon now have enough to eat and wear, returning to a state similar to 1975. As they made their way back to the old neighborhood, Bảo and Vân stopped by the home of Uncle Năm, where Bảo had once lived while working as a teacher. The neighbors told him that Uncle Năm had moved to a new economic zone over a decade ago, and no one had seen him since. Bảo gazed wistfully at the old attic, now occupied by someone else, then gently pulled Vân toward the street to catch a taxi.

"It's still early. How about we visit Tâm and Loan at the apartment?"

"Sure! I bet they'll be thrilled. I just hope they haven't moved!"

Just then, a taxi passed by slowly as if looking for customers. Bảo hurried to the curb and waved it down. The driver pulled over and got out to open the door for them. Suddenly, Bảo recognized the driver's face. He racked his brain for a moment before it hit him: "Ah, isn't this Tư Thẹo, the former ward chief during the 'liberation'?"

He spoke up:

"You look familiar. Weren't you the ward chief here back in the day?"

Tư Thẹo hadn't paid much attention to his passengers until now. But hearing Bảo's question, he turned and recognized him.

"Yes, back when we first liberated this area, I was the ward chief. Are you visiting from abroad?"

"I'm Bảo, the teacher. I used to live with Uncle Năm in this neighborhood."

"Ah, yes, I remember you now. It's been a long time, so I didn't recognize you at first."

Bảo quickly changed the subject.

"How's business? Is driving the taxi enough to make a living?"

"Ever since the government allowed private businesses, driving the taxi has been enough to get by. And you? Where are you and your wife visiting from?"

"We're visiting from the U.S. for a few weeks to see family and friends."

Tư Thẹo asked no more questions and focused on driving. They turned a corner and headed toward the apartment building where Tâm and Loan lived.

After almost 25 years, the area hadn't changed much, though it looked older due to the lack of fresh paint. The once-white walls had turned a yellowish-green from the rain.

The taxi stopped in front of a house with an open front window. Knowing it was Tâm's house, Bảo asked the driver to stop. After paying, Bảo handed Tư Thẹo an extra hundred thousand đồng, saying it was for his kids. Tư Thẹo accepted the money with both hands, thanking him profusely.

Loan, who was sewing inside, noticed the unfamiliar visitors stepping out of the taxi. She moved closer to the window for a better look. A moment later, she recognized Bảo and Vân and exclaimed loudly:

"Oh my! Is that really you, Bảo and Vân?"

Vân glanced quickly through the window and shouted back:

"Yes, it's us! Is that you, Loan?"

The front door swung open as Loan rushed out to greet them, grabbing Vân's hands.

"Where have you two been? It's been almost 20 years since we last saw each other!"

Before Vân could answer, Loan called out to her husband:

"Tâm! Come quick! Bảo and Vân are here!"

Tâm, who had been mixing cement to repair the kitchen, quickly washed his hands and came upstairs.

"Did you say Bảo and Vân...?"

Before he could finish his sentence, he froze in surprise, staring at Bảo.

"Wow, it really is Bảo!"

Bảo smiled and stepped forward to shake Tâm's hand.

"Yes, it's me, Bảo, the literature teacher. Who else could it be?"

After a moment of stunned silence, Tâm regained his composure and shook Bảo's hand firmly.

"My goodness, it's been so long since we last saw you! After your wedding, we figured you'd left to find better opportunities because you stopped coming around. Where have you been all these years? We never got any letters from you!"

Bảo sat down slowly and offered Tâm a cigarette.

"It's a long story. Let's smoke and talk."

Loan, who had been chatting with Vân, chimed in:

"Alright, you two stay here and catch up. Vân and I will go to the market to get something for dinner."

"Sure, you two go to the market. Bảo and I will head to Aunt Tư's cafe for a bit, then come back."

Once the women left, Bảo asked Tâm:

"So, Loan doesn't sell coffee anymore?"

Tâm laughed.

"You've got a good memory! My wife quit selling coffee over ten years ago. When people were starving, the government finally allowed a bit of private trade so folks could survive. But since we didn't have a proper shop for people to sit and drink, she stopped selling. Let's go to Aunt Tư's cafe down the alley for a drink."

Inside the now-quiet cafe, Tâm chose a table by the window. Bảo lit a cigarette and pushed the pack toward Tâm.

"Earlier, you asked where we've been. Well, after the wedding, Vân and I couldn't survive here, so we decided to flee the country. We eventually made it to the U.S., and that's why it's taken us so long to see you again."

Tâm lit his cigarette and exhaled slowly, his mind seemingly lost in distant memories. After a moment, he spoke:

"I had a feeling you had fled. My wife and I thought about it, too, but we didn't have the means to go, so we stayed. Over the years, it felt like being in prison — not enough to eat, not enough to wear, and always being summoned by the police for questioning."

At this point, Tâm raised his voice slightly in frustration. Bảo gently tapped his arm.

"Keep your voice down. It's better to stay out of trouble."

Tâm lowered his voice but continued:

"I'm just speaking the truth! I'm not making anything up. Remember how they used to vilify the people who fled by boat, calling them traitors who ran after American imperialism for a piece of steak? But now, those same boat people have returned, wealthy from their hard work abroad, bringing in dollars to help their families. Suddenly, the government is singing a different tune, calling them "patriotic overseas Vietnamese" and "an inseparable part of the nation." It's all about money. Without the remittances from overseas Vietnamese, how would the government get the hundreds of millions in foreign currency they need every year?"

"How are things now? Are people living better?"

"Yes, things are much better now. People and even the officials have started to wake up to the lies they were fed. Back when they were sending troops to fight in the South, they blinded and deafened the people. They said they were fighting American imperialism to save the country and that afterward, the nation would be prosperous, and everyone would have enough to eat and wear. But years after the Americans left, people were still dirt poor, even worse off than before. So, they started making excuses — blaming foreign capitalist sabotage and the lingering effects of the old regime, etc. They tried to cover up the failure of their centrally planned economy and corrupt state-owned enterprises. But now, things are much better, thanks to the collapse of the Soviet Union and the Eastern European communist regimes. That's when they finally opened their eyes, loosened the market, and allowed private businesses to attract foreign investment and create jobs."

Tâm paused to take a sip of coffee, glancing at Bảo, who was quietly listening while smoking.

"I've heard that it's easy to make a living in America. Is that true?"

Bảo stubbed out his cigarette and replied calmly:

"Compared to Vietnam, it's easier. But if you want a comfortable life with all the modern conveniences, it can be quite difficult, not as easy as people think. You need to study and have specialized skills. Otherwise, you'll just earn enough to get by, but not enough to save."

"I've also heard that there are a lot of Vietnamese festivals and gatherings over there. Is that true?"

Hearing this, Bảo chuckled and shook his head.

"America is a free country. Anyone can form an organization or a political party as long as it's legal. That's why there are so many associations and groups that it's become "inflationary." They compete for influence, hurl accusations at each other, and sling mud. In the end, you don't know who to believe anymore."

Tâm seemed puzzled by this.

"I thought in a free country like that, there would be strong political strategies and theories to guide the people in the right direction."

Bảo scratched his head and looked around the cafe to see if anyone was listening to their conversation.

"From what I've seen, the people who set up these organizations have two essential tools: an "Anti-Communist" label and a few "Communist hats.""

Tâm looked at Bảo in confusion.

"Huh? What do you mean?"

"Well, if an organization or individual opposes communism differently from the mainstream, they get labeled as "pro-communist" and a few "Communist hats" get thrown their way. You get the picture."

Tâm nodded in understanding, shaking his head in disbelief.

"Fighting like that won't defeat the communists anytime soon."

"In capitalist societies, things are different. There's no need for the kinds of convoluted tactics the communists use. You just slap a label on someone and deal with it later."

With that, Bảo decided to end the conversation. He checked his watch, then stood up to pay for the coffee.

"Let's head back. I'm sure the women have returned by now."

After more than 20 years apart, Bảo and Tâm finally had the chance to reconnect in their homeland — a place once so despised for the oppressive grip of the communist regime. Back then, people used to joke that even the street lights in Saigon would flee if they had legs.

Later that evening, after a delightful dinner filled with lively conversation about current events and life, Bảo and Vân said their goodbyes and returned to their hotel.

By this time, Saigon's streets were still relatively bustling, so Bảo asked the taxi driver to take them to Bạch Đằng Wharf. The park by the pier was more crowded with people out for a stroll compared to years past. Couples occupied the stone benches and lined the fence, gazing out over the river. The cool breeze from the water made Vân pull her coat tighter as she walked close beside her husband.

"It feels like Saigon is coming back to life, like it was in 1975."

"Yes, I've noticed that too."

The Saigon River, both then and now, remained unchanged. Its muddy waters rose and fell with the tide, like the love of youth.

As the couple walked past a stone bench by the river, Vân grabbed Bảo's hand, pulling him to a stop.

"Do you remember this bench from long ago?"

Bảo glanced at a young couple, perhaps in their twenties, sitting quietly together, their heads resting against each other, gazing out over the river. He smiled and brushed a lock of hair from Vân's face.

"They remind me of us, more than 20 years ago."

The lights of the Saigon pier shimmered on the water. Bảo and Vân had rediscovered their happiness — a simple, heartfelt happiness that they had lost long ago in their war-torn, poverty-stricken homeland, a land burdened by so much hatred.

Section 3: Reunion and Reflection

Memory, Legacy, and Healing - This final section brings the narrative into a more reflective tone, focusing on the long-term effects of war, the process of aging, and the importance of memory and reunion. It explores how the characters reconcile their pasts with their present and the enduring impact of their experiences.

Overview: Ch. 8 - "Early Season Rain"

"Early Season Rain" (Mưa Đầu Mùa) is a reflective narrative that captures the essence of life in the South Vietnamese countryside during the early monsoon season. The story unfolds through the eyes of Nam, a man who returns to his rural roots after nearly fifty years, evoking a deep sense of nostalgia and connection to his past.

The Peaceful Rural Setting: The story begins with a vivid description of the South Vietnamese countryside at the onset of the rainy season. Nam, the protagonist, sits quietly on a stool under the eaves, lost in thought as he observes the approaching storm. The heavy clouds, gentle breeze, and the first drops of rain transport him back to his youth, when life was simpler and nature played a more central role in daily life.

Memory and Change: As the rain begins to fall, Nam reflects on the changes that have taken place over the decades. He remembers the vibrant garden of fruit trees where he grew up, the large ancestral home with its red-tiled roof, and the simple joys of childhood, such as collecting mushrooms and catching fish in the rain. However, time and war have altered the landscape and the lives of those who remained. The once-grand house has been reduced to a smaller, more modest dwelling, and the bustling garden is now a shadow of its former self.

The Impact of War: The narrative also touches on the impact of the Vietnam War on Nam's family and community. His cousin Hiền, a former

South Vietnamese Army officer, recounts his experiences during and after the war, including the injuries he sustained and the hardships he faced upon returning to civilian life. The war left deep scars, not only on the land but also on the people, as Hiền's life was irrevocably changed, and the community was forced to adapt to new realities under the communist regime.

The Passage of Time and Nostalgia: Throughout the story, Nam grapples with the passage of time. He is acutely aware of how much has changed, yet finds comfort in the familiar sights, sounds, and smells of his childhood. The narrative emphasizes the transient nature of life, with the early rain serving as a metaphor for the fleeting beauty of youth and the inevitability of change. Despite the physical and emotional distance created by time, Nam's return to his roots allows him to reconnect with his past and find solace in the memories that shaped him.

"Early Season Rain" is a poignant exploration of memory, change, and the enduring bond between a person and their homeland. It beautifully captures the delicate interplay between past and present, offering a reflective and nostalgic view of a world that has evolved but still holds the echoes of a simpler, more peaceful time.

Overview: Ch. 9 - "Rain Bubbles"

"Rain Bubbles" (Bong Bóng Nước Mưa) is a poignant and real love story set during the Vietnam War, exploring themes of love, loss, and the harsh realities of a war-torn country. The narrative revolves around Tâm, a former South Vietnamese Air Force officer, whose life is marked by the struggles and heartbreaks of the war.

The Tragedy of War: The story opens with a reflection on the Vietnam War, portraying it as a destructive conflict that tore families and communities apart. The war is described as a conflict driven by foreign ideologies and political greed, leading to widespread suffering and the breakdown of normal life in South Vietnam after April 30, 1975.

A Childhood Marked by Hardship: Tâm's childhood is depicted as one of great hardship, growing up as an orphan in a challenging environment. He lived with his uncle in a poor neighborhood, where he had to work hard from a young age to survive. Tâm's early years were filled with struggles, but he showed resilience and determination, working various jobs, including delivering newspapers and selling goods, to support himself while pursuing his education.

Love and Romantic Aspirations: As Tâm grows older, his life takes a significant turn when he joins the South Vietnamese Air Force. It is during this time that he meets Loan, a beautiful and intelligent young woman who captures his heart. Their love story is filled with innocence and hope, but it is also marked by the uncertainties and dangers of the ongoing war.

The narrative captures the delicate and blossoming romance between Tâm and Loan despite the looming threat of separation due to Tâm's military duties.

The Cruelty of War and Its Impact on Love: Tâm's life is further complicated by the war, which eventually leads to his separation from Loan. After suffering an injury during a military operation, Tâm loses contact with Loan. The story poignantly describes how their love is tested by the distance and the harsh realities of war. Loan, left in despair, struggles with the uncertainty of Tâm's fate, while Tâm, isolated and injured, longs to reunite with her.

The Pain of Separation and Loss: The story reaches its emotional peak when Tâm learns that Loan has moved away, and all efforts to reconnect with her fail. The tragic separation and the unfulfilled love between Tâm and Loan highlight the deep personal losses that the war inflicted on many individuals. Tâm's sorrow and the eventual death of his wife, whom he marries later, further emphasize the theme of loss.

"Rain Bubbles" is a powerful narrative that intertwines personal love with the broader historical context of the Vietnam War, offering a reflection on the enduring impact of war on individual lives and relationships. The story is a poignant reminder of the emotional toll that war takes on those who live through it, leaving behind memories of love, pain, and lost opportunities.

Overview: Ch. 10 - "Wings of Vietnam"

"Wings of Vietnam" (Những Cánh Chim Việt) tells the story of Tuấn, a former South Vietnamese Air Force Captain, who reflects on his past, particularly his relationships with old friends like Tâm and Trung. The narrative unfolds as Tuấn unexpectedly reunites with these friends after many years, recalling their shared experiences during the Vietnam War and its aftermath.

Friendship and Reunion: The story begins with Tuấn's surprise reunion with Tâm, whom he hasn't seen in over twenty years. This meeting stirs deep emotions, bringing back memories of their time together in Pleiku, where both served as Air Force officers. Their bond was forged during the war, with Tâm sharing stories of his tragic love life and family struggles with Tuấn.

War and Separation: The narrative recounts how the Vietnam War tore apart these friendships. After the fall of Saigon, Tâm was captured and imprisoned by the communist forces, while Tuấn managed to escape to the United States with his family. Over the years, Tuấn lost touch with Tâm and feared the worst for his friend, who seemingly disappeared after the war.

Love and Loss: Tuấn reminisces about Tâm's past, particularly his ill-fated relationship with Lệ, a woman he loved deeply. Lệ tragically died in a bomb explosion, a loss that deeply scarred Tâm. This event is one of the many tragedies that haunt the characters, illustrating the profound

347

impact of war on personal lives.

Survival and Resilience: The story also highlights Trung's harrowing escape from a communist reeducation camp and his subsequent journey to the United States. Trung, a former pilot, endured significant hardship during his escape, including organizing a perilous sea journey to flee Vietnam. His reunion with Tuấn and Tâm symbolizes the resilience and survival of those who fled the post-war communist regime.

The Legacy of War: As the characters reminisce, the story delves into the lasting legacy of the Vietnam War on their lives. The memories of loss, love, and survival continue to shape their identities and relationships. Despite the passage of time, the war's impact remains deeply embedded in their lives.

"Wings of Vietnam" is a poignant exploration of friendship, love, loss, and survival against the backdrop of the Vietnam War. It captures the emotional and psychological toll of war on those who lived through it and their attempts to rebuild their lives in a new country.

Overview: Ch. 11 - "Last Leaves of the Season"

"Last Leaves of the Season" (Những Chiếc Lá Cuối Mùa) is a reflective and nostalgic narrative by Nam, capturing the experiences and emotions of Vietnamese veterans and immigrants as they navigate life in the United States, decades after the end of the Vietnam War. The story revolves around the theme of aging, the passage of time, and the bittersweet memories of a bygone era.

The Passage of Time and Nostalgia: The author opens with a philosophical reflection on the inevitability of change and the natural cycle of life, likening it to the falling of leaves at the end of the season. He muses on the futility of resisting these changes and the acceptance that comes with age. This sets the tone for the narrative, which is deeply embedded in memories and the passing of years.

Reunions and the Bonds of Brotherhood: Nam recounts his experiences attending reunions with fellow members of his military class, 63A, and other veterans from his former Air Force squadron. These gatherings are filled with camaraderie as old friends reconnect, share stories, and reminisce about their shared past. The reunions are not just social events but also a way to maintain the bonds forged during the war and in the years following their resettlement in the United States.

The Immigrant Experience: The narrative also delves into the immigrant experience, particularly the challenges and triumphs of starting over in a new country. Nam reflects on his journey to the U.S., his efforts to build a new life, and his sense of pride in having raised a family and contributed to his community. He emphasizes the importance of resilience and adaptability, qualities that have helped him and his peers thrive in their adopted homeland.

Aging and Reflection: As the story progresses, the author reflects more deeply on aging. He describes the physical and emotional changes that come with it, such as the slower pace of life, the increased tendency to reminisce, and the sense of loss as friends and comrades pass away. The metaphor of "the last leaves of the season" poignantly captures the sense of an era coming to an end, as the veterans confront the reality of their mortality.

Legacy and Continuity: Despite the melancholy that pervades the narrative, there is also a sense of fulfillment and peace. Nam expresses satisfaction with the life he has lived and the legacy he leaves behind. He urges his peers to continue cherishing their shared history and to support one another as they navigate the final stages of their lives.

"Last Leaves of the Season" is a poignant exploration of memory, aging, and the enduring bonds of friendship among those who have shared profound experiences. It captures the complex emotions of nostalgia, loss, and acceptance, offering a deeply human perspective on the passage of time.

CHAPTER 8

Early Season Rain

Original Title: "Mưa Đầu Mùa"

In the southern countryside, whenever the first rains of the season come, it seems as though everything around begins to change. The heavy black clouds on the horizon, beyond the rows of tram bau trees that grow along the village pond, stand between vast, empty rice fields where dry rice stubbles rise in ghostly white rows stretching all the way to the horizon. The rice paddies are divided into grids, creating lines like a student's notebook. The sun, which had been shining brightly, has hidden away, taking with it the scorching heat of the summer days. Gentle, cool breezes from the South are enough to make the bamboo leaves in front of the house sway.

Nam quietly sat on a small stool in front of the porch, gazing out toward the end of the alley. He leaned back against the window frame as if immersing himself in some distant void. Nearly fifty years had passed, and today, Nam was experiencing this brief, comfortable feeling once again. People change, and all things that exist change; even human hearts transform with time. There seems to be only one thing that does not alter—the mystery of creation.

The first rain of the season today was exactly like the first rains fifty years ago—dark clouds gathered thickly, thunder and lightning followed with deafening booms as if the heavens were in a rage. Then, the cool breezes gently gathered into strong storms, bending the bamboo stalks, which bowed and rose like waves on the sea. The "kut-kit" sound of bamboo rubbing together, the "raoo-raoo" of the bamboo leaves hissing in rhythm, and then the heavy raindrops began to fall.

The dark square of land in front of the house used for drying rice during the harvest season had cracked under the summer sun for months, creating a fine layer of dust on the surface. It hurriedly absorbed the first raindrops that rolled down the ground like millipedes.

The wind blew strongly into the porch, and Nam stood up, carrying the stool inside. He reached out to close the front double doors. The wind and rainwater had seeped through the cracks, and Nam felt cold on his feet. He quickly walked to the back and sat down in a hammock made of burlap tied to two poles near the porch window. The rain had grown heavier and more dense, and Nam could no longer hear the "lop-bop" sound on the tile roof as he had a few minutes before. Instead, there was a continuous "ao-ao," like the sound of a waterfall cascading from above.

Nam curled up in the warm, thick burlap hammock, occasionally feeling the chilly mist of rain blowing through the window slats. These rare sensations took him back to memories from nearly half a century ago.

Nam was born and grew up in a garden full of various fruit trees that bore fruit in nearly all seasons, such as oranges, tangerines, mangoes, and star apples, among others.

The old front house, with its red tile roof, had three large compartments with a front house, a back house, and a straw pile in the corner of the garden. Nam still remembers how every morning, he would take a basket to the straw pile, dig through the decayed straw to cut mushrooms, and bring them back for his mother to cook soup or fry with duck eggs. But time and war spread and the old red-tile house gradually shrank—from three compartments to just one. The sturdy wooden walls were replaced with cheap tin sheets.

During the dry season, the air in the house was hot, like sitting inside a furnace. The dense green bamboo hedge that served as a fence outside the moat surrounding the garden was ordered to be trimmed in 1951 by the French army, fearing it might provide cover for the Viet Minh guerrillas. The mango and guava trees, which provided shade during scorching summer afternoons, where Nam and his sister would pick chilies to crush with salt and eat with bitter guava and green mangoes, were also gone, disappearing with time.

The rain had stopped completely outside. The large dark clouds hanging in the sky were blown away by the wind to another place, and beams of refracted light in the rising steam created beautiful, multicolored rainbows. Just then, Nam's cousin and his wife arrived, pushing the door open and entering the house.

"Brother Nam, have you been sitting at home all day waiting for us? Were you bored?" Nam sat up in the hammock.

"I've been here all day; I didn't go anywhere to wait."

Le Van Hien, Nam's cousin, was a former officer in the Army of the Republic of Vietnam. In 1966, Hien was a platoon leader in the Infantry Regiment stationed at Bong Son, Qui Nhon province. He was ambushed by the Viet Cong while leading his platoon on a road-clearing mission to Tam Quan. A vicious communist bullet pierced through his left forehead and exited through his throat. Hien was transported back to Qui Nhon for surgery and then spent a year recovering at the Republic Hospital in Saigon. He narrowly escaped death and was discharged with a blind left eye.

After returning to civilian life, Hien was helped by the Veterans Administration to find a job with a relatively decent wage at a textile factory in the Go Vap area.

Then, on April 30, 1975, disaster struck. Hien was fired from the textile factory because he was a former officer of the Southern army and was sent to a re-education camp despite having long laid down his weapons and being disabled by the war. The communists accused him of having "blood debts with the people." Could it be that a communist guerrilla, self-proclaimed as a representative of the southern people, had once decided to end Hien's life, but the bullet, out of mercy, veered off by a hair, leaving him to live and now continue paying this "blood debt" to the people, to those like him?

After returning from brainwashing, local communist cadres forced him to go to a new economic zone, leaving behind the house he had struggled to buy a few years earlier in Cho Lon to return to the countryside and farm.

Fortunately for Hien, his paternal ancestral home in Long Hoa village, about thirty kilometers from Saigon, still had a few acres of ancestral farmland left by his grandparents for worship. This place, Hien's paternal homeland, was also Nam's maternal homeland. Nam's mother was the eldest daughter of the family, and Hien's father was the third son. In 1946, he joined the Viet Minh resistance against French colonial rule and was shot dead by the French expeditionary army while crossing a river in Long Dinh village. Nam's maternal grandparents were working at the Cho Lon Post Office at that time, so Nam's mother, the eldest daughter, had to stay behind to manage the ancestral home where Nam was born. It wasn't until 1951, when Nam was eleven years old that his grandfather retired and returned to this home, allowing Nam's family to move to Cho Lon to start a new life.

After being stripped of his household registration, Hien was sent back to the countryside to farm and eventually started a family there.

Now Nam noticed that Hien had truly changed from the appearance and demeanor of a platoon leader to a genuine farmer: black shorts, a conical hat, and a checkered scarf around his neck. Tu, Hien's wife, was the epitome of a hardworking Vietnamese woman. She wore a black bà ba outfit with the pants rolled up to her knees, took off her conical hat as she walked straight to the kitchen in the back, and hung it on a nail in the column.

Hien reached out to pour tea from a teapot encased in a dry coconut shell into a cup and handed it to Nam:

"We were stacking straw into piles in the field to use as firewood for the next season, but then the heavy rain came, so we stopped at Mr. Nam's house for shelter. We were worried you'd be waiting, so we came back right after the rain stopped."

"Well! I was sitting here watching the rain, and it reminded me of when I was young and still lived here. Whenever it rained, I would strip down and run outside to bathe or go to the ditches in front to catch fish as they swam up from the moat. Life was so easy back then; sometimes it felt like playing around still got you a meal."

"That's true, but staying here for too long can be disheartening, brother! We work hard all day, sometimes not even earning enough to eat. It's not as easy as when you were still living here. I remember back then, when we wanted to eat bánh xèo or bánh đúc, we just soaked the rice, ground the flour, and then went out to the fields in front with mosquito-net baskets to catch shrimp. Within an hour, we'd have half a basket of shrimp to feed the whole family, so there would be no need to work any harder than that. Then we'd pick some young bach tuoc leaves, perilla, basil, and mustard greens, and that was it. But now, they overuse pesticides so much that even the leeches can't survive, let alone shrimp, fish, or crabs!"

Nam nodded in agreement:

"This morning, when you both went out to the fields, I took a fishing rod and dug up some worms from under the plant pots in front of the house to use as bait. I sat by the moat for an hour but didn't catch a single perch or catfish. I remember there used to be so many perch and catfish here!"

"People here have caught them all, brother! They've caught everything—mothers, babies, all of them are eaten. We never used to eat barra fish, but now you can't even find tiny needlefish; they're almost extinct. Now, we only eat fish that others raise and sell. Moats and ponds that we use for water during the dry season are rarely drained to catch fish anymore. Oh! I got so caught up in talking that I forgot to ask, when are you heading back over there?"

"I'm staying in Saigon for about four weeks."

"Oh! Then after a few more rains, we can go frog hunting. This month, the season is changing, and the water is rising."

When Hien mentioned this, Nam's face lit up as he eagerly jumped into the conversation:

"Yes, frog hunting at night is so much fun! I remember when I was about eight or nine years old, I snuck out to go frog hunting with Mr. Bay Hoanh, Mrs. Tam's son. Bay Hoanh carried a carbide lamp ahead, lighting the way. The water in the fields was ankle-deep, and with the light, we could clearly see the perch and snakehead fish swimming below. I followed behind, in the dim, flickering light, when suddenly, I saw something black, about the size of my big toe, swimming in circles in the water. I thought it was a snakehead fish and quickly grabbed it with both hands. But when my fingers touched its body, it was soft and slimy. I pulled it up, and it turned out to be a leech. I shook my hand so hard it almost flew off my wrist and ran straight home. From then on, I quit sneaking out to frog hunt at night."

Tu, who was standing by the stove preparing dinner, overheard Nam's story about his unfortunate frog-hunting experience and couldn't help but laugh. She ran up and joined the conversation:

"Oh my! People go frog hunting, and you go leech hunting? No wonder you were scared! Brother Nam, yesterday, when we heard you were coming, we bought two yellow catfish at the market. Tonight, I'll fry them until golden, dip them in ginger fish sauce, and serve them with bitter herbs and watercress. I hope you don't find it strange."

Before Tu could finish, Nam interrupted:

"That's the best! I haven't had a meal like that in over thirty years!"

Hien looked at his wife and added:

"Nam's wife is a civil servant and was born in Nha Trang, so her cooking style might be a little different from ours here in the South. Besides, in America, they don't have bitter herbs, watercress, or live yellow catfish like here. Even if they do, they're shipped from elsewhere, frozen for a long time, so they don't taste as good."

Nam nodded in agreement.

"Hien is right! Most of the food over there is frozen for a long time, so it loses its fresh flavor."

As if suddenly remembering something enjoyable, Nam patted Hien on the shoulder:

"Hey Hien! Do you still have young tamarind leaves and sesbania flowers around here?"

Hien grinned:

"I thought you were asking for something hard to find, but that's easy! With rain like this, young tamarind leaves will definitely sprout by next week. As for the sesbania flowers, there are a few trees blooming in Mrs. Hai's garden. You can use a bamboo pole to pick them. Nobody here likes to eat that stuff anyway."

Nam still remembered how, in the early days of the flood season, the fields in front of the house seemed to change entirely—from the gray color of dry earth to the light green of newly sprouted rice plants, emerging from grains of rice left behind from the previous harvest. The rice paddies, which had been bare, were now refreshed with a lively green hue.

The cucumber rows, the gourd and luffa trellises in the back garden, all donned new attire. The scorching midday sun no longer blazed as fiercely as it had during the summer months that had just passed. Nam often invited his younger brother to go to Mr. Hai Quoc's garden to pick young tamarind leaves for their mother to make sour soup. Nam's favorite dish was sour soup made with young tamarind leaves and dried snakehead fish, the kind as large as a hand. After cooking, they would take the dried fish out and dip it in a dish of fish sauce with chili. Nam loved the fish's clear, fatty belly the most—it was so delicious.

If they didn't have dried snakehead fish, they'd use perch with roe or snakehead fish with its firm white meat, dipped in fish sauce with chili, and served with rice cooked over a straw fire—nothing could beat that!

Or occasionally, the two brothers would take a bamboo pole to Mrs. Hai's garden to pick sesbania flowers to either boil or make sour soup with prawn or tiger shrimp. Sesbania flowers are ivory-colored, slightly yellowish, and you only need to remove the long, slightly bitter stamen inside before rinsing and adding them to the pot for boiling or making sour soup. Dinner was served on a mat spread out in front of the yard, with the whole family gathered around, enjoying boiled sesbania flowers dipped in salted fish sauce with mắm nêm, forgetting all about being full.

Nam had spent his childhood in the countryside, so he never forgot the simple meals his mother used to cook for the family daily.

With the clear distinction between the rainy and dry seasons, the dishes also varied slightly. The rainy season brought floods to the fields, turning them green with endless stretches of rice plants. It was the season for catching shrimp, crabs, and snails, the delight of children in the village. The rice-field crabs were as large as a big toe, and kids would catch them for fun or roast them over a fire. But adults in the South never used crabs to cook for their families.

Nam still remembers a woman in the village, Mrs. Năm, who sold flowers at Chợ Lớn market. One day, she must have struck a good deal because, on her way back to the village in the evening, she rounded up some kids to catch crabs to sell to her, buying them for five cents per dozen. Back then, money was still worth something—five cents could buy five popsicles or a few bánh cam or bánh còng.

All the kids loved catching crabs to earn money for sweets. Crabs would make their burrows along the edges of rice paddies, often damaging the young rice plants, so kids could freely wade along the paddy banks to catch them without worrying about being scolded by the field owners. Later, Nam found out that Mrs. Năm took the crabs to Chợ Lớn to sell to the northern people, who would smash them up to make bún riêu (crab noodle soup).

It's worth mentioning that around 1950, not many people from the North and Central regions had migrated to the South, so there was a significant difference in customs and accents among the three regions. This was also part of the French colonialists' strategy to divide Vietnam into three regions: Bắc Kỳ (Tonkin), Trung Kỳ (Annam), and Nam Kỳ (Cochinchina), almost like three small separate countries. The French deliberately created divisions among the people to make it easier to control them.

In the village pond neighborhood down below, Mr. Hai had a daughter who also sold flowers at Chợ Lớn like Mrs. Năm Gấm. She married a man from the North, and the two moved back to her parent's house and built a home next to Mr. Hai Nhiều. The whole village gossiped about "Hai's daughter married a Bắc Kỳ man." Curious folks would pretend to walk by to see what "Bắc Kỳ people" looked like.

Then they would gossip in amazement, saying things like the accent was hard to understand, the house's thatched roof was too low, not like the ones here, and so on. Most people in the village viewed him as a foreigner rather than a fellow Vietnamese with yellow skin, sharing the same ancestors and history.

After the planting season, when the rice plants began to develop sheaths and were about to become mature rice plants, children would use small mosquito net baskets to trap shrimp in the spaces between the rice plants. They would toss a ball of mashed bran the size of a big toe into the trap. After about ten minutes, they would use a hooked stick to quickly lift the basket so the shrimp and prawn underneath wouldn't escape.

Each time, the basket would yield more than ten shrimp with roe as large as half a pinky finger and sometimes even tiger prawns as big as a thumb. The most exciting part was when they lifted the basket just above the water; the shrimp would jump wildly, causing great anticipation. In bountiful years, people would dry the surplus shrimp or make fermented shrimp paste or pickled shrimp with shredded papaya to eat during the dry season or to give as gifts to relatives in the city.

The farmers here only cultivated one rice crop per year. After the final harvest, which coincided with the approach of Tết (Lunar New Year), the fields would begin to dry up, and the fish in the rice paddies would follow the receding water to the ponds, lakes, and moats surrounding them. That was the season for draining the ponds, moats, and lakes to catch fish for Tết. This task required strong, healthy young men because, at that time, there were no water pumps like there are now.

Early in the morning, around four or five o'clock, as the rooster crowed at dawn, these men would carry wooden buckets to the fields to drain the ponds. Some large ponds required two or three buckets to be drained by noon. When the water was reduced to a trickle over the surface, fish of all kinds would start squirming and swimming back and forth, making it hard to resist. The transparent white flesh of the featherback fish, flat as a hand, swam against the current. Snakehead fish, perch, and catfish would struggle up the pond's sides, and the men would collect them in metal buckets, then take them home to sort them by type. The live ones were kept in water jars to be eaten later.

The dead ones would be filleted and dried or made into fish sauce. Fish sauce made from snakehead fish and fish intestines was a prized product in the southern countryside. People rarely sold it, reserving it for gifts to important individuals, benefactors, or close friends. Sometimes, large ponds would yield hundreds of kilos of fish, more than enough for the family. The surplus would be sold at the market to buy other necessary items.

As Nam thought about this, he suddenly remembered something even more enjoyable. He turned to Hien and said:

"Hien! Next time I come back, I'll bring my wife and children to stay here for a few days so they can experience the countryside and enjoy the local delicacies. I've told my wife about sesbania flowers and young tamarind leaves cooked in sour soup with dried snakehead fish, but she just stared blankly because it seems Nha Trang doesn't have that stuff. They mostly eat fresh sea fish rather than freshwater fish like us."

Tu agreed enthusiastically with Nam's suggestion:

"That's right, brother Nam. The last time your wife visited, she didn't stay long, so we didn't have a chance to invite her for dinner. As for your children, they probably don't know anything about it. You should bring them here to learn about their homeland and their roots, especially the place where you were born."

Hien chimed in:

"But if you want to treat your wife and kids to those delicacies, you have to come at the right season. Otherwise, there won't be any sesbania flowers or young tamarind leaves because these plants only bloom seasonally."

Nam bowed his head, seemingly deep in thought:

"It's easier said than done; bringing the whole family back is quite challenging. When they were little, it was relatively easy to take them anywhere, but now they're all grown up, with jobs and their schedules. Planning a trip takes at least a year of preparation. Moreover, there's the issue of whether they even want to come back. Back when we were escaping from the communists, they were only two or three years old. Apart from being attached to their parents, they don't have any memories to cherish, and they have no old village roads, old schools, or childhood friends to make them want to come back like we do. Vietnam, to them, is just a legend, stories told by their parents and acquaintances. That doesn't mean we, as parents or uncles, should hastily label them as unpatriotic Vietnamese. It's a bit unfair, given that the circumstances and history of our generation have made it this way!"

Hien looked down at the ground, a hint of sadness surfacing on his face as he spoke with regret about the past:

"It's our fault, our generation's fault, not theirs. How can we blame them now? People get used to where they live. Besides, if your family had stayed, who knows what would've happened? Would your kids have been able to study, or would they have been condemned to ignorance and suffering for life because of the selfishness and discrimination of a heartless regime?"

Nam replied with a bitter laugh:

"Thanks to Uncle Hồ and the Party for coming down South; they despised and cursed us, so our families had to risk their lives escaping by boat. Now, our kids have the opportunity to advance, study broadly, and see the light of civilization and freedom in a foreign land. If Uncle Hồ and the Party had cared for us, treating the people like their children, we'd probably be stuck with official rations of moldy bo-bo and damp rice from Russia and China! Has there ever been a time in our nation's history, from the era of King Hùng to the present, when the people of Giao Chỉ have fled their country in such massive numbers?! The Vietnamese people are known for their perseverance and resilience, always enduring and accepting the harsh challenges of natural disasters, floods, and foreign oppression, yet they remain determined to cling to the land and roots of their ancestors."

Today, they're leaving their country not because they crave delicious food or fine clothes or to curry favor with American capitalists. But surely, there's one very simple and singular reason—they want to live a life that's relatively worthy of being human.

"It's better now, brother! Back then, even bringing a liter of rice to give to your parents somewhere else would get you questioned by the police as if you were smuggling contraband. Sometimes, because of food and the unreasonable and harsh laws of the Party, Vietnamese people have become mean, ashamed, lowly, and cunning. Even with chickens and ducks that we raised ourselves, we didn't dare to butcher them openly. To bring them to family in the city, we had to hide them in our clothes, our stomachs, or under our shirts, and the police at the checkpoints along the way would search and inspect even more thoroughly than the French colonialists did in the old days. Looking back on how people lived then, it's terrifyingly bleak. Everyone cursed the heavens and blamed the earth, but no one could do anything against them!"

The conversation was starting to become disheartening. Nam got up from the burlap hammock and urged Hien:

"Let's change the subject. I'm tired of hearing about them! Let's go to the moat and pick some Malabar spinach and pennywort to cook with okra for a cooling soup."

Hien stared at Nam in surprise:

"Wow, after more than forty years, you still remember that special soup?!"

"I used to eat it all the time when I was a kid. That soup, poured over rice and eaten with fried catfish dipped in ginger fish sauce, is unbeatable."

After a simple but delicious dinner, Hien was still busy helping his wife clean up the dishes and take them out to the moat for washing.

The sun was starting to hide behind the bamboo hedges at the end of the village. Nam slowly made his way out to the bamboo gate, intending to relive memories from nearly half a century ago. In his childhood days, when he was five or six, he would sit by this gate for hours, waiting for his mother to return from the market. He would help her carry the basket into the house and then enjoy bánh cam, bánh còng, or bánh neo, bánh bò, which would make him happy and content.

When he was nine or ten, Nam began to daydream, feeling a gentle excitement at the sight of the sunset in the distant horizon, where the multicolored clouds appeared like beautiful paintings against the sky. He would wish for wings to fly to those places, to see the splendid palaces and temples of Heaven because he believed that up there, there must be many beautiful fairies in magnificent attire, dancing and singing for the Jade Emperor!

A feeling of lightness and euphoria suddenly swept over him, as if time were running in reverse, bringing Nam back to the past, reliving the days when he used a sardine can with clay wheels to pull along the bamboo path leading to Mr. Ba's house, racing against the makeshift cars of Bay Hoành and Mười Xù.

The clattering sound of the toy cars on the dirt road seemed to still echo faintly. The water-apple tree growing by the moat in front of the gate, where Nam would climb to pick the large yellow fruits the size of his fist, had long since died, leaving only a blackened, decayed stump. The young girl at the end of the alley named "Quyên," who was so charming, the youngest daughter of Mr. Quản Năm, whom Nam used to tease by calling "Quyên with the dimples, looking so awkward," was also gone.

A few white egrets flapped their wings on the distant horizon, seemingly hurrying back to their nests as twilight fell. The sound of a frog grinding its teeth in the corner of the garden could be heard, signaling the arrival of the rainy season. Nam was about to turn and head back into the house when Hien approached:

"Brother Nam, do you see anything strange around here? It's so dull in the evenings, isn't it?"

"Yeah, but if you stay long enough, you'll get used to it."

"We heard from sister Nam that you're about to retire, but are you planning to stay there or come back here?"

"At first, I intended to come back because I still have my elderly mother here. But after thinking it over, it seems too difficult. Besides, my children and grandchildren are all over there—how could I come back alone? So I'm planning that if everything goes well, I'll try to come back here for a few months each year in the winter, partly to help take care of my aging mother so she'll be happy, and partly to avoid the cold during the snowy season over there."

"Your aunt is over eighty now, but she's still relatively healthy. Who knows what will happen in a few years? If you were here with her, she'd be so happy..."

Behind the kitchen, Tú had already started lighting the lamp. The flickering light of the oil lamp from its wick barely illuminated her face as she held it in her hand, walking up to the front. When she saw Hien and Nam entering from the gate, Tú asked her husband:

"Were you and brother Nam at the gate earlier?"

Hien nodded and half-jokingly replied:

"Brother Nam is planning to build a house down here for retirement!"

Tú stared at Nam with both disbelief and a special kind of delight:

"Really, brother Nam?"

"Hien is joking. I don't plan to stay here for good!"

Tú turned to Hien, giving him a sideways glance:

"This guy loves to tease too much, doesn't he? Oh! I brewed a pot of tea earlier and put it on the table. Brother, invite brother Nam to have some tea and peanut brittle while I go get a few more cups."

Outside, everything was quiet, thanks to the rain earlier that cooled the earth, prompting frogs and toads to emerge from their burrows, calling to each other incessantly, "en-ét," "quền-quệt," almost endlessly.

A few gusts of wind from the South blew through the cracks in the door, bringing a slight chill into the house. The old-fashioned pendulum clock hanging on the wooden post near the bedroom wall was slowly ticking toward nine o'clock. The outside had now become pitch black.

Nam stood up from the small stool and turned to Hien:

"Let's go to bed. Tomorrow, I have to go back to Saigon early to buy a plane ticket to visit my wife's hometown in Nha Trang for a few days. Then I'll return here to go frog hunting with you one night."

Hien nodded, standing up and reaching for the oil lamp to check the doors and windows in the house one last time before going to bed.

It seemed like the rain had started again, with soft "pitter-patter" sounds hitting the tin roof as the wind blew. The sound of insects in the garden continued to buzz without stopping. Nam pulled the woolen blanket up to his neck, curling up on the cold wooden plank, suddenly remembering that it was on this very bed that he used to sleep with his great-grandmother and sister.

His great-grandmother, who was over eighty years old at the time, would lie in the middle, with the two siblings lying on either side, fighting to hug her for warmth. Back then, Nam used to think that if he counted from one upward, there would eventually be a day when he ran out of numbers to count. So, he would challenge his sister to a counting competition to see who could count the most. Every night, after climbing onto the bed and lying next to his great-grandmother, the two siblings would count, arguing over who had counted more until their great-grandmother scolded them into silence before they could finally close their eyes and sleep.

Now, his sister was still in Vietnam, her hair turning white. Sometimes, when Nam brought up the counting game from their childhood, his sister's face would turn slightly sad, as if regretting the beautiful innocence of their youth that had passed. She would look at Nam as if seeing her little brother again at three or four years old! "Wow, you still remember? I stopped counting after grandma passed away."

Later on, Nam no longer wanted to bring up those memories with his sister.

Fifty years had passed like a dream. Today, Nam had returned to this house to count backward those old numbers that the two siblings had competed to count, to hear the crickets chirp, the frogs grind their teeth, signaling the coming rain, and to see the shimmering clouds at dusk beyond the gate.

The rain had grown heavier, and Nam drifted off to sleep in the dreams of his childhood.

Written in memory of the Autumn of 1997 in Seattle (USA), after Nguyễn Văn Ba ("Nam") came back (post-1975) to Vietnam for the first time in 1992.

CHAPTER 9

Rain Bubbles

Original Title: "Bong Bóng Nước Mưa"

Raindrop Bubbles is a true love story that took place during the Vietnam War. The reflections and grievances of the characters in the story may also be the reflections of sincerely patriotic Vietnamese people. The names of the characters in the story do not match their real names in life.

The Vietnam War was a fratricidal conflict that rekindled the internal strife reminiscent of the TRINH-NGUYEN civil war, though it differed in appearance, with contrasting political views and incompatible ideologies between communist and capitalist regimes. Ultimately, it stemmed from the greed, anger, and selfishness of a group of political leaders who were seduced and incited by foreign ideologies, driven by the power and interests of foreign powers. The end of the war on April 30, 1975, marked the beginning of a period of upheaval and destruction of the normal way of life for families in South Vietnam, with some weeping for their sons who were imprisoned and others sorrowfully parting ways to escape by boat.

These were the confessions of an H.O. friend, Lê-Văn-Tâm, a former captain and pilot in the Republic of Vietnam Air Force. After many years of being sent for brainwashing and subjected to harsh labor in communist reeducation camps, he and his second wife had been in the U.S. for nearly two years.

One afternoon, by chance, I met him in a coffee shop and listened to him recount his thoughts on his life.

Having lost his parents at a young age, Tâm grew up in the bitter circumstances of an orphaned boy in a city filled with material temptations. He was taken in by a cousin's family in an alleyway in the Bàn Cờ neighborhood. Tâm's childhood was a long chain of hardships, lacking not only material comforts but also emotional warmth. He had to fight with his young mind, heart, and small hands to find a way to make a daily living.

In the eyes of everyone, he was a poor orphan boy. Despite his still fragile, youthful, and naive reasoning, Tâm understood that in this society, who would waste their time trying to understand the thoughts of a lonely, orphaned child like him? He accepted and was ready to endure his daily life. No matter how stormy or rainy it was, he had to get up very early, ride his bike to deliver newspapers to over fifty homes in the neighborhood, and then stop by a street vendor on the way back to quickly buy a "baguette" stuff it with a handful of yellow sugar, and head off to school.

In the afternoon, after throwing his bag onto the table, he had to rush out to the ice cream shop to pick up "cà-rem" (ice cream) to sell around the neighborhood or sometimes collect lottery tickets, balloons, and cheap toys for children from poor families in the laboring neighborhoods. These were his daily jobs when he was about fourteen or fifteen years old.

The days passed by as the rainy and dry seasons alternated. His peers gradually joined the army to fulfill their military duties, while Tâm, luckier than others, was exempted from service to continue his final year of high school. By then, Tâm's work was somewhat easier. He had found a job tutoring a few children at their homes, so he no longer had to struggle as he had in the past. However, he still maintained his newspaper delivery job for the loyal customers he had served for many years.

Heaven does not disappoint those who persevere; his efforts, endurance, and determination were rewarded handsomely when, at the end of that year, he passed the high school graduation exam with honors.

The Vietnam War had started to spread, and news of intense battles was broadcast repeatedly on Saigon's radio stations, military stations, and in the newspapers daily. Tâm knew he could no longer continue his studies at the university level, so he had to decide on a path for his future.

After many sleepless nights torn between two choices—whether to find a job immediately to ease the hardships of life while waiting for the draft to Thủ Đức Reserve Officers School or to volunteer right away for the Navy, Army, or Air Force—he finally made his decision.

Not long after, on a Saturday morning, as usual, he was still riding his bike to deliver newspapers. While he was busy rolling up the newspapers to throw in front of customers' doors, he was startled to hear someone calling his name loudly:

"Hey, Tâm! Are you still delivering newspapers? You're a Bachelor now, and you haven't changed jobs yet?"

Startled, he turned around and, with a wide-eyed look, immediately recognized the young man standing in front of him, dressed in a military uniform, looking heroic—it was TRUNG, yes, Nguyễn-Văn-Trung, who had been in Class 12B two years ago. Trung lived in the upper neighborhood, in an area where Tâm often delivered newspapers, while Tâm lived in the lower neighborhood. The two only knew each other from school and hadn't had the chance to become closer.

"Oh, Trung, you've joined the military? No wonder I haven't seen you around here. You look so gallant! You must be a student officer in the Air Force, right?"

"Yes, I joined the Air Force almost five months ago. Today is my first day of leave to visit my home. Have you not found a job yet?"

"I'm still undecided, whether to find a job or join the military. I don't know which branch of the military would suit me. What field did you choose in the Air Force?"

"I chose the pilot field."

Hearing the words "pilot," Tâm suddenly remembered something in his subconscious, and his eyes seemed to light up.

"Oh, I've heard many friends talk about the book Life of a Pilot by Toàn Phong. Most young people dream of such a heroic life!"

After a brief chat about the stages of training that Trung recounted, the two school friends temporarily parted ways.

###

Two months later, Tâm was at the Nha Trang Training Center. Trung was now his senior, having just received orders from the government to go to the U.S. for flight training. After three months of basic military training, Tâm was awarded the "ALPHA" insignia and granted special leave for three days to visit his sick cousin in Saigon.

By now, Tâm was no longer the thirteen or fourteen-year-old boy delivering newspapers or selling "cà rem" around the laboring neighborhoods in the rain or enduring pitiful looks from passersby. Gone were the days when he struggled to blow up balloons that sometimes seemed bigger than him, trying to sell them to children in the neighborhood. Today, Tâm was a tall, handsome, and mature young man. His future was bright, and he was on track to become one of the officers in the Republic of Vietnam Armed Forces. Looking back on his life, it was a dark tunnel, but ahead was a radiant sun full of light and hope.

As he walked into his old neighborhood where he once lived, the children, seeing him in his military uniform, which was so different from before, ran up to him, laughing and asking questions. The neighbors, who had known him for a long time, praised him for being ambitious, and he felt very proud, being warmly and affectionately welcomed back.

"Tâm, you look so gallant!"

A sweet, gentle voice, just loud enough to hear, came from a girl standing near the front door. Still busy answering the curious children in the neighborhood, Tâm quickly turned around and was immediately met with a warm smile from a girl who shyly stood half-hidden behind the front door of a nearby house.

"Oh, Loan, how are you? Still attending school at Gia Long?"

"Yes, I'm at home today because it's the weekend. How long are you on leave, Tâm?"

"Just a few days, Loan!"

Loan was a senior student at Gia Long High School. Her family had moved to a brick house at the entrance of the alley about two years ago. It was said that her father was a building contractor and doing very well in business. In the past, Tâm often rode his bike past this house on his way to deliver newspapers or go to school, sometimes sneaking a glance through the window and catching Loan brushing her hair and getting ready for school. They would smile at each other in passing, but that was all. Because of his poor and orphaned status, Tâm didn't dare to approach Loan.

Tâm's love life was, in essence, empty. He had many dreams, like any other young man, and when he stood in front of a girl, he felt his heart beat faster. But to him, love was a luxury that only the rich could afford. Perhaps it was because of this harsh reasoning that he had dedicated all his passion and energy to climbing out of material poverty and emotional deprivation to reach the relatively stable future he now faced. Hearing the girl's voice and the children laughing in front of the house, Loan's mother stepped out to see what was happening.

"What's going on, dear?"

"Mom, this is Tâm from the inner alley. He used to deliver newspapers past our house every day. Remember? He's on leave today, so the kids are excited."

"Yes, I've heard the neighbors talk about him, but I've never met him."

Tâm politely stepped forward to greet Loan's mother.

"Hello, ma'am, I'm Tâm from the inner alley."

After a brief look of surprise, she stepped outside to greet him.

"Oh, I've heard the neighbors say you're a talented young man. It's nice to meet you today. Have you left school to join the military?"

"Yes, ma'am, I've left school and volunteered for the Air Force a few months ago."

As soon as Tâm finished speaking, Loan chimed in with a hint of admiration:

"Mom, he didn't join as a conscript; he's a pilot!"

After a few minutes of polite conversation, Tâm said goodbye to Loan and her mother to visit his cousin further inside the neighborhood.

Thanks to the money he had saved up before joining the military, Tâm didn't have to spend much on himself now, so he used that money to pay for his cousin's medical bills and doctor's fees. Two days later, his cousin's condition improved significantly, and it was time for Tâm to return to the Nha Trang training center for English language training and to prepare his security clearance for overseas deployment.

This morning, Saigon's sky was exceptionally clear and high, with a few light breezes mixed with the last remnants of early morning dew, bringing a coolness as refreshing as the untouched skin of a young girl coming of age.

In his crisp, freshly ironed yellow khaki military uniform, with shiny black shoes, kepi, and a brand-new tie, Tâm's strides were firm and confident as he walked toward the end of the alley to catch a ride to the airport. His life was like a precious gem, newly unearthed from the muddy depths of the orphaned and impoverished society of the capital known as glamorous Saigon.

As he approached the end of the alley, Tâm suddenly saw Loan stepping out of her house. Knowing for sure that he would return to his unit today, she had been waiting for this moment for several days, so she spoke first:

"Are you heading back to Nha Trang today, Tâm?"

"Hello, Loan. Yes, my leave is over, so I have to return to the unit. Take care of yourself!"

She lowered her gaze to the ground, pouting as she replied:

"You're so formal in how you address me; it scares me! Why don't you just call me Loan? Doesn't that sound better?"

Feeling a bit awkward, Tâm replied just loud enough for her to hear:

"Alright, I'll call you Loan. If you ever get a chance, why not ask your parents for permission to visit Nha Trang? The scenery there is beautiful; the sea is much cleaner and clearer than in Vũng Tàu."

Loan didn't answer immediately. Her cheeks suddenly flushed a rosy hue, making her look even more charming. Her hands were tightly clasped in front of her chest as she glanced around, checking to see if anyone was watching or listening, then quietly said just loud enough for Tâm to hear:

"Write to me, okay?"

"Yes, it's quite lonely out there since I don't know anyone. I'll write and tell you about Nha Trang, alright?"

A bridge of affection was beginning to form. A flower of love was about to bloom. Tâm felt that life was beautiful and full of happiness; he wanted to shout out loud to thank the heavens. For the first time, he had sensed and felt the sweet taste of love, though it was only in its early stages.

Back at the Nha Trang training center, he felt invigorated and couldn't believe how quickly his love life had changed; the innocent, childish dreams of the girl from the beginning of the alley had become a reality.

Tâm had noticed Loan ever since her family moved into the new house at the entrance of the alley, but he had always kept his feelings hidden, fearing that his unstable situation would prevent him from pursuing her. Meanwhile, Loan only knew Tâm through the neighborhood children and from seeing him ride his bicycle past her house on his way to school or tutoring a few students in the upper neighborhood. She admired his patience and perseverance but had only smiled politely when they crossed paths, never having the chance to get to know him better.

In the first few weeks, Tâm's letters to Loan were mostly about the beautiful natural scenery of Hòn Chồng, Tháp Bà, Xóm Bóng, etc., as he was too shy to express his true feelings to her. After several months of exchanging letters, Tâm received orders to go to the United States for flight training, so he prepared his belongings to report to the Air Force Command and complete the necessary paperwork for his departure. He didn't forget to send an urgent letter to Loan, informing her of his departure from the training center and advising her to stop sending letters to Nha Trang.

The Dakota C-47 transport plane of the Republic of Vietnam Air Force had just landed at Tân Sơn Nhất airport. As soon as it taxied to the parking area, Tâm, with his "ba lô" (backpack), was ready at the main door to step outside.

Today, the sunny Saigon weather reminded Tâm of the scorching summer days at the training center, where he had to carry a rifle, crawl on the sand, or practice marching to the correct rhythm on the training grounds at the end of the runway at Nha Trang airport during basic military training.

The Air Force pickup truck that took the passengers to the military air terminal at the Phi Long gate had just come to a stop. Tâm, somewhat absentmindedly, picked up his backpack and followed the crowd out to the main road to find transportation home when suddenly he heard a voice calling from behind:

"Tâm! Tâm!!"

Caught off guard, Tâm quickly turned around, almost in disbelief. He recognized Loan and her friend standing by the entrance of the air terminal. Loan was smiling at Tâm, her white áo dài (Vietnamese traditional dress) fluttering in the breeze, making her look as beautiful and innocent as an angel!

"Oh, Loan, when did you get here? How did you know I was here?"

"How could I not know? Didn't you write to me about it, remember?"

Turning to her friend, Loan introduced her:

"Tâm, this is Hương, my classmate."

Tâm politely nodded to Hương, then turned back to Loan and suggested:

"How about we go out for breakfast together, Loan and Hương?"

Hương looked at Loan and politely declined:

"No, Tâm, you two go ahead. I'll head home; I already had breakfast earlier this morning."

Then she winked at Loan, speaking just loud enough for her to hear:

"My mission ends here, my friend!"

Tâm pretended not to hear Hương's remark and said:

"I'm really hungry; I haven't eaten since this morning. If you don't join us, I'll be disappointed!"

Knowing she couldn't refuse, the three of them went to a nearby restaurant on the main road near Lăng Cha Cả.

Afterward, Hương took a xích lô (cyclo) home first, leaving Loan and Tâm to plan their secret meetings for the days to come.

Tâm stayed in Saigon for nearly a week, taking care of tailoring and paperwork, health checks, vaccinations, etc. Despite being busy all day, Tâm still made time for Loan. In the mornings, he would take her to school, and in the afternoons, she would find ways to sneak away from her mother to spend time with him. Their first love was gentle, sparkling like silk, and pure like the moon and stars. Though they never once explicitly declared their love, their trust and affection for each other were stronger than any flowery words.

The days in Saigon with Loan were sweet and fragrant like ripe fruit, sometimes making Tâm feel like he was living between a dream and reality. Perhaps it was nature's way of compensating for the hardships of war, bringing destruction and separation, but also nurturing beautiful love stories for those in love.

If not for this ongoing war and the general mobilization order by President Ngô-Đình-Diệm, Tâm's life might have followed a typical path, working as a public or private employee somewhere, going to work with an umbrella in hand each morning, and returning home each evening, saving up to marry, and then having children like millions of others on this earth. But because of the war and the government's mobilization order, "when the nation is in danger, every man has a duty," Tâm decided to change his life in search of a more meaningful and heroic existence.

Loan, an intelligent and gentle girl, was the most beautiful student in the neighborhood, a fresh and radiant flower, a dream for any young man. But Loan's dreams were not like those of many other girls who knew they were beautiful. Her dream was simple and ordinary: the man she loved had to be brave, heroic, and resilient in the face of adversity. Tâm came to her as a fated match.

After two years of studying abroad and graduating, Tâm was transferred to the 1st Air Division in Đà Nẵng to take on a new assignment.

Loan had graduated from high school and was now working as an accountant for a bank in Saigon. During this time, they continued to exchange letters faithfully and promised to save up for their engagement ceremony the following year.

The Vietnam War grew increasingly intense, with fierce battles in Quảng Ngãi, Tam Kỳ, and the border regions of Lower Laos.

Second Lieutenant Tâm, a newly commissioned officer, was enthusiastic about every mission in the I Corps Tactical Zone. As a helicopter pilot, he witnessed and felt the extreme suffering and misery of the Vietnam War during medevac and civilian evacuation missions from contested areas. In his childhood, Tâm had thought he was the most unfortunate orphan in Vietnamese society, but now he realized there were hundreds of thousands of children living even more wretched lives in the impoverished mountain villages, where food was scarce, clothes were insufficient, and homes were nothing more than simple thatched roofs without walls. These children lived in constant fear, caught between the warring factions of the Nationalists and the Communists. Tâm had often choked back tears at the controls of his helicopter when he saw children clutching milk bottles, sitting beside the bodies of their parents, who had been killed by artillery shelling that very morning. Their innocent, naive eyes couldn't comprehend that their parents were gone forever, still thinking that their loved ones were merely lying there.

War had turned people into cruel and selfish beings. Who would take the time to explain to the child that their parents were dead? The bodies would be buried in a shell crater, and the child would never see them again. In the training center in Nha Trang, the psychological warfare officers occasionally lectured about the communists, and Tâm had imagined them as hardened, battle-scarred, and sinister, armed with lethal weapons from the Soviet Union and China.

But on the battlefield today, he saw only children, around fifteen or sixteen years old, with fresh, innocent faces, or thin, frail people wearing black shorts, bare-chested, with bony chests, unkempt hair, and dirty like beggars in Saigon. People told him they were communists, but to Tâm, they were ignorant farmers, poor rural children living in contested areas without the protection of the Republic of Vietnam government, easily swayed, coerced, or threatened by communist political officers into joining the local militia. They were victims of both the Nationalists and the Communists. Tâm felt more pity than hatred for them.

About two years after being assigned to Đà Nẵng, an unexpected accident occurred. During a mission to deploy troops to clear an enemy base in the mountains of Quảng Ngãi, his helicopter was shot down as the formation was about to land at the rendezvous point. Tâm was seriously injured in the left knee and had to be evacuated to the Đà Nẵng military hospital.

In Saigon, Loan continued to send and receive letters from Tâm regularly, but today, when she received a returned letter, she was shocked to see that her letter had been returned with a note on the envelope: "Second Lieutenant Lê-Văn-Tâm is no longer at this KBC." Her world seemed to collapse, and she couldn't understand what had happened. So many questions and doubts flooded her mind: "Could it be that Tâm is trying to avoid me? Impossible! He loves me so much!"

She quickly rushed to find Tâm's cousin's house to inquire about him, but his cousin had sold the old house to pay for medical bills last year, and the neighbors didn't know where the family had moved. Loan was thrown into near panic because her family was about to move to Cà Mau, as her father had retired more than a year ago. Their house in Saigon had already been sold, and they were set to move out by the beginning of the month. Being the only daughter and unmarried, she had to quit her job in Saigon to follow her parents to Cà Mau, where the bank had promised her a similar position.

Everything was ready, just waiting for the moving day. The returned letter was a crucial piece of information because it contained the news that she was moving to a new province and other important details. She wanted to ask her parents for permission to go to Đà Nẵng to find out what had happened, but aside from Tâm, she knew no one else there. As a young, unmarried woman, she had never left her family's side; where would she stay, and whom would she ask for help?

Thinking of this, she lowered her head in despair. Suddenly, two tears rolled down her cheeks. Loan cried. This was the first time she had known love, and it was also the first time she shed tears for an unfulfilled love. She had loved Tâm with the purest first love of a schoolgirl, as pristine as the white of her student áo dài, still fragrant with the scent of fresh ink and paper.

###

More than a month after being treated at the Đà Nẵng military hospital, Tâm had not written a single letter to Loan due to the severe pain in his leg, making it difficult for him to walk. He preferred to stay in bed and didn't want to send bad news to her too soon, intending to inform her after a few months when he had recovered. Today, feeling somewhat better, Tâm called the squadron to check if there was any mail from Loan, but the staff informed him that his unit had been transferred to another KBC, so no mail was being sent to the squadron anymore.

Completely surprised, Tâm hurriedly wrote a letter to Loan to inform her of his new address. After waiting for two weeks without a reply from Loan, he grew anxious and sent a telegram to Saigon, asking for an urgent response. A few days later, the telegram was returned, indicating that the recipient could not be found. He read the words in the telegram: "Recipient not found." Tâm's heart was completely numb. Hundreds of questions raced through his mind: "What has happened to Loan?"

Tâm wanted to return to Saigon to find Loan and explain that he had been injured, hospitalized, and unable to write to her due to the excruciating pain from his wound. He feared that Loan might have misunderstood and thought that Tâm had intentionally forgotten her.

That evening, when the senior doctor visited, Tâm requested permission to be discharged and return to Saigon, but his request was denied due to his still-healing wound. Any premature movement could cause the wound to reopen and become dangerously infected.

Knowing that he couldn't leave the hospital as he had hoped, Tâm called the squadron to see if any of his friends were flying to Saigon in the coming days. The duty officer reported that a lieutenant would be flying to Saigon for a routine medical check-up on Friday. Overjoyed, Tâm wrote a long letter to Loan, clearly stating her home address, and asked the lieutenant to deliver the letter directly to her.

After nearly a week in Saigon, the lieutenant returned to the unit and visited Tâm at the hospital, reporting that Loan's house had been sold to someone else last month. The new owner had no information on where Loan's family had moved, only that the neighbors said they had moved to another province. Tâm was devastated, taking back the letter from the lieutenant and collapsing onto the pillow with a deep sigh of despair.

War is truly cruel and bitter. Tâm had narrowly escaped death, a bullet fired by another Vietnamese, sharing the same bloodline, the same race, and a common history of four thousand years of building and defending the nation. It was absurd to engage in such fratricidal slaughter, driven by a few individuals who had sold their souls, constantly justifying themselves as being more righteous than others, using foreign weapons to destroy their ancestral graves, tearing each other apart! Who were the victims of this war, of this destruction today? As Tâm pondered this, he felt a sting in his eyes. In a moment of weakness, he cried. With these tears, he couldn't determine whether he was crying for his homeland or for himself—crying for a love that had just begun but was destroyed by the turmoil of war. Moments later, he fell into a restless sleep, filled with unanswered questions and concerns.

Six months later, with his wound mostly healed, Tâm was discharged from the hospital and returned to the Air Force Command in Saigon. After a short period of rest to regain his strength, he was assigned to the 2nd Air Division in Nha Trang. Due to his injury, the flight surgeon grounded him.

When he reported to the personnel office as a former pilot, he was assigned to be a liaison officer for an infantry division stationed in the Central Highlands.

Pleiku, a city often described as eternally sad, with red dust swirling and rainstorms blocking the way, was a poor town with simple houses and streets, almost forgotten, lying near the Vietnam-Cambodia border. People in Saigon thought of Pleiku as a frontier town, often mentioned in connection with battles like Đức Cơ, Play Me, and as the location of the 2nd Tactical Zone Command.

Tâm had been stationed there for over two months. Aside from his daily routine duties, he mostly stayed in his room, reading newspapers or renting books to read. The half-body portrait of Loan that she had given him was enlarged and hung on the wall next to his metal bed. On weekends, alone in his empty room, he would often take out the letters Loan had sent him, read them again, carefully put them back into their envelopes, and then store them in the drawer. He would then lie back on his bed, rest his hand on his forehead, and gradually drift off to sleep as if he were retreating into the memories of his old love.

Loan's state of mind was no different. It had been over eight months since she moved to Cà Mau. Her family now lived with her grandmother in a grand brick house on the outskirts of the province, as her father, being the eldest son, was responsible for taking care of his elderly mother and maintaining the ancestral shrine.

Loan had resumed her job at the bank in the town center, and everything in the family was running smoothly. Her parents were comfortable and happy in the quiet countryside setting. In contrast, Loan felt a sense of emptiness. She missed the bustling scenes of Saigon, the sound of children laughing and playing at the end of the alley, but in reality, she missed Tâm the most. The other memories were just images of her old neighborhood in the Bàn Cờ district. To put it more accurately, she was in love with Tâm, placing all her hopes and trust in her first love.

Sometimes, during work, she would suddenly become lost, as if in a trance, because her instincts told her that Tâm was still alive and still loved her deeply. Then, countless questions and doubts would flood her mind, ultimately leading her into a world with no way out.

Today was the 23rd day of the lunar month, the day of the Kitchen Gods' departure to heaven. The market and neighborhood activities seemed busier than usual. Tâm was tidying up his room, organizing his belongings neatly. Just then, Tuấn, who had just finished his shift in the division operations room, opened the door and walked in.

"Hey, Tâm! When are you going back to Saigon on leave?"

"Who told you I was going to Saigon for Tết?"

"I heard they've already divided up the leave slots!"

"But I'm not going anywhere. I gave my slot to Vũ."

Tuấn scratched his head, puzzled, thinking to himself:

"Huh? Is this guy crazy? Why wouldn't he go home to visit his family and girlfriend and have fun?"

Tuấn and Tâm were two Air Force friends, temporarily assigned to work with another unit, and since both were single, they stayed together in the officers' quarters on the Pleiku military base.

"If you're not going to Saigon, why don't you come to my aunt's place for Tết? My parents said they'd come up to visit the coffee plantation and spend Tết here."

Tuấn walked up to Tâm and gently patted his shoulder, raising his eyebrows as he whispered:

"Hey, I've got a cousin who's really pretty, the belle of the town! I'll introduce you to her."

The weather in December on the highlands brought a special kind of cold, with mountain winds blowing from Laos, drying the skin and nearly cracking it during the early morning hours when the sun was still behind the hills or during the twilight hours after the sun had set, with thick, low white clouds hovering halfway up the mountains.

In his dark green combat uniform, Tâm sat with his arms crossed next to a steaming cup of black coffee, quietly staring out into the distance, seemingly listening to the slow passage of time outside the window. The days leading up to the year-end Tết were indeed somber for Tâm.

His friends were all preparing to go home on special leave to reunite with their families for the three days of the Lunar New Year, while he had voluntarily chosen to stay behind in this desolate, windy outpost. To be honest, Tâm had no one to return to! His cousin, who had once taken him in, was now dead. His friends had all joined the military or gotten married, each going their own way. Even his one and only love, Loan, had unwittingly left him and moved away. He was once again living a lonely existence, with the same feelings of an orphaned child from years ago. At times, he blamed himself, wondering if the heavens and gods were unfair to him.

Just then, a military Jeep screeched to a stop outside. Tuấn jumped out of the vehicle and hurried into the room:

"Hey, are you ready yet, Tâm? We need to head to Cù Hanh airport to pick up my parents. The AIR VIETNAM flight is landing in about ten minutes."

"I've been sitting here waiting for you, practically bored to death."

"Sorry, sorry, I got held up there for a bit and had to sweet-talk my way out to get here early."

The two friends quickly hopped into the Jeep and drove towards the civilian airport in Pleiku.

As Tuấn was searching for a parking spot, he noticed a young woman with fair, smooth skin and shoulder-length hair waving and walking toward the Jeep as it tried to park.

"Brother Tuấn, my mother is waiting inside!"

Turning to Tâm, who was sitting in the front seat, Tuấn said:

"Lệ, my cousin!"

When the Jeep came to a stop, Lệ approached. She spoke to Tuấn, slightly pouting as if scolding him:

"My mother has been waiting for you in there, thinking you'd forgotten!"

"How could I forget? My parents told me about this a week ago."

Turning to Tâm, Tuấn introduced:

"This is my friend, Tâm, and this is Lệ, my cousin."

Tâm stepped forward with a polite smile and nodded to Lệ:

"Hello, Lệ, that purple áo dài you're wearing looks beautiful!"

With a charming, slightly shy smile, she looked at Tâm and gently bowed:

"Hello, Tâm, thank you for the compliment."

The three of them quickly entered the airport terminal where Lệ's aunt and Tuấn's parents were waiting.

It was the 30th day of Tết (Lunar New Year's Eve), and according to our traditional customs, Lệ's aunt and Tuấn's parents had risen early to prepare offerings to welcome the ancestors to enjoy the New Year with their descendants for the three days of Tết. Tuấn's father was a veteran official of the Saigon Postal Service, known for his kind and virtuous nature. After hearing from Tuấn about Tâm's orphaned and lonely life, he was deeply moved. He suggested to his sister-in-law:

"Ba, I think it would be good for Tâm, Tuấn's friend, to come here for Tết. He's all alone in the barracks, and it must be quite lonely during the holiday."

Before he could finish, Lệ's aunt joyfully interrupted:

"Oh! I was thinking the same thing, that Tuấn should invite Tâm to come and spend Tết with us, but I was hesitant, fearing that you and your wife might not be comfortable having a stranger in the house."

She paused for a moment, then stepped closer to her brother-in-law and spoke softly:

"Brother Hai, I find Tâm to be a very likable young man, and it seems to me that Lệ has taken a liking to him as well!"

"Well, Lệ is growing up—she's almost 21 this year. If she finds someone respectable and decent, I wouldn't mind them becoming close."

As they discussed this, Tuấn's mother arrived, having overheard part of the conversation, and she chimed in:

"I think it's a good idea, Ba. Let them get to know each other. Besides, since your husband passed away so early, your home must feel lonely with just you and Lệ. Having a good son-in-law nearby would be like having a son!"

That Tết, Tâm experienced a nearly complete New Year, though the image of Loan, his first love, still occasionally flickered in the empty spaces of his mind.

In the weekends that followed, Tâm, Tuấn, and Lệ often went out together, enjoying each other's company. Eventually, Tuấn began making excuses to leave them alone, giving Lệ and Tâm more opportunities to get to know each other.

Lệ was an only child. She had lived in Saigon with Tuấn's parents while attending school. After graduating from junior high school and preparing to enter her first year of high school, her father fell ill and passed away that same year. With no one else at home, she returned to Pleiku to live with her mother and continued her studies there. Thanks to the coffee plantation and the assets her father left behind, Lệ and her mother, despite their widowhood, lived a comfortable and well-off life.

It was now coffee-harvesting season, and Lệ wanted to take Tâm to the plantation to enjoy some fruit and watch the coffee harvest. So, early that Saturday morning, she prepared lunch and enough drinks for both of them. Lệ dressed in the style of the local ethnic girls, wearing a long-sleeved blue shirt and a "sarong" wrapped around her waist, giving her a charming yet wild look, like a wildflower from the countryside.

Since meeting Tâm, Lệ had become more cheerful and filled with dreams. Tâm often invited her to go out for ice cream or to listen to music in the quiet, nearly empty cafes at the end of the street. Occasionally, during rainy afternoons, the two would walk together under the eaves of buildings in the quiet streets. Tâm would point out the rainwater bubbles floating and about to burst, whispering to Lệ:

"I don't want our relationship to be like those rain bubbles."

She would lean her head against his chest, wrapping her arms around his back:

"I know what you're thinking, and I'm praying we won't end up like that."

Lost in these memories of her time with Tâm, Lệ was suddenly startled out of her daydreams by the sound of a Honda motorbike screeching to a stop in front of the house. Knowing it was Tâm, she grabbed the food basket and headed to the front to ask her mother for permission to go out. As Tâm parked his bike and entered the house, he greeted and asked Lệ's mother for permission to take her to visit the plantation. Her mother was pleased and reminded them:

"You two, be careful on your trip. The roads up there are steep and slippery. Come back early and have dinner with us before heading back to the barracks."

Tâm bowed in gratitude:

"Yes, Aunt, we'll be back early."

After securing the food basket on the back of the bike, Lệ climbed on behind Tâm. As the Honda sped away from the gate, Lệ no longer held onto his shoulders lightly as she did in front of her mother. Instead, she wrapped her arms tightly around him, just like a newlywed couple. The warmth and rhythmic breathing from her soft, full chest transferred through Tâm's thin shirt, giving him an indescribable, floating sensation.

Lệ leaned forward, whispering close to Tâm's ear:

"Once we're out of town, drive slowly so we can enjoy the scenery. I brought a camera, so if we find a nice spot, let's stop and take pictures, okay?"

Tâm quickly turned his head to smile at Lệ, showing his agreement. He gently stroked her smooth, cool wrist resting on his back:

"I'll do whatever pleases you."

Playfully, Lệ pinched Tâm's thigh:

"You're just trying to flatter me, aren't you?"

"No, I'm serious! And even if I am flattering you a little, what's wrong with that?"

A few minutes later, Tâm stopped the bike by a large rock on the roadside, telling Lệ to climb up so he could take some pictures.

Earlier that morning, with Lệ's mother present, Tâm hadn't paid much attention to her attire. But now, he realized how beautiful and captivating she looked in her ethnic outfit. To him, she embodied the beauty of a wildflower, the perfect blend of the highlands' simplicity and charm. He couldn't find any other words to describe her than to elevate her to the queen of the universe that day.

"Tâm, why are you staring at me like that? Take the picture already!"

Startled from his trance, Tâm fumbled:

"Oh... right, I'm taking it now!"

After taking a few shots of Lệ, Tâm slung the camera over his shoulder and helped her down from the rock.

"Lệ, you know, I don't know what the real 'Phà Ca'—the mountain girl from those Saigon theater shows—looks like, but I'm certain of one thing: my 'Phà Ca' is stunningly beautiful. You made me so mesmerized that I forgot to take your picture."

Blushing at his compliment, Lệ pushed him playfully away:

THE LAST FLIGHT OUT

"There you go, flattering me again!"

A little over a year later, Lệ and Tâm were officially married. The couple lived with Lệ's mother in town. Their daily life was peaceful and happy. Lệ didn't need to work, as her mother's family was wealthy, so she chose to devote all her time to caring for her husband. In return, Tâm was very fond of his wife and treated her with great respect, as well as honoring her mother.

Time passed swiftly and mercilessly. Since Loan had moved with her family to Cà Mau, her mind remained entangled in unresolved memories. Loan couldn't believe that Tâm had suddenly abandoned her so abruptly. Sometimes, news reports about intense battles in the northern region would leave Loan anxious and fearful. Occasionally, a foolish thought would briefly cross her mind: "Could Tâm be dead?"

When she thinks this, she panics and immediately dismisses the thought: "No! No! Tâm could never be dead!" Tears would stream down her face, not for the first time, but many times whenever she thought of Tâm. She cried for the cruel fate of her first love and for her weakness in the face of adversity as a woman.

Today was Armed Forces Day in the Republic of Vietnam. The province of Cà Mau was holding a parade for the local militia and officials. Loan decided to go watch with a friend from work. The two of them rode a Honda "dame" scooter towards the main street, and Loan suggested to her friend:

"Let's stop at that shop over there for some phở first, then we can go out and have fun!"

"OK! Whatever you want."

As they entered the shop, Tuyết, Loan's friend, lightly tapped her on the back and whispered:

"Hey, Loan! Look at that table in the corner; there are some Air Force pilots eating over there. See if you recognize any of them as your Tâm?"

Loan quickly turned around to take a closer look at each person, but she didn't recognize anyone. She whispered back to Tuyết:

"Besides Tâm, I don't know many Air Force officers."

At this point, Tuyết interrupted:

"If you don't know, just ask them. Go ahead and ask if they know anything about your Tâm."

"I feel too shy."

"Shy about what? You're just asking, not asking for a favor!"

Loan realized Tuyết had a point. She adjusted her posture and boldly approached the table where the Air Force officers were sitting. After politely nodding in greeting, she said:

"Excuse me, gentlemen, my name is Loan. I have a question to ask you if you don't mind."

The youngest officer, a lieutenant, stood up and pulled out a chair for her:

"Please, have a seat. If we can help you, Ms. Loan, we'd be happy to."

After regaining her composure, she sat down, trying to ease her discomfort in front of these unfamiliar men:

"Gentlemen, I have a brother, Lieutenant Lê Văn Tâm, a helicopter pilot in the 1st Air Division in Đà Nẵng. I haven't been able to contact him for years, and I don't know what has happened to him."

Before she could finish, Captain Mỹ, seated at the far end of the table to her left, suddenly stood up:

"I know Lieutenant Lê Văn Tâm..."

He moved closer to Loan and asked:

"Excuse me, are you Ms. Loan, who used to live in Bàn Cờ and later moved to the province with your family?"

A bright light seemed to shine on her face as Loan quickly stood up and replied:

"Yes, yes... I used to live in Saigon, in Bàn Cờ..."

Feeling as though they had known each other for a long time, Mỹ grasped Loan's hand and led her to an empty table nearby, where he told her everything that had happened to Tâm, including the last letter Tâm had asked him to deliver to her house, only to find out that Loan was no longer living there. Mỹ also mentioned that after being wounded, Tâm was no longer in the aviation unit, and around that time, Mỹ himself had been transferred to a new unit in Cần Thơ, so he had lost contact with Tâm and didn't know where he was or what unit he was in now.

The old love ignited once more in her heart as Loan blamed herself for unintentionally causing Tâm so much pain while he was suffering from his severe wounds. Unable to hold back her tears, she thanked Mỹ and bid him goodbye before returning to her table and asking Tuyết to take her home immediately, as she was no longer in the mood to eat or go out.

Loan was overjoyed to know that Tâm was still alive and that he still thought of her with love, dispelling all the doubts she had harbored before. Many young men in the province were very handsome, well-educated, and from wealthy families, including the young, single manager at the bank where Loan worked. They all wanted to get to know her and asked to visit her home. Out of politeness, she always received them graciously but never allowed anything beyond formal social gestures. Loan still hoped, and her instincts told her that she would reunite with Tâm one day.

The war between the two ideologies of communism and capitalism had reached a fever pitch. The American ally had been politically defeated at home. After spending billions of dollars and sacrificing tens of thousands of American lives in the Vietnam War, they realized there was no further profit to be gained. The powerful American ally, faced with an indigestible situation, was forced to negotiate with the enemy, preparing to pack up and leave, leaving their reluctant Vietnamese ally to die like a terminally ill patient.

The North Vietnamese Communists, taking advantage of the withdrawal of U.S. and allied forces, launched a massive influx of troops and modern war equipment, including advanced heavy tanks from the Soviet Union and Communist China, into South Vietnam, aiming to launch decisive battles to end the Second Republic of South Vietnam.

The Viet Cong began heavy shelling on the II Corps Headquarters and the Pleiku military airport. Tâm had been confined to the base for several days, only occasionally being granted special permission to visit his home for a few hours to check on his wife and gather additional clothes and necessities. He never forgot to remind and comfort his wife to take care of her health and secure the house.

Lệ hurriedly packed clean clothes into a bag, including half a dozen oranges and some of Tâm's favorite snacks to take to the base. Tâm felt deeply saddened as he watched his wife. This was the first time since their marriage that they had been separated for so long. He approached her, gently rubbing her back:

"I feel so sorry for you and Mom being alone at home. If the situation gets worse, I might have to send you and Mom to Saigon for a short while."

Lệ abruptly turned around, her eyes wide as she looked at Tâm:

"If you're staying here, I'm not going anywhere. If we die, we die together!"

Tâm quickly silenced her with a finger pressed to her lips:

"Don't say that, darling. It would make Mom sad. My leave is almost over; I have to get back to the base."

Tâm embraced his wife tightly, wanting their bodies to merge into one so they wouldn't have to face the separation.

The sound of the North Vietnamese Communist artillery shelling the Pleiku airport from an unknown location was very clear in the night, followed by deafening explosions nearby. Every time there was shelling, Lệ and her mother would lie flat on the floor to avoid any potential danger, as their house was very close to the airport. In those moments of crisis, Lệ would forget about herself entirely and constantly pray, asking for divine protection for Tâm.

To her, the war was a manifestation of human brutality. She disliked hearing political commentary on the radio or in newspapers, only wishing to live a normal life, to love, to be pampered, to admire beautiful flowers, and to breathe in the fresh air of early mornings. Ever since Lệ and Tâm had met, fallen in love, and married, she had been increasingly worried about Tâm. In her simple way of thinking, Tâm was the victim of today's war, with the mark of hatred still etched on his knee from the bullet wound.

Many times, Lệ shuddered in disgust when she heard about "glorious victories" where hundreds of enemies were killed in a reported battle, feeling no different than a North Vietnamese Communist soldier who, after shooting down Tâm's helicopter, would surely be rewarded for "killing the enemy," even though that "enemy" was a fellow Vietnamese.

This morning, based on intelligence reports from II Corps and reconnaissance aircraft, Tâm had heard of signs indicating that North Vietnamese regular army divisions were moving from Laos and Cambodia into Vietnam, planning to capture border provinces like Kontum, Pleiku, and Ban Mê Thuột in the coming days.

Many high-ranking civil servants and wealthy individuals in Pleiku were hurriedly preparing to evacuate to other locations. Knowing there was no time to delay, Tâm requested special leave to arrange for his family to move to Saigon so he could return to the unit to fight with peace of mind.

All civilian flight tickets had long been sold out, and the South Vietnamese Air Force planes were reserved for combat personnel and battlefield supplies, leaving no other air transport options except for traveling by bus along the Pleiku-Ban Mê Thuột-Saigon highway.

Knowing she couldn't go against her husband's final decision, Lệ and her mother reluctantly secured the house and packed some money, jewelry, and necessary clothes to follow Tâm to the bus station to head to Saigon. This morning, Pleiku's streets looked desolate and gloomy. A fine misty rain, like early morning dew, barely chilled the earlobes of passersby. Lệ clung to Tâm's side as they walked along the sidewalks where they used to stroll together during the weekends when they first met.

"Tâm, I'm so scared!"

"What are you afraid of?"

"Do you remember when we first met? During a rainy afternoon, we walked through here, and you showed me those beautiful rain bubbles floating by, only to burst suddenly. It made me so sad I almost cried. Did you know that?"

Tâm remained silent, his head bowed as if trying to suppress the intense emotions of the old days between him and Lệ. Seeing Tâm stay quiet, Lệ gently tapped him on the back:

"Did I make you sad by bringing up those memories?"

"Yes, I remember them well. But we're so happy now; don't let your mind wander into negative thoughts."

Lệ leaned her head on his shoulder, following him like a soulless body. She cried, warm tears soaking into Tâm's shoulder.

"When will you be allowed to visit me?"

"I don't know, darling. We're under a 100% lockdown. If I get a chance, I'll come right away because there's no one more important to me than you."

The conductor's hurried calls for passengers to board the bus made Tâm hasten his wife along.

"Hurry, or you and Mom won't get a seat."

After Lệ and her mother settled into their seats, the bus, filled to capacity, began to slowly roll away. Tâm stood on the sidewalk, watching them leave, as Lệ leaned out the window to look back at him one last time.

There's no more heartbreaking scene than bidding farewell to a loved one without knowing when you'll meet again! The bus drove further and further away until it disappeared into the misty rain. Tâm remained standing there, like a statue, staring blankly into the distance.

The bus station now felt cold and empty, with only a few ethnic women carrying children on their backs and bundles of firewood on their heads, quietly walking across the road. Tâm pulled his Air Force jacket tighter around his neck and trudged towards the familiar street corner restaurant where he and Lệ used to have breakfast. He felt an intense need for a strong cup of coffee to help numb the pain tearing through his mind.

A notice, written on red paper and posted on the door, read: "WE REGRET TO INFORM YOU THAT WE ARE TEMPORARILY CLOSED." Tâm read the sign again. The restaurant's apology was posted on the tightly closed steel shutters. Frustrated, Tâm ran his hand through his rain-dampened hair, muttering to himself: "They must have already evacuated the city." Glancing through the shutter slats, he saw the wooden table by the window in the corner, where he and Lệ often sat during breakfast. Lệ always said to Tâm, "Sitting here, we can see all the lively, joyful activities of Pleiku's street..."

Now, the table with its four stools sat silently, alone. Lệ had left, and outside, the streets were sparse, leaving Tâm standing there, lost in memories.

As he was about to step out onto the main road to find a ride back to the base, Tuấn suddenly arrived on his Honda motorbike, screeching to a stop in front of him:

"Tâm, where did your wife and aunt go? I stopped by the house, but no one was there!"

"Oh, Tuấn! My mother-in-law and my wife have already left for Saigon!"

"How did they travel?"

"By bus!"

Tuấn frowned, clearly displeased:

"Were there no other means of transportation than by bus?"

"All civilian and military flights were booked. But how did you know I was here?"

"I called the operations room. They said you took special leave to evacuate your family, so I came out to check. When I saw the house locked up, I drove around to find you. So, they've already gone?"

"Yes, about half an hour ago."

"Most of the shops are closed. Hop on; I'll take you back to the base."

Back at the barracks, Tâm didn't even bother removing his combat boots. He wearily dropped onto the military cot, hands resting on his forehead, eyes shut tight. Outside, the 105mm artillery guns from a nearby battalion fired every few minutes, a constant reminder of the ongoing war.

Hearing hurried footsteps approaching, Tâm opened his eyes. A panicked, frightened voice reached his ears:

"Tâm! Did you hear?"

Tâm shot up as if struck by lightning:

"Hear what, Tuấn?"

"A bus was hit by a Viet Cong mine...!"

Tâm felt like the world had collapsed beneath him. His face went pale as all the blood drained from it, and he stared at Tuấn, unable to speak. Tuấn continued:

"I heard in the operations room that a reconnaissance plane radioed in, reporting that a bus traveling from Pleiku towards Ban Mê Thuột was hit by a Viet Cong mine on Highway 14, about 70 kilometers from Pleiku. Many people were injured or killed. I was so scared!"

Finally, Tâm managed to choke out:

"Oh my God! I'm doomed, Tuấn!"

Saying this, Tâm could no longer hold back the overwhelming emotions inside him. He buried his face in his hands and wept. Tuấn approached, pulling him to his feet, trying to calm him down:

"We need to act quickly; we can't afford to wait. We don't know the situation yet. I'll call the Air Force's emergency operations room to request two medevac helicopters, and we'll go with them to see what happened."

Half an hour later, the two medevac helicopters safely landed near the overturned bus still smoking by the roadside. A squad of local militia was trying to carry the wounded passengers out of the bus and line up the dead bodies on the roadside. Tâm and Tuấn hurried out of the helicopter, running towards the wreckage. Scanning the injured passengers sitting by the roadside, Tâm didn't see Lệ or her mother. As he anxiously searched, a short, nearly breathless cry came from Tuấn, almost as if he was about to faint:

"Tâm, over here!..."

As if his worst fears had been confirmed, Tâm's face turned ashen, drained of all color. He rushed to where Tuấn was kneeling beside a bloodied body, lying on its back on the roadside. Tâm let out a short, grief-stricken sob:

"Oh God, Lệ!..."

Almost too heartbroken to cry, Tâm collapsed to the ground, cradling her in his arms, pressing her lifeless body tightly to his chest. He gently smoothed her hair and wiped away the dust from her face as his hot tears began to fall onto her unmoving face. Tâm bowed his head close to his wife's as if sharing his last words with her, believing she was still alive in his loving embrace.

Tuấn knelt beside Tâm, speaking through his own tears:

"Tâm, I'm so sorry. Lệ and your mother-in-law are gone. I'm just as heartbroken as you are!"

"Tuấn! If I could die now with my wife, it would be the happiest moment of my life..."

"I understand, but everyone has their fate. All we can do now is give Lệ and your mother-in-law a proper burial to bring them peace."

Tuấn and Tâm, along with the neighbors, carefully buried Lệ and her mother with great care. Today was the day to open the grave, and the two friends had risen early, buying fruit and flowers to take to the gravesite as an offering according to the ancient customs. Tâm let Tuấn handle the incense and prayers while he sat beside his wife's grave, head bowed, hands clutching the freshly turned earth. His body shook with quiet sobs, and his eyes were red and swollen from crying, evoking deep sympathy in Tuấn, who knew that no words of comfort could ease Tâm's pain. So he simply sat in silence, waiting.

"Tuấn, you go back to the base first. I want to stay here with my wife for a while."

"Alright."

Tuấn quietly left and drove back to the base. Now, it was just Tâm and his beloved wife who lay beneath the earth. He gently stroked the mound of earth as if it were her body. A small sparrow, lost from its flock, landed on the sugarcane stalk nearby, chirping forlornly for a brief moment before flying away.

Tâm didn't know who to blame for the tragedy that had torn them apart. He didn't believe in fate, thinking that people often turn to fate as an explanation when they can't find a solution. The gods, heaven, and Buddha don't kill people—people kill each other.

But how cruel and tragic it was for the Vietnamese people to be incited by foreign powers to turn on each other with more ferocity than wild beasts, tearing each other apart on the land that their ancestors had bled and died for, defending against foreign invaders and preserving their heritage!

The strategic withdrawal order issued by President Nguyễn Văn Thiệu, abandoning Kontum, Pleiku, and other provinces, was a key factor that led to the complete collapse of South Vietnam. Realizing the situation was beyond saving and under intense pressure from the ally, President Thiệu was forced to resign and quietly leave the country.

Then came April 30, 1975, the day when the last president of South Vietnam, former General Dương Văn Minh, officially surrendered to the North Vietnamese Communists, marking the exodus of millions of South Vietnamese people and the internment of hundreds of thousands of military personnel and civil servants by the Communists in reeducation camps, where they were subjected to brainwashing and hard labor. Captain Lê Văn Tâm was one of those unfortunate ones.

After leaving Pleiku, Tâm's unit moved to Nha Trang, then to Saigon, and finally to the IV Air Division base in Cần Thơ. There, Tâm was reunited with Captain Mỹ, his former colleague from the Đà Nẵng air unit. Mỹ reminded Tâm of their conversation about Loan, Tâm's first love, who was now working at a bank in Cà Mau.

Mỹ's story left Tâm deeply shaken as he recalled his old love. He felt as though he had betrayed his beloved wife, Lệ, who had passed away not long ago. She had taken her loyalty and love for her husband to the grave, and he knew he would never see her again. Now, by chance, Loan had reappeared in a completely unexpected situation. Tâm lowered his head, shaking it slightly:

"Thank you, Mỹ, for telling me about Loan, but my wife Lệ has just passed away. I still believe that her spirit will stay with me forever. I love her, and I don't want to hurt her spirit by having another love, even though that person came before Lệ."

Mỹ stepped closer, gently patting his friend on the shoulder in consolation:

"Tâm, you're right. But I think you should at least let Loan know that you're still alive so she can have peace of mind."

"I want that too, but I don't have the courage, Mỹ. I'd rather stay silent, as I have for all these years. I don't want to make Loan sad because of me."

Realizing he couldn't persuade Tâm in this matter, Mỹ shifted the conversation to the latest developments in the ongoing conflict:

"Tâm, I've heard rumors that some of the special treatment officers have already made arrangements to evacuate their families abroad, but I haven't heard anything about the families of the rest of us."

"I've heard those rumors too, but I don't believe in fleeing in such a cowardly manner. Sometimes, I think those officers are even more cowardly than the Montagnard soldiers in the highlands, who fought for their lives because their families were right there with them at the frontline. They didn't have any lofty ideals of fighting for freedom or nationhood; they simply fought to protect their wives and children. If their families hadn't been there, the border outposts would have been abandoned long ago, as the soldiers would have deserted to find safety for their families."

"I share a similar view Tâm. Sometimes, when I hear those officers brag about their heroic deeds of securing escape routes for their families, I get so disheartened."

"Forget it. What will happen will happen. Let's head to the squadron's club, grab some coffee, and play some billiards. That sounds better than sitting here talking about life."

About six years later, from the day Tâm heeded the Communist propaganda officers' calls through the loudspeakers in Cần Thơ and joined other officers in surrendering, he hoped that, as the Communist state had announced, after a short period of reeducation, everyone would return to their normal lives.

But then, days turned into months, and as the weather changed with the two seasons of the year, time dragged on until he could no longer keep track of the days. The long wait had turned hope into a distant memory as if it had been buried under a thick layer of dust.

After several transfers, Tâm was finally taken to the "Kà Tum" camp, where he had been for nearly a year. The camp was located deep in the dense jungle, near the Vietnam-Cambodia border in the northwest of Tây Ninh province. This morning, as usual, he was digging up roots and tilling the soil to plant cassava and sweet potatoes for the camp when he was surprised to hear the camp guard call his name clearly:

"Lê Văn Tâm, you have a visitor. Go back to the camp, clean up, and change your clothes. Someone will guide you to meet your relative."

For a few seconds, he was doubtful, wondering if they had the right person, as he had had no relatives for almost six years. His wife's family had escaped by boat after the events of April 30, 1975, and as for his own family, there was no one left who could be called a relative. He stepped forward to ask the guard for confirmation:

"I'm Lê Văn Tâm?"

"Yes, you have a visitor."

Leaving the hoe by a banana bush nearby, he hurried back to the camp, his mind filled with questions. He couldn't figure out who would have accepted him, a lonely, wretched prisoner, as their relative.

Wash your face, hands, and feet thoroughly, then choose the relatively best outfit in your possessions to change into. He picked a blue shirt that still looked new and a pair of long trousers that had just been washed yesterday. For Tâm, this was already quite a formal and fresh appearance!

The communist guide from the reception committee approached and gave him some instructions before leading him to the camp gate. At this point, Tâm was like a machine, merely accepting without voicing any opinion, as he thought that when a person's value is reduced to nothing, there's nothing left to say. He nodded at the cadre to finish the conversation and then followed him out to the visiting area.

Tâm looked far into the distance. He saw five or six women with a few children standing outside the fence, anxiously looking in. He knew for sure that they were the parents or wives and children of his fellow prisoners who had come here to visit their husbands and sons. He did not recognize any of them as his own family.

As he approached the camp gate, he suddenly stopped dead in his tracks when he saw a woman—more accurately, a girl around thirty years old—standing by a bamboo pole near the left side of the camp's visiting area. She was dressed in a white bà ba shirt and black trousers, with a conical hat on her head. He mumbled to himself like someone who had lost his mind:

"Oh my God! Could that be Loan?"

As if responding to some telepathic connection, Loan read the question in Tâm's eyes and on his lips. She widened her eyes, quickly walked toward him, and, full of tears, asked in a choked voice:

"Yes! It's Loan! And you're Tâm, Lê Văn Tâm, right?"

The communist guide swiftly stepped in, pushing the two apart, followed by a few condescending words:

"You two can't act like this; it looks weak and pitiful. Go inside and sit properly to talk."

Startled, Loan stepped back, lifting the hem of her shirt to wipe her tears. Perhaps it was only now that she truly cried, having waited so long and now faced with the bitter reality that had weighed down on his life.

Tâm had aged and was gaunt from malnutrition. Most of the other prisoners were regularly supplied by their families, so their bodies weren't in too bad shape. But Tâm was lucky if he received just a bit of leftover food that his friends kindly shared with him. His eyes were deeply sunken, and his cheekbones nearly touched each other. Sometimes, when he looked in the mirror, he could hardly recognize himself. He was no longer Captain Lê Văn Tâm, the young and handsome man with captivating eyes, once nicknamed the "women killer" of the South Vietnamese Air Force.

Now, he was a ragged, starving communist prisoner simply because he was born in the South and had served the Southern government.

After entering the area, the two had to sit opposite each other at a roughly made wooden table. The communist guide also sat nearby, seemingly to listen in on their private conversation.

Tâm sat on a bamboo chair, elbows on the table, head in hands, his face contorted as he looked down, unable to say a word. Loan, sitting opposite, still had tears streaming down her face as she gently asked in a voice full of anguish:

"Tâm! Why do you want to avoid me?"

After a few moments of trying to regain his composure, Tâm lifted his head and looked straight at Loan but didn't want to answer her question directly:

"How did you know I was here?"

"Your friend Mỹ, who was just released, came to my house and told me everything, so that's how I knew you were here. I never expected your life would be filled with so much misfortune!"

He sighed, accepting and resigned to all the painful memories that had occurred in his life:

"Yes, I got married, and my wife died because of the war in Pleiku almost seven years ago. I also told Mỹ all about my personal life, and that's when Mỹ told me you were working in Cà Mau."

As Tâm spoke, Loan interrupted:

"You knew I was working at the bank in Cà Mau, so why didn't you come to visit me?"

After a moment of silence, Tâm responded, almost in a whisper:

"Because Lệ, my wife, had just passed away, I didn't want to disrespect her spirit. Moreover, I didn't want to make you sad."

"I deeply respect your loyalty to Lệ! As for me, I have waited so long, hoping to see you again, and I'm just happy to see you now!"

"I know, but I was afraid of making you sad because I had a family."

The communist guide sitting at the next table checked his watch, then turned to them and said:

"You two have about five more minutes before visiting time is over, so wrap it up and return to the camp."

Tâm turned to look at him, nodded in agreement, and then continued his conversation with Loan:

"Now, we each have our own circumstances. I'm imprisoned and don't know when I'll be released. Loan, you traveled all the way from Cà Mau to visit, but it's such a long and difficult journey, and it's expensive too. I feel bad. Besides, I've gotten used to being alone; a little more won't hurt."

Loan reached across and gently placed her hand on Tâm's, which was resting on the table:

"Tâm! Please call me 'em' instead of 'Loan,' and don't refer to yourself as 'tôi.' If I didn't care about you, I would have gotten married long ago. Your situation now needs me more than ever, so please don't reject me!"

Tâm looked down at the table:

"Alright, I really appreciate your kindness."

Overwhelmed with emotion, Tâm turned his face away to hide the tears streaming down his cheeks:

"You should go home before it gets dark. Please send my regards to your parents."

Loan suddenly remembered the basket of food and daily necessities she had brought for Tâm. She quickly walked to the bamboo pole nearby, where she had placed the basket, and brought it back to him:

"I brought some fruits from the garden, candies, and dried food for you."

Tâm reached out to take the basket full of food that Loan handed to him:

"Thank you so much!"

The guide had already stood up, grabbed his military helmet, and was ready to return to the camp.

"Visiting time is over; Tâm must return to the camp now."

Both Tâm and Loan stood up from the table. He turned to Loan and urged:

"You should head home before it gets too late. Traveling through the forest at night isn't safe!"

Loan stood there, heartbroken, watching Tâm walk away with the guide toward the camp gate. She called after him:

"Take care of yourself. I'll come back to visit you in a few weeks!"

After Tâm had disappeared into the camp, Loan finally sat down at the bamboo chair again, overwhelmed with sadness. She had truly loved Tâm since they first met, then sacrificed and waited through the long years, season after season passing by. She met him when they were still students, in the innocence of her youth, like a blossoming flower, her heart fluttering with the excitement of spring. Eighteen...nineteen...twenty...twenty-one... Eventually, she no longer had the courage to count the passing Tet holidays and the blooming seasons of the mango trees in the garden. Today, Loan had just turned thirty, over a decade of waiting, a decade too long in one's life to test and measure the value of love.

While deep in thought, feeling sorry for her lover, who had to endure the tortures of prison life, she was startled by a light tap on her shoulder:

"Loan! Let's go home; the motorbike taxi drivers are waiting outside!"

"Ah, sister Hai, I'm just getting ready to leave!"

Time passed by, and every month or so, Loan would prepare various food, clothes, and other essentials to bring up to visit Tâm.

The name "Kà Tum" no longer seemed strange or difficult to her; it had almost become a familiar, endearing, and memorable name. A few vendors at Tây Ninh market and the motorbike taxi driver had grown accustomed to seeing her. Each time Loan arrived, they welcomed her warmly, like a close friend. Sometimes they would give her a dozen mangoes or ripe star apples to bring into the camp to share with the inmates, and other times, the motorbike driver would let her pay whatever she could. They were very sympathetic to the families of the former South Vietnamese officers imprisoned there. These gestures and acts of kindness touched Loan deeply.

The love between Loan and Tâm gradually rekindled. The memories of their past, when she was still in school, their dates at the zoo or Tao Đàn Park in Saigon, slowly resurfaced in her mind. The brief moments they spent together during visits to the camp allowed Tâm and Loan to relive the dreamy days of their youth: "Your house at the beginning of the lane, mine at the end; our love connected by a winding path through the neighborhood..."

The image of Lệ, Tâm's beloved wife from the past, gradually faded into the background of time. In front of Loan, he never mentioned his past with Lệ. Tâm's state of mind was sometimes like that of a person reading a book—wanting to turn the page to a new chapter but hesitating, afraid to lose the emotions they had felt.

More than six months after Loan began visiting and supplying him regularly, Tâm's health and spirits gradually improved. Today, he and several other prisoners were called up by the communist cadre to be transferred to another camp. The changes no longer puzzled him. He prepared to write a letter to Loan, informing her of his new address so she wouldn't have to go up to "Kà Tum" for visits anymore.

About a week later, Loan received Tâm's letter notifying her of his transfer. She learned that he had been moved to a camp in Hóc Môn, near Saigon. There, he was no longer subjected to the harsh labor he had endured in previous years, perhaps because at that time, the Eastern European communist countries and their Soviet benefactor had begun to reevaluate their dictatorial regimes and their chronic poverty-stricken economies. The growing human rights movements in the free world forced the Hanoi communist party to reconsider, leading them to ease their oppressive grip on former South Vietnamese officers. Thanks to this, the prisoners received more frequent and relatively easier visits and supplies from their families.

After the Southern government fell to the communists, the economy here completely changed, focusing on state-owned enterprises run by incompetent officials lacking expertise in management and technical knowledge. They dragged the Southern economy, which in 1975 was on par with Taiwan, Thailand, and South Korea, down to the level of the poorest nations in the world.

For the first time in history, the people of the South had to eat "bo bo" (a type of coarse grain) mixed with moldy rice or cassava. The majority of the country's elite—scholars, intellectuals, technical experts, and economists—were suspected and severely mistreated by the communists. Those fortunate enough not to be imprisoned for their capitalist mindset could not survive under the oppressive and ignorant rule of the local communist government. Hundreds of thousands of people were forced to leave their ancestral graves and lifelong savings to flee the country in search of freedom in foreign lands.

From that time, the bank in Cà Mau was shut down by the communist authorities. Loan had to quit her job and return home to help her parents by selling fruit on the streets and opening a tutoring class for children in the neighborhood. This allowed her to save a little and visit Tâm regularly in prison.

Today, as usual, after finishing her class around four in the afternoon, Loan leisurely cycled home. When she reached the gate of the garden at the front of the house, she got off her bike and walked it inside. Suddenly, she was a bit surprised to see her mother step out of the house, seeming busy:

"Honey, Loan's back!"

Hearing her mother call out to her father in a tone slightly different from usual, she paused:

"Mom, do we have a guest?"

"Yes! Your father is talking with Uncle Tâm from the old Bàn Cờ neighborhood. He's just been released from the re-education camp and came down to visit us!"

This unexpected news almost left her in shock. She asked her mother again to be sure:

"Mom, are you talking about my friend Tâm?"

"Yes, that's right! Poor thing, he looks much more haggard than before. He's been asking about you all morning!"

Before her mother could finish speaking, Loan hurriedly interrupted, leading her bike inside:

"Let me go in!"

About six months later, a very simple wedding ceremony took place in the presence of close family members and a few neighbors. Loan's parents officially accepted Tâm as a son-in-law. The decade-long wait for Loan was finally rewarded, even though she knew and accepted that her husband had been through a previous failed marriage.

After the wedding, the couple moved into a small thatched house by a canal on the outskirts of the city. Loan continued to ride her bike to teach while Tâm worked as a bicycle tire repairman at the corner of the main road near their house. Despite the ongoing hardship and lack of resources, Loan and Tâm still felt happy and content with their new life.

This morning, on a Monday, Tâm was getting ready to set up his shop under a tamarind tree by the roadside—a basin of cold water, a pump for inflating tires, and a few old tools in a box. As he finished, the rain began to fall heavily. He quickly pulled on a raincoat made from coconut palm leaves to keep warm.

Leaning against the tamarind tree to avoid the rain, Tâm suddenly recalled memories from the past, so vivid as if they had just happened yesterday. He remembered visiting the White House in Washington D.C., swimming on the beaches of Florida, then flying heroically over enemy lines, getting wounded in the military hospital in Đà Nẵng, and strolling through Pleiku with his beloved. Then came the collapse of his world—the wife he had trusted to live with for the rest of his life was lost forever on the evacuation route. Before, his eyes now were only tears and prison. Tâm wiped the rain from his face and watched the raindrops, the perfectly round bubbles that quickly burst:

"Tâm, what are you doing looking at raindrops like a child?"

Startled as if waking from a dream, Tâm looked up to see Loan, who had just stopped her bicycle in front of his repair shop under the tamarind tree.

"Oh! Why are you home so early today?"

"It's Uncle Hai's death anniversary, so he told me to let the children off early today. How about we pack up and go home? There won't be any customers in the rain anyway. We could also take this chance to visit Uncle Tư in town and see if there's any news about Mỹ."

Tâm stepped closer to his wife, speaking just loud enough for her to hear:

"Don't you know?"

"What don't I know?"

"Mỹ has escaped by boat!"

Shocked, Loan stared at her husband:

"Oh...Oh! I've heard a lot of rumors about people escaping by boat, but you need to spend a lot on the boat owners and bribe the local police, and even then, there's no guarantee. Sometimes, people lose their money and end up in prison!"

"Yes, I've heard that too. Some people have the means to attempt it, but for us, we'll just take it as it comes. We're already cursed, so we're not afraid of being further cursed! Let the police do whatever they want; I'm no longer afraid of them."

"Don't say that! If they hear you, it could bring trouble. Let's head home early and go fishing for some small perch. It's drizzling like this; the perch will be biting like crazy!"

Hearing about fishing, Tâm suddenly became more cheerful and quickly packed up his things:

"Ah! My wife has a great idea! We'll catch some perch this afternoon, grill them, and mix them with basil and Vietnamese coriander, dipping them in fish sauce with boiled water spinach—perfect for a meal!"

Time passed slowly, and the water in the canal behind their house would rise and fall with the seasons. The crabs and snails clinging to the canal's banks gradually became scarce due to the ongoing poverty and scarcity. Anything edible was nearly extinct. The communist regime had turned the vast, fertile rice fields of the South into barren wastelands because no one cared to work for free anymore. A certificate of "Labor Hero" was no longer worth as much as a full bowl of rice!

The victory of North Vietnam over the South should have been celebrated by the entire nation as the unification of the territory to reconcile the people and work together to make the country prosperous. But, unfortunately, the North Vietnamese communist regime only wanted to rule the country according to rigid, inhumane Marxist-Leninist doctrines. Hundreds of thousands of former Southern military and government personnel were sent to prison, and millions of people abandoned their properties, sweat, and tears—years of savings—to risk their lives crossing the ocean in search of freedom anywhere that wasn't communist.

Chairman Mao Zedong of the Chinese Communist Party once said and disdainfully regarded intellectuals as being less useful than manure. That lowly thinking frightened those considered the elite, the intellectual assets of the nation. If they were needed, the communists would use them like "squeezing a lemon and discarding the peel." That's why they fled the country whenever they could, which modern Vietnamese history has proven beyond doubt.

Along with the rhythm of society's poverty, Tâm's bicycle repair shop under the tamarind tree also became less busy because hardly anyone could afford a bicycle anymore, so most people would walk to avoid the costs of repairs!

This afternoon, Tâm was busy fixing a broken chair when Uncle Tư, who had just arrived, saw him sitting with his back turned and called out:

"Tâm, what are you doing?"

Tâm turned around and smiled politely at Uncle Tư:

"How are you, Uncle? I'm just fixing this chair; one of its legs broke. Oh, did you hear any news about Mỹ?"

After straightening his pants, Uncle Tư sat down on a wooden box nearby:

"Yes, I just got some news! How are you and your wife doing lately?"

"Business is slow Uncle! But please, tell me the news about Mỹ!"

"After all these years, he finally got in touch with me. He told me he's settled in the U.S., but I forget which state. He said life is tough over there, and he also sent his regards to you and your wife."

As he spoke, Uncle Tư reached into the pocket of his bà ba shirt and pulled out a thick wad of cash:

"Oh, he sent this to you two. He said it's from the support of your Air Force friends over there, sent to help you start a small business. Count it to make sure it's all there, and then sign the receipt so I can send it back to him, and they'll be happy to know."

For a moment, Tâm was almost paralyzed with shock, staring at the wad of cash and then at Uncle Tư, unable to say a word. He had never imagined such a thing would happen. How could friends who had been separated for over a decade still remember him and send him money for help?

He recalled the years he spent in the communist re-education camps, where he encountered people who claimed to be on the same side as him but, driven by hunger and the grueling labor, would stoop to flattery and sycophancy toward the camp guards—some of whom were young enough to be their grandchildren—or become informants in exchange for extra food and lighter work. Thinking about it now, Tâm felt deeply ashamed that the communists had placed them alongside him and disrespected the dignity of South Vietnamese officers.

Noticing that Tâm hadn't taken the money yet, Uncle Tư asked:

"What's wrong? Why are you standing there frozen? Take the money and sign the receipt so I can get back home. I need to look after the little ones at home; I can't stay here long."

Almost as if waking from a dream, Tâm reached out and took the bundle of cash:

"Thank you, Uncle Tư."

"Do you want to count it now or later?"

"I'll sign the receipt now and give it to you, and please send our thanks to the friends over there who remembered us."

Uncle Tư pulled up his pants, stood up, and took the signed receipt from Tâm:

"Oh, I almost forgot. Mỹ mentioned that the U.S. has a program for former South Vietnamese officers who were imprisoned, allowing them to settle in the U.S. I heard they're processing applications in Saigon or Hanoi. Maybe you should look into it?"

"I'll check to see if it's true!"

"Alright, I'm heading home now, Tâm!"

"Thank you very much, Uncle Tư!"

After Uncle Tư disappeared behind a thatched hut, Tâm checked the money, totaling about two hundred and fifty U.S. dollars—a sum of money so large that he hadn't seen anything like it in ten years. He hid the bundle in a burlap bag used for storing bicycle inner tubes.

A light and uplifting feeling came over him. Life is often unpredictable; sometimes, you think that friends who shared the same boat with you would help and protect each other in times of trouble, while those who had the opportunity to rise above would forget their friends and their bond! But that wasn't the case with him during his years of imprisonment.

Thanks to the money his friends sent him, he and his wife had the chance to return to Saigon to visit a few friends and gather more accurate information. It had been a long time since he had left the communist prison camp, and today he and Loan finally had the chance to revisit their old city. The streets were still there, with the footprints of their youth echoing in their memories. The tall tamarind trees still stood, casting their shadows in the sunlight, but why did it feel so empty and lonely, like a stranger in a foreign land? Who had changed the names of the streets once named after heroes and warriors, the pride of the nation, to something filled with the stench of blood?

Loan and Tâm's Saigon of the past was filled with laughter, with bustling streets and crowded vehicles after school. But where had those Saigonese gone now? Leaving the streets and shops almost desolate and cold! Across the street, a few cyclo drivers were sluggishly hunched over, pedaling with bent backs, their cyclos loaded with state goods. Shouldn't they be cheerful, holding their heads high, proud that their country had advanced rapidly toward socialism?

Thinking about this, Tâm shook his head and sighed in frustration, quickly pulling Loan across the street toward Tao Đàn Park.

"Do you remember this place?"

Loan gently pinched his shoulder and shot him a sideways glance, half teasing, half scolding:

"Do you think you're the only one who remembers? Do you think everyone else has forgotten? You used to ask me to come here every afternoon, and now you pretend to ask?"

"Time sure flies, doesn't it? It's been over fifteen years, and we finally get to come back here. I remember you used to wear the white áo dài of Gia Long High School..."

Before Tâm could finish, Loan jumped in:

"And you were in your second lieutenant's uniform, with those silver wings on your Air Force jacket, looking so handsome and dashing!"

"Let's not dwell on it anymore; it only makes us sad. We're lucky to be alive and to have found each other again!"

Saigon wasn't too hot today, and occasionally, a gentle breeze from the tropics would rustle the leaves of the almond trees, creating a moving shadow as if waving hands were welcoming old friends back. The hedges of "bùm sụm" trees, once neatly trimmed, now looked withered and tattered. Where was the gardener now? Who had let the plants grow wild, uncared for? Had the gardener also fled with the so-called reactionaries, refusing to stay behind to line up for a few kilos of rice mixed with bo bo, instead of chasing after a slice of beefsteak? As they walked past a few stone benches in the park, Tâm suddenly reached out to stop his wife:

"Let's sit down and rest for a bit. We've been walking in silence, lost in the memories of a time worth remembering and cherishing."

It was only now that Loan snapped out of it:

"Ah! I remember this bench!"

Tâm looked at his wife, smiling secretively:

"If you remember, tell me about it."

Loan gazed into her husband's eyes with love and tenderness, playfully scolding:

"Who tricked me into looking up at the frangipani tree over there, then stole a kiss on my cheek?"

Tâm's open and carefree laughter made Loan blush slightly. She leaned her head against his chest, and Tâm kissed her lightly on the forehead:

"It's because you're so beautiful and charming! It wasn't my fault!"

Loan explained:

"You know, there were many people passing by, and I was so embarrassed. By the way, did it hurt when I pinched you back then?"

Tâm shook his head, playing the role of the tough guy:

"Not at all! A pinch from a beautiful woman never hurts!"

Loan seemed even more mischievous:

"Well, let me pinch you again now and see if it hurts!"

Tâm quickly raised his hands in mock defense:

"Oh, no...no!"

The couple burst into laughter, feeling completely at ease. At that moment, it was as if time and the memories of their youth had returned to them.

After a few days of staying in Saigon to gather information, today Tâm and Loan had to return to Phú Lâm to catch a bus back to Cà Mau. As the pedicab carrying the couple pulled up at the bus station, it began to rain. The harrowing image from nine years ago at the Pleiku bus station flashed back in Tâm's mind, as if it had happened just yesterday. It was also a rainy day, at a bus station, with muddy puddles everywhere. He had bid farewell to his first wife, Lệ, and her mother as they evacuated to Saigon on the last bus, not knowing that it would be their final farewell.

Noticing her husband walking silently with his head down, looking unusually distant, Loan quickly followed behind, grabbing the hem of his shirt to stop him:

"Why are you walking so fast? Wait for me!"

"Oh...Oh! I forgot."

"You seem so sad. Is there something bothering you?"

Tâm remained silent, not saying a word, as he took her hand and led her toward the bus ticket counter nearby:

"Let's buy our tickets, or we might not get a seat."

The express bus left Phú Lâm station, following the highway toward the Mekong Delta, passing through flooded rice fields. On both sides of the road, newly planted rice seedlings, still yellow and weak, struggled to take root in the nutrient-poor soil. A few blackwater buffalo grazed leisurely by the roadside.

Tâm remained silent, staring out the window, lost in memories of the early days when he first met Lệ. They used to ride a Honda motorcycle through the empty fields on the outskirts of Pleiku, visiting her coffee plantation on weekends. The warmth of her embrace, the love emanating from her full chest, made him feel intoxicated, like the rice wine of the local tribes.

Tâm often called her the "phà ca" mountain girl, his own queen of the highlands. They fell in love, got married, and hoped to stay together for the rest of their lives. But who could have guessed that in this life, there would be moments of union followed by separation, like the raindrops and bubbles he had once mentioned to Lệ.

Seeing her husband deep in thought, staring out the window since they boarded the bus, without saying a word to her, Loan began to feel uneasy. She gently rubbed his shoulder:

"What are you looking at out there? You haven't said anything since this morning!"

Tâm turned around, putting his arm around her shoulder, trying to hide the silent turmoil in his heart:

"Oh, I was just admiring the scenery, nothing more! The breeze is so cool it makes me sleepy."

"And here I was thinking you were upset about something!"

Back in Cà Mau, after a few months of hesitation, partly due to his wife's family and also because of their fear of falling into the communists' treacherous trap, Tâm didn't dare to submit an application to emigrate to the U.S. right away. Besides, they were living in such poverty—how could they survive if they went to the U.S.?

The rumors about the U.S. government or friends over there helping out were just unconfirmed hearsay. But if they stayed here, his life would wither away under the old tamarind tree, where he made a living repairing bicycle tires, constantly harassed by the local police.

This evening, after dinner, Tâm noticed that his wife seemed troubled. He walked over, took her hand, and pulled her to sit with him on the hammock strung between two pillars of the house:

"Are the kids being too naughty? You don't look happy."

"It's not that…. I'm going to have to stop teaching."

A hint of surprise appeared on Tâm's face:

"But you said you love teaching!"

"Yes, I do! But this morning, the local police came and demanded all sorts of documents and permits. They're forcing me to close the class by the end of the month."

Another misfortune was about to befall them. It was thanks to Loan's regular teaching income that they could scrape by day after day. Tâm's work was inconsistent; there was no way they could survive on it alone.

Tâm gently rubbed her back to comfort her:

"Don't worry! We'll find another way to make a living. I'll teach you how to repair and pump bicycle tires so you can help me."

"And what will you do?"

"I'll rent a cyclo and pedal around delivering state goods or transporting goods for people. I hear it's a decent way to make a living!"

Loan quickly turned to look him in the face:

"Oh my God! With your frail body, how could you possibly pedal and pull that load?"

Tâm wrapped his arms tightly around her, smiling as he tried to reassure her:

"Are you underestimating me? No matter how weak, I'm still a former Air Force captain, remember? I might look skinny, but I'm strong!"

Though he spoke to comfort Loan, Tâm knew himself better than anyone else. After nearly seven years of imprisonment in the wild forests, working harder than an ox every day for just two or three handfuls of rice mixed with cassava, even an iron body would deteriorate over time, let alone human flesh and bones!

Loan's eyes suddenly brightened as if she had just discovered something new. She quickly glanced toward the front door, checking to see if anyone was passing by or paying attention:

"How about you go to the town and submit an application? We're already cursed, so we have nothing left to fear! I heard that some people have already submitted theirs!"

After a moment of silent thought, Tâm turned to look at his wife:

"If you agree, then I'm not afraid anymore! Tomorrow morning, I'll go to town to get the application and see how the police handle it."

The rooster from the neighboring garden flapped its wings noisily and crowed at the break of dawn. Outside, the sky was just beginning to lighten. Loan had been up for a while, preparing a few essential items and some clothes, packing them into the only suitcase that her mother had given her when the couple moved out on their own. Once everything was in order, she came over to lift the mosquito net and wake Tâm:

"It's morning, dear! Get up and change. Afterward, let's visit Mom and Dad one last time before we catch the bus."

Startled, Tâm clumsily crawled out of bed, stretched, and, still half-asleep, asked his wife:

"What time is it?"

"It's nearly six in the morning! Go wash your face and get dressed; it's time to go!"

The bus from Cà Mau to Saigon was less crowded than usual today. Tâm and Loan sat quietly in the front row. Perhaps the couple was deeply absorbed in thoughts about the journey ahead. By all accounts, Tâm should have been happy since soon he would be able to see the light of freedom and enjoy the respect for human rights in a progressive democratic country. But his current state of mind was that of someone who had lost too much.

His homeland was still there, his people were still there, the villages and neighbors remained, but why did he feel as if they were a thousand miles away? He suddenly remembered his dear late wife, with her lonely grave in the outskirts of the province, and felt like a heartless man. Over the past ten years, he hadn't had a single fortunate opportunity to return to Pleiku to visit her grave. He felt as though he had betrayed Lệ. Silently, he prayed that her soul would forgive him given the circumstances today.

To him, the Vietnamese communists were not true nationalists in the purest sense; they were not like the warriors of Hưng Đạo Vương or Quang Trung Nguyễn Huệ of the past. Instead, they were mercenary soldiers of Marx, Lenin, and Chairman Mao, who had come to occupy and impose their colonial rule over this homeland. The communists had labeled him with the derogatory term "Mỹ Ngụy."

In truth, though he was just a small individual, Tâm never once thought of accepting or volunteering to be a pawn for any foreign power. He didn't like the presence of any foreign military on his ancestors' land. He was sure that most of the people in the South, like him, had a very simple mindset: when the homeland calls, young men must stand up to answer.

In reality, today's wars—and perhaps those in the future—are not simple conflicts that can be resolved internally between two parties or on a local level as they were many centuries ago. The world has become much smaller due to the extraordinary advancements in communications, transportation, and modern weaponry.

Small, underdeveloped countries can no longer live in isolation as they did in the past. Whether they want to or not, they are forced to rely to some extent on more powerful nations. The terms used in new strategies are often beautifully and eloquently gilded with words like allies, comrades, or, more gloriously, freedom fighters, and so on.

So, we must ask: "Who is a puppet, and who is not? Who is a lackey, and who is a patriot?" Often, these are just the words of the victors, imposed upon those who were less fortunate.

Tâm's true feelings today were those of a citizen in a subjugated nation. Whether directly or indirectly, he and his wife still felt as though they were being expelled from the place where they were born and raised.

For over two years now, Tâm and Loan had been living happily in a small house in the southern part of Seattle, in the Pacific Northwest of the United States. Autumn had now arrived, and the weather outside was starting to get cold. The trees lining the streets were beginning to change their colors, and a light rain was falling, just enough to wet the hair of pedestrians on the road in front of the park.

Since arriving here, Loan had been doing piecework, sewing items at home due to her limited English skills. Tâm, who had some English from before, was fortunate to find a job sorting and delivering mail at a post office in a nearby district. Although they were getting older and had no children, their life was relatively comfortable.

Today, being the weekend, Tâm woke up earlier than usual. He was sitting in a chair by the window, looking out at the front of the house. On the table was a hot cup of coffee, with steam still rising from it. Suddenly, he remembered his homeland, recalled Ban Mê Thuột, Pleiku, Kontum, Đà

Lạt... during those days when the drizzle seemed to fall endlessly. At that moment, Loan had also woken up and came over to sit beside him:

"The rain here looks just as gloomy as it does back home, doesn't it?"

Tâm pulled his wife close:

"Yes, rainy days like this remind me... " Then, as if he suddenly remembered something, he stopped mid-sentence and changed the subject:

"Alright! Let's wait until the rain stops, and then we'll go out for some pho, okay?"

Loan looked at him, bewildered, her eyes wide with surprise:

"Huh? Why did you suddenly switch topics? You were saying the rain reminded you of something?"

Tâm tried to dodge her question, but Loan was determined to find out what he was thinking. Knowing he couldn't keep it a secret any longer, he lowered his voice and began to tell her about his past love with Lệ, filled with memories from when he was still in Pleiku.

"My dear, I'm going to tell you about something that happened in my life over twenty years ago, something I've kept hidden because I didn't want to make you sad."

Loan rested her head on his shoulder:

"I won't be sad if it's about the memories you shared with Lệ."

"Yes, every time it drizzles like this, it reminds me of those old memories. The day I met Lệ, and the day she left this world forever, both happened on days like this. Fate can be so cruel! She was always afraid that her life would be like a raindrop, here one moment and gone the next, and in the end, she couldn't escape that fate."

"Please don't talk about it anymore; it's too sad for her. Now I understand why I sometimes catch you in moments of solitary sadness. I hope that in this new life, with me and good friends around, you'll gradually forget the painful past!" Loan suddenly changed her tone and switched to a different topic.

"Oh! Isn't there a meeting with your friends from the old unit this afternoon?"

"Yes, a few of us from the old unit want to get together to catch up and also to plan a small fundraiser to send some money back to Vietnam to help friends who are struggling."

"I've heard that there are dozens of associations or political groups here. Some people take advantage of these groups and fight fiercely over them. Some individuals are completely clueless, lacking both talent and virtue, but they're vain and competitive, often boasting about their expertise in organizing. They bribe others to spread lies, using dirty tactics to cover up their deceitful actions and lure in the gullible. I think you should stay away from such people."

"I know! Being a soldier and losing the war is already a shame. Being indirectly or directly driven out of our homeland, away from our ancestors' graves, is a second shame. Here in the U.S., some politicians, including former U.S. Secretary of Defense Robert McNamara, have made malicious statements, unfairly blaming and undervaluing the fighting spirit of the South Vietnamese military. Now we're like renters in someone else's home. Sometimes the homeowner says whatever they want, and the tenant just has to turn a deaf ear, staying silent and resigned—this is the third shame! The fourth shame is seeing our own people fighting each other in a foreign land for personal or factional interests, or due to extreme, rash thinking, slandering and attacking without cause. They intentionally or unintentionally sow division within the community, making us a laughingstock to the natives!"

"As you can see, African Americans here are often thought to be poor and less educated compared to other ethnic groups. However, if you look closely, they have a strong sense of unity and solidarity, standing up for each other more than any other group in this country. Thanks to that unity, their community has a powerful voice in American politics and economics today."

Hearing him say this, Loan interjected:

"Lately, the newspapers and television have been talking a lot about human rights abuses in Communist China and the human rights activist Harry Wu. What do you think about that?"

Tâm took a slow sip of his coffee before calmly responding to his wife's question:

"We haven't had the chance to discuss current events in a long time— or rather, we didn't dare to back home, where the communist police seemed to have ears everywhere. Even a small slip of the tongue could easily land someone in jail."

Loan jumped in again:

"That's precisely why we left the country. If they had allowed even a little bit of freedom, we could have stayed back home, working the land with our parents. That would've been a good life!"

Tâm nodded in agreement with his wife:

"But let me tell you my thoughts and reasoning. If you disagree, let me know, okay?"

Loan playfully pinched his thigh:

"You always preface things so cautiously—just say it! If I disagree, I'll let you know right away. In this civilized country, you have the freedom to speak your mind, not like under the communist regime where you have to be careful!"

Tâm continued, more relaxed:

"In principle, I agree with Harry Wu. We need people like him to let the world know about the dictatorship and human rights abuses of the Chinese Communist government. Human rights are as essential as food and drink for survival. Without human rights, people become slaves—or worse, they live like animals under the control of an individual or a dictatorial party. No one can deny that.

However, in the realm of international politics, things aren't always as simple as they seem. Many countries use human rights as a tool to criticize and distract international attention while pursuing their own insidious agendas that benefit their nation. As you know, today's world is not like it was decades ago, where powerful nations would send troops to invade and colonize weaker nations, competing for economic benefits and exploiting the natural resources of the conquered country.

Now, they use more cunning strategies. One of the new strategies is to create these beautifully crafted, heroic-sounding terms like 'freedom fighters' or 'human rights warriors,' among others. But in reality, their goal is to incite dissatisfaction, create internal turmoil, and overthrow governments to install puppets that benefit them in any country that refuses to bow down or poses a threat to their economic interests."

Hearing this, Loan sighed in frustration:

"So the world isn't fair at all, is it?"

Tâm looked at his wife, agreeing, then gave a wry smile:

"If everyone thought as honestly as you, this world would be a paradise! The principle of 'the strong do what they can, the weak suffer what they must, the cunning live with honor, the foolish live in disgrace as beasts of burden' has existed since the dawn of humanity, and it will likely continue until this world explodes. Only then might it end! The important thing is to remain calm, think critically, and discern what's real and what's false, so we don't fall into the trap of blindly following unjust causes."

As Tâm spoke, Loan grew weary of the topic. She stood up, shifting to a more enjoyable and practical subject:

"Listening to all that is exhausting! Now that the rain has stopped, let's get dressed and go out for breakfast. While we're out, I'll stop by the store to pick up some vegetables and bean sprouts so I can make sour shrimp soup for dinner tonight."

Tâm stood up, carrying his coffee as he followed his wife to the kitchen:

"Okay! Shrimp sour soup with caramelized fish sounds perfect!"

This afternoon, the Seattle sky suddenly became clearer. The soft golden rays of autumn seemed to be bidding a wistful farewell to the glorious summer that had passed. The sun wasn't warm, but just enough to make the neighborhood a little less dreary as summer faded and autumn's chill began to take hold.

Tâm and I sat watching the coffee slowly drip into the glass in front of us. I had met Tâm during a troop deployment mission more than twenty-five years ago in Pleiku.

Today, by chance, we met again here and sat together as he recounted his life story—a story that felt like a tragedy, a tragedy written in his own tears and the tears of many Vietnamese who had either left or stayed behind, sharing similar fates as his.

Tâm, I'm writing this not just for you but for many others like you, for the soldiers of the Republic of Vietnam who, despite the agony of defeat, still proudly fulfilled their duty as men during a time of national crisis and who have always remained loyal and steadfast to their people. I also write to honor the gentle Vietnamese women of our homeland, who have always been patient, enduring sacrifices for their husbands and children and answering the call of the nation. - Nguyễn Văn Ba ("Nam")

CHAPTER 10

Wings of Vietnam

Original Title: "Những Cánh Chim Việt"

The season had shifted to autumn over a month ago. With persistent, heavy rains that lasted all day, tiny raindrops drifted softly in the gusts of wind.

The large "Alder" tree in front of the house, which typically had lush foliage during the summer, had shed its last yellow leaves the week before. The little squirrels with their beautiful long gray tails, curled up like feather dusters, charmingly ran around the yard, seemingly sensing the approaching cold of winter, and had since disappeared.

Wrapped in a thick warm coat, Tuấn sat on the rocking chair on the porch. The steaming cup of coffee beside him, along with the morning cigarette in his hand, made him feel as though time had slowed down. Since moving here, most of the bad habits associated with the four pleasures had been temporarily discarded—not just by him, but by almost everyone he knew, who seemed to have fallen into the same unique routine.

Tuấn no longer smoked at work or inside the house as he used to in Vietnam. However, on weekends when friends came over to chat, he liked to enjoy a few cigarettes to sweeten the conversation. Perhaps the old saying of our ancestors was very accurate: "A cigarette and a betel nut are the beginning of a conversation." With a hot cup of coffee on a cool autumn morning like today, or a cold beer with foam in the scorching summer, a cigarette between the fingers seems to complete something essential in life.

Tuấn did not intend to glorify or promote the image of smoking, thinking it might make a man appear more elegant or romantic, because his wife often nagged him about lung disease, the stink of smoke on clothes, and the house, and so on. But old habits die hard, and her words were like water off a duck's back—he would sneak out to have a puff to avoid the nagging.

After finishing the last sip of coffee, Tuấn stood up, intending to turn around and open the door to go back inside when he heard the sound of a car turning toward his house. Tuấn paused and looked out toward the alley, squinting through the drizzle to see if the visitor was familiar or a stranger. Before he could clearly identify the person, the car had already stopped in front of the garage.

The driver's side door opened, and a man in his fifties stepped out. Tuấn easily recognized the visitor. Overjoyed, he called out:

"Hey, Trung! Today you're driving a new car—I thought it was some stranger! Last Saturday, my wife prepared some snacks waiting for you, but you never showed up."

"Oh! Last week I was busy with some errands in Seattle. When I got home, my wife told me you had called. Never mind, let's put that aside. Today I have a new friend with me—I'm not sure if you remember him."

Tuấn was still unsure when the passenger side door opened. A man around the same age stepped out and nodded in greeting. Just as Tuấn was about to step outside to greet the guest, he froze in surprise, unable to say a word. Trung approached and patted his shoulder, hinting at something:

"Do you recognize him?"

Tuấn didn't answer Trung's question directly. Instead, he continued to gaze at the guest, mumbling:

"Is that really Tâm?"

The guest stepped forward, seemingly able to read the surprise in Tuấn's eyes:

"Yes, it's me, Tâm!"

Overwhelmed with emotion, Tuấn extended his arms to embrace Tâm:

"Oh my God! It's been over twenty years since we last met...!"

To respect the emotional and surprising reunion, Trung quietly stepped inside the porch.

Hearing the sound of the car stopping and the unfamiliar voices outside, Phượng opened the door to see the three men still chatting excitedly. She quickly called out:

"Tuấn! Why don't you invite the gentlemen inside? It's cold out there!"

As if snapping back to reality, Tuấn quickly urged:

"I was so happy to see Tâm that I forgot about the cold! Let's go inside!"

Turning to Phượng, who was waiting by the door, Tuấn introduced his wife:

"You already know Trung, but do you recognize this gentleman?"

Phượng widened her eyes, trying to recall if she had met this guest before. She shook her head apologetically:

"You have so many friends, I can't remember them all!"

Tuấn smiled mischievously at his wife:

"Of course, you wouldn't know! This is Tâm, an old acquaintance of ours. Do you remember the story I once told you about my cousin's younger sister, Lệ, in Pleiku?"

As Tuấn mentioned this, Phượng's face lit up, and she eagerly interrupted:

"Oh! Is this the Tâm who was Lệ's husband?"

Tâm had been standing quietly since Phượng opened the door. He now stepped forward and bowed slightly:

"Yes, that's me, ma'am."

Phượng's expression suddenly turned sad, and she bowed her head, sighing:

"Poor Lệ..."

Tuấn didn't want to dwell on the past sadness between Tâm and Lệ, so he intentionally changed the subject:

"It's too cold out here! Let's go inside and have a little drink to celebrate Tâm's arrival in the States!"

###

Tuấn was a former Captain in the Air Force of the Republic of Vietnam. He was assigned as an Air Liaison Officer to the II Corps Tactical Zone Headquarters, stationed in Pleiku. It was at this new unit that he met Captain Lê Văn Tâm, who had been transferred from the 2nd Air Division in Nha Trang to work together in the joint air operations office, coordinating between the army and the air force.

At that time, both Tâm and Tuấn were still single, so the two friends lived together in the bachelor officers' quarters. Occasionally, on days when the persistent rain in Pleiku kept them indoors, Tâm would recount to Tuấn the bittersweet love story between him and Loan, as well as the hardships of his orphaned childhood. The more they lived together and got to know each other, the closer they became, like brothers. One such bond led to an occasion during the quiet New Year's holiday in the dreary town, where Tuấn introduced Tâm to his cousin's younger sister, Lệ, who lived in Pleiku. That spring in the border town of Pleiku seemed warmer and more joyful than usual. The bright yellow apricot blossoms appeared to bloom in celebration of a new love that had just begun.

Spring passed, summer arrived, and then autumn came. On some weekends, during the long, drizzly rains of the highlands, people would see a couple walking closely together under the eaves of the street. She would often stop to watch the rain bubbles floating on the surface of the water, and a gentle sadness would suddenly come over her as those bubbles would soon burst and disappear into the flow of water.

Fate is the law of God that all living beings must implicitly accept as an undeniable reality that cannot be explained or understood. No matter how intelligent or exceptionally talented humanity may be, each individual must ultimately submit to the divine arrangement of fate.

###

After President Nguyễn Văn Thiệu ordered the abandonment of strategic positions in the provinces of Ban Mê Thuột, Kontum, and Pleiku, Tâm was transferred to the 4th Air Division (Cần Thơ), while Tuấn was reassigned to the Biên Hòa Air Base.

The day they parted ways when leaving Pleiku was also the final farewell between the two.

About six months later, Tâm was imprisoned by the communist forces to pay the price for being a defeated national soldier.

More fortunate, Tuấn and his family managed to leave Saigon on the afternoon of April 30, 1975, aboard a naval ship at the Bạch Đằng River port. After many days of transition during their time as refugees, his family eventually settled in Portland, Oregon.

In the early years, Tuấn made inquiries about Tâm, but none of the old friends he was able to contact knew whether Tâm was still alive or had perished after the communists took control of the South.

Over time, the name Lê Văn Tâm gradually faded from his memory, only occasionally brought up as part of the sad love stories that had occurred within Tuấn's extended family.

###

After everyone had settled into the living room, Phượng turned to her husband and whispered:

"They probably haven't had breakfast yet. Can you help me by going to the garage and getting some packets of instant noodles from the box on top of the refrigerator? I'll make some shrimp noodles for them, and then I'll go to the market to buy beef to make 'shaking beef' with bread for later."

Before Phượng could finish her sentence, Trung stepped in and said:

"Hey, Phượng! Don't go through all that trouble. I actually planned to come early to invite you and Tuấn, along with Tâm, out to the street to have some pho."

Phượng didn't seem to agree:

"Are you saying my cooking isn't good enough, Trung?"

Trung shook his head, trying to explain:

"That's not what I meant at all! Everyone here knows that you're a great cook, especially when you make appetizers—they're the best! I just didn't want to trouble you."

"You guys only come over occasionally; it's no trouble at all. Besides, today we have Tâm here, so you should stay and chat. Going out in the cold rain will ruin the fun."

As if she suddenly remembered something, Phượng added:

"Oh! I forgot to ask, why didn't you bring your wife, Thảo, with you today, Trung? Why are you going 'solo'?"

"My wife was a bit busy at home today, so I came to bring Tâm over to visit you both."

At that moment, Tuấn came in, carrying the box of noodles, and placed it on the table. Phượng turned and went to the kitchen to prepare breakfast.

Tuấn and Trung had only known each other for about fifteen years, and their friendship began by chance at a mutual friend's house.

Trung had escaped from a communist prison after being transferred from the North back to the South during the Vietnam War. He later fled by boat and made it to the United States.

When Tuấn learned that Trung was a former Major and L-19 reconnaissance pilot in the Air Force, the same branch as him, he frequently took his family to visit Trung and help him get settled in the early days. From then on, they became close friends.

After handing the box of noodles to Phượng, Tuấn returned to the living room and, with a slightly puzzled expression, asked Trung:

"Hey, Trung, I'm a bit confused. How did you meet Tâm? And how did you know that Tâm had a connection to my family and bring him here?"

Trung didn't answer right away; instead, he smiled as if keeping a secret:

"What's so hard about that? Don't you remember? The world is round like a ball. You walk around, and eventually, you meet at some point, just like ants circling the rim of a bowl!"

Tuấn was a bit annoyed by Trung's roundabout way of starting the conversation:

"I've heard your 'ant circling' philosophy before, but it's dragging on too much. Just get straight to the point already!"

Trung, still maintaining his calm demeanor, laughed heartily:

"Slow down, man! Why are you in such a rush, like the girls in the old Cầu Hàng neighborhood? My wife loves me because of my easygoing nature. But if I tell the story, it won't be as interesting. Let Tâm tell it; he'll do a better job."

Trung then turned to Tâm and said:

"Go ahead and tell your brother-in-law. He's eager to know!"

Tâm, who had been sitting quietly, finally spoke up:

"Trung just wants to tease you, Tuấn! The truth isn't that complicated. I've known Trung since our school days in the Bàn Cờ neighborhood of Saigon. It was Trung who introduced me to the Air Force pilot officer training program in Nha Trang."

"And after that, as you know, President Thiệu ordered the tactical withdrawal from Pleiku. I was transferred to the 4th Air Division in Cần Thơ, while you were reassigned to the 3rd Air Division in Biên Hòa. Since then, we lost contact. A few months later, the situation changed rapidly, and the country unfortunately fell into communist hands. The entire ARVN dissolved overnight like foam on water. Trung and I happened to meet again under bitter circumstances in a communist prison. We were in the same camp, where they classified us by rank. About a month later, they moved Trung to another place; I later found out they took him to the North, while I remained in the South. Around that time, I tried to inquire about you and Uncle and Aunt Hai, and I was told that your family had left long ago. Later, I came to the United States under the H.O. humanitarian program and have been renting a place in the southern part of Seattle for over four years now. Just yesterday afternoon, by chance, I took my wife grocery shopping and ran into Trung, who was also shopping with his wife in Seattle."

As Tuấn listened intently to Tâm recount the old story, his eyes suddenly widened in surprise, and he interrupted the conversation:

"Wait! Tâm, you're married?"

"Yes, I'm married now," Tâm replied.

He paused for a moment, lowering his head as if trying to hide a sad memory resurfacing from the past. Tuấn, more than anyone, understood the beautiful love story between Lệ and Tâm during their time in a poor border province.

Tuấn had also witnessed a horrifying scene on the road between Pleiku and Ban Mê Thuột—a mine exploded, flipping a bus onto its roof, its wheels still smoking. Many bodies were laid out on the road, including a young woman who had just died, her eyes still wide open as if in regret or waiting for a loving hand to gently close them so she could peacefully depart. Tâm arrived just in time, and Lệ finally closed her eyes to say goodbye to everyone. She ascended, higher and higher, like a cloud drifting toward an unknown sky, where surely there would be no war and no human hatred.

After a moment of emotion, Tâm composed himself and looked directly at Tuấn:

"Do you still remember Loan?"

Still puzzled by the news of Tâm's marriage, Tuấn was caught off guard by Tâm's question:

"Loan? Which Loan?"

"The Loan whose photo I used to show you when we were living together in the bachelor quarters in Pleiku!"

As if suddenly remembering events from over two decades ago, when they were young Air Force officers assigned to II Corps and lived together in the camp, Tuấn responded:

"Oh! Oh! I remember now! Loan had long, silky black hair and beautiful dove-like eyes. She was the neighbor girl you mentioned, right?"

"Your memory is still pretty sharp!" Tâm replied.

"But you said back then that she had moved away, lost contact, and you couldn't find her anymore, didn't you?"

Tâm's face lit up with happiness as he turned to Trung, who was sitting nearby:

"You were right, Trung. The world may be big, but you keep circling around, and you'll meet again. It proves your 'ant circling the rim of a bowl' philosophy is a hundred percent accurate!"

Seeing Tâm side with him, Trung chimed in:

"See, Tuấn? I'm never wrong! Tâm is my witness."

Knowing Trung's love for jokes and teasing, especially now that he had an ally, Tuấn pretended to concede:

"Alright, 'senior officer,' I admit defeat to your superior 'ant circling' philosophy! Now, please shut up and let Tâm talk. Stop jumping in and spoiling everything!"

Tâm continued the story about himself and Loan, explaining how a series of coincidences led to them meeting again and, ultimately, how fate brought them together as husband and wife.

Although busy cooking in the kitchen, Phượng was still listening to the conversation between the three men in the living room next door. When Tâm finished recounting the difficult journey he and Loan had been through, from their first meeting in their school days to the many years of suffering and sorrow that followed, Phượng stepped into the living room and sat down next to her husband:

"Tâm's story is so sad, it made me want to cry!"

Tâm turned to Phượng:

"I'm sorry for making you sad! I didn't want to bring up old memories, but since Tuấn and Trung, who are like family, wanted to know, I shared my personal story with them."

Sensing the room had become quiet and somber, Tuấn decided to change the subject:

"Hey! Have you finished cooking the noodles? We're hungry here."

"They're ready! I've set them on the table, still hot. Now, come join us for breakfast. Oh! By the way, Tâm mentioned earlier that Loan is currently at home with Thảo, right?"

"Yes, that's right, Phượng. The reason I didn't bring Loan here to visit you and Tuấn is that she requested that I come alone first to have some freedom. She'll come visit later."

Tuấn gently patted his wife's shoulder:

"How about we go to Trung's house this afternoon to visit them and get to know Loan?"

"I agree, but we should ask Trung and Tâm if it's okay."

Tâm remained silent, but his glance at Trung indicated there would be no issues for him or Loan. Sensing his friend's intentions, Trung cheerfully said:

"You're both very welcome to visit. I'll call my wife and let her know to prepare."

Phượng playfully teased:

"Trung, you're making it seem like we're strangers, having to notify your wife in advance, like preparing for a presidential visit!"

"No, no, it's not like that! I just want to remind her to pick up some cilantro, okra, and fresh fish from the market so Loan can make sour fish soup and caramelized fish in a clay pot for you two to enjoy some authentic Cà Mau dishes this evening!"

"You're going to trouble them again!"

Hearing about sour fish soup and caramelized fish, Tuấn's face lit up with excitement, and he winked at Trung:

"Stop being difficult! Don't make Trung change his mind now!"

Pretending to reassure his wife, Tuấn then turned to Trung and whispered:

"Hey! I love that dish. Call home and tell them to get ready. I've got a brand-new bottle of Hennessy I haven't opened yet. I'll bring it over to your place this evening, and the three of us can drink and unwind."

Cà Mau has long been famous as a source of fresh fish and crabs in southern Vietnam. The jars of fragrant fish sauce made from snakehead fish after the draining of ponds, creeks, or fields have become one of the well-known specialties in Vietnam.

The girls of Cà Mau, with their long, silky black hair, graceful sun-kissed skin, and gentle demeanor, dressed in áo bà ba and tilted conical hats, hiding their half-smiles, have captivated and enchanted many city boys. Often, Tâm's old friends, knowing he married a girl from Cà Mau, would tease him:

"Cà Mau is easy to go but hard to return,

Boys go with a wife, girls return with a child."

Loan was born and raised in a well-off real estate business family in Saigon. As an only child, she was dearly loved and spoiled by her parents. Her mother, a smart and exemplary woman in the family, did not let her affection for her daughter detract from her teaching Loan the virtues of "công-dung-ngôn-hạnh" (work, appearance, speech, and conduct), which she believed were important virtues, especially for women.

Her mother's view on being a wife and mother was simple: to her husband, she was a companion, a comfort, an encouragement, and a support whenever needed. To her child, she was like a towering tree providing much-needed shade throughout life.

Aside from her regular work at school or the office, Loan often helped her mother with cooking and various household chores. So after Loan and Tâm got married and moved out on their own, she, like her mother when she still lived with the family, paid close attention to her husband's favorite meals and ensured he had restful sleep after tiring workdays. Tâm had noticed this many times and deeply cherished his wife, silently thanking heaven and earth for granting him such an intelligent, kind, and virtuous partner.

After a delicious dinner cooked by Loan herself, with a pot of sour soup made from fresh snakehead fish heads, fragrant with rice paddy herbs and basil, along with a pot of caramelized catfish in pepper, and a bowl of fish sauce with chilies—everything appeared simple and rustic, yet the art of cooking and seasoning the soup and the meat was nearly perfect, everyone praised it heartily.

Tuấn was the first to speak:

"Loan, back in the day, when I was still in the military, I would occasionally stop in Rạch Giá for lunch during operations because it was rumored that the sour snakehead fish soup there was the best. But now, I see that the sour fish soup and caramelized fish from Cà Mau are truly number one, and Rạch Giá has to take second place!"

Tuấn's flattering words made everyone laugh heartily. Phượng, sitting next to him, chimed in:

"Tuấn is right, Loan. He almost had a wife in Rạch Giá, but when he returned to Biên Hòa, he married me because he couldn't resist the sweetness of Biên Hòa pomelos and forgot his way back!"

Tuấn playfully patted his wife's shoulder:

"Hey, you can say that again. Was it sweet pomelos, or did someone eat a sour one and end up with a numb tongue?"

Then he turned to Trung:

"Hey, Trung! Tâm fell for the sour fish soup and caramelized fish from Cà Mau, I was seduced by the pomelos of Biên Hòa, so what got you?"

Trung pretended to think, then looked at his wife:

"Do you remember what got me, honey?"

Thảo looked at Trung and gave him a sideways glance:

"Was it something that 'got' you, or something you 'got'? Those two verbs have different applications, you know!"

Tâm, who had been sitting quietly, noticed that Trung was about to get into trouble, so he interjected to help his friend:

"Trung was almost right, Thảo. See, I 'got' thrown into prison, and that's how I 'met' my wife. If I hadn't been imprisoned, who knows what I would have 'gotten' in the future?"

Everyone burst out laughing at Tâm's light-hearted reasoning. Thảo joined in the laughter and shook her head:

"You Air Force guys have a way with words; you can talk your way out of anything."

Tâm didn't miss the opportunity and continued while Thảo was still pondering:

"Everyone here knows each other's secrets, so what are you still hiding, Thảo? Share your story to make it fair!"

At this point, Phượng and Loan also started teasing Trung:

"See, Trung? Loan got Tâm with sour fish soup and caramelized fish, and I hooked Tuấn with Biên Hòa pomelos. You two must have had a very romantic relationship, so why are you still keeping it a secret from us?"

Realizing they were being cornered, Thảo finally gave in:

"Alright, alright, I surrender. If we had had a golden time to get to know each other before the war, like you all did, that would have been the best. But Trung and I met during the hardships of escaping by boat, with nothing to spare."

The truth is that after the communists transferred prisoners from the North back to the South, during a moment of negligence by the communist guards, Trung managed to escape from the group of prisoners. He hid and eventually made his way back to his family home in the Bàn Cờ neighborhood of Saigon. Fearing local police and nosy neighbors, Trung's mother arranged for him to stay temporarily with an old friend who was now a fish wholesaler in Gò Công, while they planned his escape by boat.

During this waiting period, Trung disguised himself as a laborer, helping to unload fish from the boats when they returned from the sea and distributing the fish to buyers who came to take them to the market.

At that time, Thảo was the youngest daughter of Mr. Hai, the owner of the fish wholesaler, and she handled the financial aspects of the business. Thảo was just twenty years old, and she had two brothers who had served in the ARVN: one was a Navy Lieutenant who had escaped to the United States by ship on April 30, 1975, and the eldest brother was a Battalion Commander in the Rangers who died in action at Đồng Xoài during the fierce summer of 1972.

Mr. Hai was old and often ill, compounded by his grief over the loss of his eldest son. Not long after, he passed away. The fish wholesaling business was left to Mrs. Hai and Thảo to manage. Several times, local administrative officers requested that she sign over the business to the state cooperative, but she resisted, knowing that sooner or later, the business would be taken over by the communist government—the assets her husband and she had built through their hard work since the birth of their first son.

A little over a month ago, Mrs. Hai and Thảo secretly received a letter from her second brother, sent from the United States through Europe. He advised his mother and sister to sell all their assets and hire someone to build a boat to cross the sea to Malaysia or the Philippines, and he would arrange for them to be sponsored to the United States.

Mrs. Hai was hesitant, uncertain about the method and the route, so she procrastinated. Now, with Trung's help, she saw this as a great opportunity for the family. Mrs. Hai and Trung's mother had been very close friends when they were in school and later in life, so she treated Trung like her own son, and Thảo saw Trung as an elder brother in the family.

Mrs. Hai trusted Trung and gave him full authority to secretly arrange for the construction of a boat for their escape. Not wasting any time, Trung found and connected with Mr. Bảy Hiền in the Cầu Nổi neighborhood. Mr. Bảy was a skilled craftsman who specialized in building large offshore fishing boats.

After negotiating the price, dimensions, and load capacity, Mr. Bảy promised to expedite the work within three months. The boat would be powered by a new G.M. three-block engine from the United States. The remaining tasks for Trung included purchasing fuel, drinking water, and provisions for about seventy people for a fifteen-day journey—this was the number of people planned for in Mrs. Hai's family to take on this trip.

Although Trung was a former Air Force Major and reconnaissance pilot, navigating maps and compasses was not a problem for him. However, he was concerned about potential engine trouble, so he carefully recruited a former Navy mechanic, Sergeant Nguyễn Ngọc Oánh, who had served in the coastal patrol fleet in the IV Corps Tactical Zone and was very familiar with the local seas. Trung promised to take Oánh's entire family on the trip for free. With everything arranged and kept in strict confidence, the plan was considered solid.

The fish wholesaler continued to operate as usual, with no signs of any impending departure. Trung, still shirtless in shorts, continued to help load and unload fish at the dock.

To gradually liquidate valuable real estate, Mrs. Hai often mortgaged or sold properties to wealthy individuals in the province. Sometimes, after selling, she had to lease back the property for a few months to continue using it temporarily, leading many to believe her business was in decline and she was selling off assets to pay debts. With the proceeds, she discreetly purchased gold bars, storing them in fish sauce jars buried underground.

The "N-7" day arrived, and Mr. Bảy Hiền sent word to Trung that the boat would be ready in about a week. That evening, Trung and Oánh secretly went down to Cầu Nổi to inspect the engine installation. If anything was unsatisfactory, Mr. Bảy would fix it before the boat was launched into the small river in front of his house.

After carefully inspecting every detail, Trung was very satisfied, and Oánh praised him for selecting a powerful engine, even though Trung was not a professional in the field.

All plans and schedules for boarding the boat were meticulously arranged by Trung. The large boat was camouflaged as a professional fishing boat, while the smaller boats carrying people and provisions were to be hidden under the mangrove and water coconut trees near the riverbank, close to the sea.

After the boat was launched, Trung had everything—fuel, essentials, and provisions—loaded onto the boat, fully prepared. To familiarize themselves with the waterways, Trung, Oánh, and Mr. Bảy Hiền frequently pretended to be fishermen, taking the boat in and out of the estuary a few times. The border police occasionally stopped them for inspections, but finding nothing suspicious and being familiar with Mr. Bảy Hiền, they let them pass without issue.

The "N" day had arrived, coinciding with the communist authorities in Gò Công celebrating the liberation of the South. The provincial officials, engrossed in feasting and festivities, were a bit more lenient than usual, and Trung took advantage of this to move people with less fear of being watched by the local population or police. Around 9 PM, Mrs. Hai informed Trung that everyone had boarded the small boats and was ready to depart for the rendezvous point as planned. Oánh, the helmsman and mechanic for the trip, had already taken his family aboard the large boat since 6 PM.

Now, Oánh was at Mrs. Hai's house, waiting for the signal to set sail at the "G" hour. Thảo, appearing anxious, clung closely to her mother. The suitcases of clothes and necessary items had been moved earlier in the day, leaving only a small handbag containing valuable gold and jewelry, which Thảo was holding tightly.

After a final check of the maps and nautical instruments, Trung looked at his watch and told Oánh:

"It's time. Take Mrs. Hai to the boat. Be calm and watch out for the police. I'll bring Thảo along in five minutes."

Oánh placed Mrs. Hai on the back of his bicycle as if nothing unusual was happening. A short while later, he disappeared down the road.

Trung folded the small map, tucked it into his waistband, carefully locked all the doors in the house, and looked at Thảo:

"It's our turn now, Thảo!"

At that moment, Thảo was like an obedient little girl, following Trung's instructions like a student. She quickly stepped out and sat on the back of the bicycle, wrapping her arms around Trung's waist for support. The two of them rode together as if they were a couple out for a leisurely evening ride, unnoticed by anyone.

About half an hour later, Trung and Thảo had to dismount and lead the bicycle along the rice fields toward the riverbank. It was dark, but the faint starlight provided some guidance. Thảo often had to hold onto Trung's arm tightly to avoid falling into the fields. Occasionally, they encountered irrigation ditches, broken by water buffalo, where the rising river water flooded in. Trung had to carry Thảo in his arms to cross the ditches to keep her clothes dry. Despite the difficulties, they remained silent to avoid attracting attention. Soon after, they reached the riverbank, where Mrs. Hai and Oánh were already waiting under a clump of nipa palms. Thảo was overjoyed to see her mother again.

Trung told Oánh:

"Now, let's board the boat and head out!"

All four of them climbed into a small tam bản boat that had been hidden there for several days. Trung took the bow, Oánh the stern, while Thảo and her mother sat in the middle. The boat quietly glided out of the canal, heading toward the river where the larger boat was anchored.

It was past 10 PM, and everything around them was silent, except for the gentle splashing of the oars in the water. A few herons, startled from their sleep in the mangroves along the canal, flapped their wings and flew away.

A minute later, the small boat approached the side of the large boat, which was camouflaged with nipa palm leaves. Hearing the commotion, Oánh's two eldest sons, who were guarding the boat, came out with flashlights. After recognizing their family members, they lowered the rope ladder for everyone to climb aboard.

Trung had arranged a special spot on the boat for Mrs. Hai and Thảo, where they could lie down and rest relatively comfortably.

The newly built boat was quite large, with a powerful engine and fewer passengers than its capacity. The escape plan had been carefully studied, so Trung and Oánh were hopeful that they wouldn't encounter any problems when navigating the estuary.

The "G" hour for departure had arrived at precisely 1 AM. Trung ordered the anchor to be pulled up and the mooring lines untied from the tree. Oánh started the engine, which purred quietly. With just a quarter throttle, the boat smoothly glided across the calm water like a leaf floating along with the tide. Trung sat at the bow, watching for obstacles and keeping an eye on both sides of the river. Having practiced the route a few times, Oánh was fully in control of the boat's direction.

About five minutes later, on the right bank of the river ahead, a signal light blinked four times. Trung responded with four blinks of his flashlight and signaled Oánh to stop the engine. The boat slowed down, and Trung lowered the rope ladders as six small boats, hidden among the mangroves, emerged, loaded with people.

Children and the elderly were the first to be brought aboard, followed by young men carrying bags, suitcases, and food. Everything proceeded according to plan and schedule. At that moment, Trung, like a ship's captain, ordered everyone to sit still and remain as quiet as possible. Even the children's parents had to do everything they could, including covering their mouths, to prevent them from crying out, as this journey was a life-or-death escape between freedom and imprisonment for everyone on board.

In no time, the boat reached the mouth of the river, where the tide was at its highest. Oánh steered the boat close to the left bank, knowing that the right side of the river was lined with houses and a police checkpoint.

It was nearly 2 AM. The river, full of water, reflected the dim light of the distant stars, rising and falling with the boat's gentle waves. Fireflies flickered among the trees along the riverbank like ghostly lights.

To maintain maximum silence, Oánh kept the boat at half-throttle. The engine hummed softly as the boat, carrying over seventy people and their provisions, glided steadily and smoothly. The bow of the boat sliced through the water like a sharp knife, splitting the surface into two equal parts. Now, only the darkness surrounded them, and the only sound was the soft splashing of the water.

The women held their children close, while the men, anxious, craned their necks to observe the surroundings on both sides of the river. Trung sensed the heavy, tense atmosphere weighing down on everyone. He left the bow and went to the stern to give Oánh final instructions:

"Keep the pace steady. When we pass the checkpoint at the estuary, if they don't notice us, we'll speed up afterward. But if they spot us and order us to stop, push the throttle to full and keep going."

Trung also reminded everyone on board that when they passed the police checkpoint, they should lie flat on the deck because the boat would not stop, and the police might shoot at them.

Returning to the bow, Trung saw the lights from the police checkpoint about two hundred meters ahead. He signaled everyone to lie flat on the deck. Oánh calmly kept the boat moving at the same speed as if nothing unusual was happening. As they closed the distance—100 meters... 50... 20—suddenly, a loud voice from the police at the checkpoint ordered the boat to pull over for inspection.

Trung quickly jumped to the back of the boat, lying flat against the deck. Oánh pushed the throttle to full speed, causing the boat to lurch forward like a dinosaur, its bow rising high as it cut through the water like a naval speedboat from the old days. The bright searchlight from the checkpoint tracked them, and the sound of AK-47 gunfire rang out like New Year's fireworks.

As a former soldier, Oánh knew well that from this distance, small firearms were only being fired in hopes of hitting the target by chance, not with precision. He calmly steered the boat, avoiding obstacles as they headed straight out to sea.

A few minutes later, after they had put a good distance between themselves and the checkpoint, the AK gunfire ceased. Trung got up and, seeing no police boats in pursuit, checked that everyone was safe. He instructed Oánh to maintain their speed, setting a course east (90 degrees) to avoid Coast Guard patrols, planning to change course toward Malaysia once they reached international waters for safety.

The Vietnamese coastline gradually faded into the distance, becoming a faint black line on the horizon. The vast ocean stretched out before them, with waves gently rising and falling, making the boat seem small and fragile, like a bamboo leaf bobbing on a lake.

Now, there was no trace of land—only water and sky surrounded them. Trung estimated that they had reached international waters, and Oánh reduced the speed to half to ensure the safety of the engine. Occasionally, they saw large ships in the distance, brightly lit, passing by, but it was still too dark for Trung to determine which countries those ships belonged to.

The first nerve-wracking hours of their journey had passed, and everyone, exhausted, found a spot to rest their heads and quickly fell asleep to pass the time until morning.

With everything proceeding safely as planned, Trung entered the boat's cabin to check on Mrs. Hai and Thảo, who were resting inside. As soon as she saw Trung crouching to enter, Thảo quickly stood up, relieved, and took his hand:

"Are you okay, Trung? I was so scared when I heard them shooting at us!"

Before Trung could respond, she continued:

"Mom and I are fine in here, but we were so worried about you out there!"

Trung, trying to appear calm, smiled gratefully. He gently pressed her shoulder:

"Thảo, sit down—standing can make you seasick. He turned to Mrs. Hai:

"We've reached international waters now. How are you feeling?"

Mrs. Hai gestured for Trung to sit next to her:

"I have a bit of a headache, but Thảo brought a bottle of Nhị Thiên Đường oil for me, and it helped. Everything out there went well, didn't it?"

Trung replied:

"Yes, everything went as planned."

Mrs. Hai continued, praising him:

"You did well! Thảo kept asking about you—she's really afraid of gunfire. Now that our family is all we have left, consider Thảo as your sister. Don't call her 'miss'—it feels too distant."

As soon as Mrs. Hai finished speaking, Thảo eagerly chimed in:

"See? You heard what mom said, right? Calling me 'miss' all the time sounds weird! Can you just call me Thảo?"

Trung, feeling slightly bashful, lowered his head:

"Okay, from now on, I'll call you Thảo. It's still dark outside—why don't you and your mom try to sleep a little more? I'm going back out to relieve Oánh so he can rest."

The sky began to lighten as dawn approached, with the stars fading and the horizon seeming to rise. The ocean breeze blew gently, enough to lull those lying together on the deck to sleep, as if they were in a cradle suspended by an invisible thread stretched between the two horizons.

The splashing of long fish jumping out of the water, like dancers in a festival, was the only sound breaking the silence.

In this vast emptiness, Trung suddenly thought of God. His God could be the Sky, Buddha, or some other supreme being he couldn't quite imagine or identify, but he always believed that his God had the miraculous power to save everyone in their time of need. Just as when he was in the military, his mother would light incense and pray every night for God to protect him from harm. God was his mother's hope and his as well.

When the communists took over the South, they rejected God, creating a divide between his belief in a higher power and their atheism. The communist ideology, based on the materialistic theories of Marx and Lenin, aimed to build a real paradise on earth, but they confused "people" with "objects."

People are living beings capable of diverse creativity beyond control or imposition, while objects are passive entities governed by specific rules created by humans. Communism is not a religion because true paradise cannot exist on earth, as the communists believe. For religions, paradise is only attainable after death, which is the ultimate hope for the human soul—hence, religion endures.

A new day was dawning. The sun, as large and red as a tray, rested on the water's surface, casting the first golden rays that sparkled brilliantly on the sea. The faint murmuring of an old man reciting prayers echoed from the cabin.

Though exhausted, Trung fought off sleep to relieve Oánh and monitor the engine, speed, and course to ensure the boat's safety. But now, he could hardly resist, as sleep seemed to pull him down, his eyelids heavy as if weighed by large lead balls. He struggled to stay awake and reached out to shake Oánh, who was sleeping nearby, to take over.

Oánh woke up, still groggy, as Trung threw himself down to sleep. He quickly fell into a deep, peaceful slumber.

The boat's compass still pointed at a heading of 180 degrees south. The weather was good, with only a light breeze, so the course remained steady.

The sun had fully risen, and the boat was now more lively, with people cooking rice and boiling water for breakfast. Thảo brought her mother to the back of the boat to find Trung and ask for news. When she saw Oánh, she asked:

"Where's Trung, Oánh?"

He pointed to where Trung was sleeping on the deck:

"He's over there, sleeping!"

Thảo turned to look at the small corner where Trung lay sleeping peacefully like a child. Without saying a word, she quickly returned to the cabin. A moment later, she came out with a blanket in hand and carefully covered Trung with it, her gesture full of concern, like a wife caring for her husband.

Mrs. Hai, feeling a bit anxious, asked Oánh:

"Do you know how much longer it will take to reach Malaysia or some other island?"

After a brief thought, he replied calmly:

"If the weather stays like today, we should reach the Malaysian coast in a few days, Mrs. Hai."

Hearing the sounds of conversation and the morning sunlight, Trung woke up. He was surprised to find a thick blanket covering him. Thảo, standing nearby, saw him rubbing his eyes and smiled:

"Trung, you're still tired—go back to sleep! I covered you because it's cold out here."

Trung groggily pulled the blanket off and stood up:

"Thanks, Thảo, but I wasn't that cold!"

Mrs. Hai added to her daughter's words:

"There's still room inside—if you need to sleep, go lie down in there. It's easy to catch a cold out here. We're in the middle of the ocean, so we have to be careful!"

Trung didn't directly respond to Mrs. Hai's advice. Instead, he turned to Thảo:

"Have you taken care of Mrs. Hai's breakfast?"

With a playful tone, she responded:

"Stop worrying! I asked her, and she said she's feeling a bit seasick and doesn't want to eat right now. I also made some hot coffee with milk for you, but you were still sleeping, so I didn't bring it out. I'll go get it for you now!"

With that, Thảo snatched the blanket from Trung's hand and quickly disappeared into the cabin. Mrs. Hai, watching her daughter, laughed easily:

"She may be a bit short-tempered, but she's very caring, Trung. I pray to God that this journey goes smoothly and safely, because I've heard that encountering Thai pirates can be very dangerous."

Trung sighed but tried to reassure her:

"I've heard about the unfortunate encounters others have had, but Oánh and I plan to steer the boat as far away from Thai waters as possible."

On the fourth afternoon of their journey, the weather began to change. Dark clouds gathered on the southwestern horizon, and strong gusts of wind started to blow. Large waves crashed against the side of the boat, making ominous sounds. Despite using all his skills and experience from his time in the Navy's coastal patrol, Oánh couldn't prevent the boat from being tossed by the waves.

Everyone on board was exhausted, vomiting all over their clothes and the deck, the stench becoming unbearable. Mrs. Hai and Thảo couldn't take it anymore and lay down on the wooden floor. Despite being weary from the waves, Trung ran back and forth, caring for Mrs. Hai and helping Thảo up to take her medicine. He managed everything for the two of them.

Occasionally, they encountered foreign merchant ships of various nationalities passing along the international shipping route. The people on board desperately waved, raising white flags for help, but as the ships approached and saw them, they quickly sailed away.

Trung felt a deep sense of humiliation for the fate of those from weaker nations. People often speak of humanity and the value of human life, but now, to Trung, those words mean nothing.

Humanity can sometimes be more savage than demons, only seeing their own greatness. It's as if by rescuing people like us, especially yellow-skinned people from poor, backward countries, they believe they are only bringing trouble upon themselves. Conversely, if we were white people from wealthy, civilized countries, rescuing us would surely bring them benefits and good publicity.

However, their abandonment was still more humane than encountering Thai pirates—people born in a land known as the Buddha's land, yet whose hearts were more cruel and savage than wild beasts.

In the end, Trung didn't resent or hate those merchant ships for ignoring them, because even his own leaders, of the same race, shared the same ancestors, and had the same four-thousand-year history of building and defending the nation, had intentionally sought to annihilate those who loved freedom and opposed dictatorship. How could he expect anything from foreigners? Every commodity, whether abstract or concrete, has its price. Today, Trung and everyone on this boat might have to pay the ultimate price—human life.

By the sixth day, the weather had improved; the wind had calmed, and the sea was only gently rippled. Thảo and Mrs. Hai felt better and asked Trung to take them to the bow for some fresh air since the cabin still smelled bad and damp.

Thảo gripped Trung's shoulder tightly as she stood up. Trung wrapped his right arm gently around her slim waist to steady her as they slowly made their way to the bow.

A light, airy feeling passed between their gazes.

The offshore breeze was still blowing gently, the salty smell of the tropical sea was pleasantly refreshing, and Thảo and Trung stood close together, like a couple on a leisurely stroll or honeymoon cruise. Thảo looked far into the distance, as if searching for something to anchor herself, only to be met with disappointment, which she expressed with a sigh:

"It's so desolate and lonely here, isn't it? If you weren't here, I'd cry a lot!"

Trung pulled her closer, tightening his embrace.

"People fear loneliness, so they come together and love each other. Just as everyone fears death because it's a lonely departure, all alone, not knowing where to go, it scares them. Like our boat drifting on the vast ocean, surrounded by imaginary horizons limited by the curvature of the earth and the range of our vision, missing familiar landmarks like a coastline or an island to remind us that we still exist, it makes everyone uneasy. On the other hand, if all these people were on a bus traveling across provinces, they would feel completely different."

Before Trung could continue, Thảo playfully interrupted:

"So, are you afraid on this journey?"

"I'm not afraid because I have you here, so I'm not alone!"

Thảo didn't ask any more questions. She reached out and playfully pinched Trung's side, silently acknowledging his meaning. After a brief silence, with a distant sadness in her eyes, Thảo asked:

"I wonder if we'll see each other again once we reach the United States? I've heard it's a huge country, with each state being as big as Vietnam. I'm worried! Oh, didn't you go there before to learn to fly? What was it like?"

Trung didn't directly answer her question. Instead, he pointed to a thin white cloud drifting alone in the blue sky:

"Do you know where that cloud will go?"

Thảo turned and gave him a sideways glance:

"How would I know? Why are you asking such a strange question?"

Trung laughed lightly:

"Let me explain! That cloud is carried by the wind, so the wind and the cloud are like traveling companions, leading each other to some destination. Or, to put it another way, it's the predestined fate that unites the wind and the cloud. Similarly, even though the United States is vast, if the desire and determination to find each other are strong enough, even a country a million times bigger would seem small in comparison. Moreover, their country is far more civilized and free than ours. Here, traveling from one province to another to visit a friend can be more difficult than a cross-state journey in the United States."

Thảo lowered her head and spoke softly:

"Let's not talk about those miserable things anymore. It's because of those wicked people that we're risking our lives like this. I pray and hope that one day, Vietnam will change so that everyone can have more freedom. Oh, we've been out here long enough; let's go back inside so Mom won't worry!"

Perhaps thanks to the help of God and Buddha, by the evening of the seventh day, despite the hardships caused by the sun, wind, and storms of the sea, everything went smoothly. Fortunately, the boat did not encounter any Thai pirates.

Once again, dusk was about to envelop the sea. The setting sun was a bright red orb, perfectly round like a golden tray, and the last golden rays of sunlight shimmered beautifully. The clouds on the western horizon seemed to be changing into new clothes, with vibrant colors forming a semicircular arch, resembling a triumphal arch bidding farewell to the day.

Darkness quickly descended, turning the deep blue sea into a menacing, eerie black. The boat continued to drift like a lonely leaf on the vast water in the endless void.

Trung checked the map and compass heading once more, murmuring to Oánh:

"According to our course and speed, we should be on track, but why haven't we seen anything yet? Do we have enough fuel to keep going for a few more days, Oánh?"

Now, Oánh looked more thoughtful. He, too, was worried because his entire family was on the boat, so he was determined to bring them safely to shore. He reassured Trung:

"The engine is new, so the boat is fuel-efficient. With the fuel reserves, we can go for another five days or so, no problem. I'm only worried about bad weather or currents we might not have accounted for slowing us down. You should get some sleep now; I'll wake you up around midnight to take over."

It was past eleven at night, and everyone was asleep. Oánh walked over to trim the wick of the kerosene lantern hanging by the wheelhouse. Under the glow of the phosphorescent compass, the needle still pointed at 250 degrees. The engine continued to hum steadily. A few small fish, attracted by the light, jumped out of the water, thinking the sun had risen early, only to splash back into the sea around the boat. Oánh poured more coffee from the thermos his wife had prepared earlier in the evening.

Far ahead, there seemed to be a tiny light flickering like a firefly, bobbing with the waves. Rubbing his eyes, Oánh squinted to get a better look at the mysterious light. A mixture of joy, anxiety, and suspicion quickly overcame him as he realized the light was definitely from an electric lamp on a ship heading toward them.

He rushed to where Trung was sleeping and shook him awake:

"Trung, wake up! There's a strange ship coming toward us!"

Trung jolted awake quickly, and Oánh suggested turning off the light to avoid detection, but Trung stopped him just in time:

"Don't worry, Oánh. They've probably already seen us. If we turn off the light now, they might not see us and collide with us."

Trung then went to inform everyone on board to hide their money and jewelry and lie down as if they were gravely ill. Thảo and Mrs. Hai were visibly frightened and panicked. Trung smeared black oil all over their faces, arms, and hair, making them look like filthy beggars. Five minutes later, the scene was set as Trung had planned—a ghost ship carrying corpses in the night.

Oánh shut off the engine and lay down on the deck, pretending to be severely ill, leaving the boat to drift freely.

The time dragged on, feeling almost like a century, as the strange ship drew closer. The electric lights became brighter and clearer. Trung crouched on the deck, anxiously observing. He feared for Thảo, as he had heard many stories about Thai fishermen who had turned into pirates, robbing, raping, and killing people before throwing their bodies into the sea in a most savage manner. He thought he might have to risk his life if they tried to harm Thảo.

A bright searchlight from the distant ship swept back and forth, piercing through the night over the sea. Trung thought to himself: with such a powerful spotlight, it could only be a large ship—pirates wouldn't have something like this. His suspicion was nearly confirmed when, just a few minutes later, he saw the brilliant lights on the ship's deck, revealing it to be an international merchant vessel.

Overjoyed, he shouted to everyone that it wasn't a pirate ship. Oánh quickly sat up, grabbed a white shirt, tied it to a stick, and waved it back and forth as a distress signal.

The enormous ship slowly approached and stopped next to Trung's small boat, which seemed like a mouse lying next to a giant elephant. Then, ropes and ladders were lowered, and the children, women, and elderly were helped aboard first. Finally, Trung and Oánh climbed up.

With his basic English, Trung learned that the ship belonged to the Indonesian Merchant Association. The captain, a European who spoke fluent English, promised to take them to "Bulau-Tanga" Island in Malaysia.

Six months later, with sponsorship from relatives in the United States, Mrs. Hai, Thảo, and Trung arrived in California. In the first few days, Trung stayed with Thảo's brother, who had already settled and bought a house there. About a week later, Trung made contact with some old friends from the South Vietnamese Air Force who were living in Oregon.

Today was Saturday morning, and Trung intended to take Thảo on a final visit to the Vietnamese district in San José and to say goodbye before he moved to Portland (Oregon) to live with his old friends and look for a job. As Trung was folding and packing some freshly washed clothes, Thảo approached:

"What are you doing, Trung? Let me iron those clothes and fold them properly."

"Thanks, Thảo, but these are just old clothes that someone gave me—no need to iron them. I can wear them as they are. By the way, are you busy later?"

"No, not really! I've been cooped up inside all week and don't know anyone here to go out with."

"If you're not busy, I thought I'd take you to San José. I heard there are a lot of Vietnamese restaurants and shops there!"

With a mix of excitement and playfulness, she replied, "But we don't have a car! The roads are so wide, and there are so many cars—if we walk around aimlessly, we could get run over!"

Trung glanced around to make sure no one was nearby, then stepped closer to Thảo and whispered in her ear:

"There are no fields full of water here like back home, but if there were..." Before he could finish, Thảo seemed to catch on and smiled with a natural blush on her cheeks. She playfully asked:

"If there were fields, what would you do?" Trung leaned closer to her ear and whispered:

"I'd carry you so you wouldn't get your clothes wet!" Thảo quickly pinched Trung's arm and said:

"You're making me so embarrassed!"

The sky over San José was clear and blue, with no clouds in sight. The sunlight was bright and soft, reminiscent of the highlands of Đà Lạt. Thảo, a young woman just over twenty, was charming, innocent, and beautiful like a blossoming flower. Although she was from Gò Công, a coastal province in the South, she grew up in a business family and didn't have to toil under the sun. Her skin was as white and smooth as someone from Đà Lạt. Trung often jokingly called her the girl from the Land of Cherry Blossoms.

In contrast, Trung, a man in his early thirties, had a tanned, weathered complexion, a former Air Force Major in the South Vietnamese Air Force, with deep black eyes that sparkled with the experience of life. He had traveled abroad many times and fought in the Vietnam War for nearly twelve years, from the 17th parallel to the southern tip of Cà Mau. He had set foot in many places.

With a gentle and open personality, his smile was always a powerful tool for winning over those he met for the first time. It was this warm smile and expressive eyes that often left Thảo feeling flustered and shy during their conversations.

The two walked silently beside each other through the shops in town. The sun was still shining brightly, and people around them were chatting and laughing casually. In this busy and joyful atmosphere, Trung didn't want to create a sense of parting while Thảo's mind was still lost in a beautiful dream.

As they walked, lost in thought, Thảo suddenly saw a wedding dress hanging in a shop window. She tugged at Trung's hand, stopping him.

"Look, Trung! This wedding dress is so beautiful in white. It looks so elegant, don't you think?" Trung smiled, leaning down to tease her:

"Do you like it?" Thảo shook her head, pouting slightly:

"You're being silly! Who would marry me that I'd need a wedding dress?"

"Someone does! You just haven't noticed."

"Who?"

"If you pay a little attention, you'll know right away!" Thảo playfully pushed Trung away, and they both laughed, seeming to understand each other well.

By four in the afternoon, the sun had softened, no longer as harsh as before. Trung held Thảo's hand and led her to a stone bench under a tree in the park. With a touch of sadness on his face, he tried to lighten the mood as he looked at Thảo:

"Thảo, I'm going to leave this state soon." Before Trung could finish, Thảo suddenly stood up, her eyes filled with surprise:

"What did you say? And where are you going?" Trung gently pulled her back down to sit beside him:

"Sit down, Thảo. I wanted to tell you that today is both our first and last outing together. Next Monday, I'll be heading to Portland, Oregon, to stay with some old friends and look for work. They want me to move up there for company."

As Trung spoke, Thảo didn't reply or ask any more questions. She lowered her head, staring at the ground. Trung lifted her chin gently, feeling her warm tears fall onto his hand:

"Why are you crying?"

With a soft, almost choked voice, she replied:

"We've only been here for less than a week, and now you're leaving?" Trung felt a lump in his throat, struggling to find the words as he spoke in broken sentences:

"Thảo... I'm not leaving you... I'm just going up there to find work... It's hard to stay here, living off others. I promise I'll come back to visit you often."

Her faith in Trung's promise sparked hope in her heart. Thảo looked up and rested her head on his shoulder:

"I love you, Trung. Did you know that?" Trung kissed her forehead:

"I love you very much too!"

And time kept moving... like the shadows of trees stretching across the porch, growing longer until the sun set behind the western hills, signaling the end of another day. Every day, Thảo continued attending community college to study English and accounting. Trung had found a stable job at a private company in Portland.

More than a year later, during the warm spring of the Pacific Northwest, Trung returned to San José (California) to officially ask for Thảo's hand in marriage.

The pure white wedding dress that Thảo had once dreamed of while walking with Trung in town was now the dress she wore for her wedding. That dream had come true — Thảo was now the bride of the Land of Roses and the rightful queen of former Air Force Major Nguyễn Văn Trung.

After listening to the full story of the twists and turns and the suspense between Trung and Thảo, everyone felt as if they had just watched a romantic film with a peaceful and happy ending. Phượng was the first to speak:

"And to think that all this time, neither Thảo nor Trung ever hinted at their romantic past to anyone!" Trung laughed sheepishly:

"Well, now you all know everything!" Thảo joined in, adding to her husband's words:

"I guess Trung and I were meant to be from a previous life. He's older than me, has been a soldier, and has known so many people from Huế, Quảng Trị to Cà Mau—he's been everywhere. But in the end, we met in Gò Công, fell in love on the boat during our escape, and here we are. Isn't that amazing?"

Loan, sitting nearby, felt that Thảo had left something out, so she added:

"Thảo, you told the story well, but it's missing something. You should have said, 'We met in Gò Công, fell in love on the boat, and got married in the Land of Roses' to make it complete!" Loan's witty remark made everyone burst out laughing. Tuấn joined in:

"Hey Trung! I've read many of your short stories and memoirs in the papers, but I've never come across anything like your love story with Thảo. You're missing out!" Tâm was surprised to hear Tuấn mention that Trung often wrote for newspapers, so he asked:

"Trung, are you writing for the papers now? What pen name do you use? What papers do you write for so I can read them for fun?" Trung laughed heartily:

"I write just for fun, to express what I can't say out loud, so I rely on pen and paper. I'm no professional writer or journalist—I just write for friends. That's why I don't bother with pen names or fancy titles. I just use the name my parents gave me and let people think whatever they want!"

"No wonder I've read a few short stories, reports, and memoirs under the name Nguyễn Văn Trung—I thought it was someone else. Turns out it was you! I suggest you be careful, especially when expressing political opinions in the papers, or you might end up being labeled a communist sympathizer." Trung continued laughing as if it wasn't a big deal:

"You've only been here a few years, but you already know and understand how things work. I've already been labeled a communist sympathizer by some of these armchair warriors. The truth is, I do get mad at them, but I've decided not to pay attention to their childish nonsense, and eventually, they feel ashamed and leave me alone."

Then, as if recalling something that annoyed him, Trung blurted out:

"Damn it! I was imprisoned in the far north, escaped, and nearly died at sea to avoid those bastards. Now, here I am, and these armchair warriors dare to label me as one of those damn communists. Don't you think that's infuriating? Sometimes, I think their way of fighting communism is no different from a cheap contractor who takes the money and then does all the work with his mouth. Just think about it—if you write for a newspaper they don't like, sooner or later, they'll label you a communist. It seems they want all Vietnamese writers abroad to be nothing but 'literary slaves,' like Tố Hữu, the Central Committee's propaganda chief who wrote poems praising Stalin:

'Long live Stalin

Forever great deeds

A shadow of peace

At the forefront of the storm!'

Or:

'Love my father one

Love Uncle Stalin ten!'"

As Trung spoke, Tuấn, who had been listening, felt a bit uneasy and jumped into the conversation:

"You're lucky you have a record of being imprisoned and nearly dying at sea. If you were labeled as pro-communist or a sympathizer for national reconciliation, people like me, who were lucky enough to escape on the afternoon of April 30, 1975, would have to keep our mouths shut and follow the 'four no's' policy of former President Nguyễn Văn Thiệu. That policy of tying our hands and blindfolding ourselves ultimately led to painful and humiliating results because we deceived ourselves and failed to assess the political and military situation of both sides accurately. The enemy had taken nearly all the land, and their tanks were at the gates of Independence Palace, yet the Saigon Radio and the Army were still loudly broadcasting President Thiệu's golden words: 'We will not negotiate with the communists... We will not give up a single inch of land...' etc. President Thiệu was right—South Vietnam didn't give up a single inch of land on paper because it was too insignificant. Instead, they gave up entire provinces, whole towns, like Phước Long, Ban Mê Thuột, Kontum, Pleiku, etc., following his strategic withdrawal plan."

And then, around 11:30 AM on April 30, 1975, in Saigon, the brave soldiers of the South disbanded like a headless snake. President Dương Văn Minh and his staff, dressed in full regalia, sat in Independence Palace, waiting for the invading North Vietnamese armored brigade 203 to break down the palace doors and rudely tell the last leader of the South: 'There's nothing left for you to hand over—just unconditional surrender.'

We have to ask ourselves, before 1975, President Thiệu still had land and military power, so he came up with the 'four no's' strategy. But now, what do the true, non-communist Vietnamese patriots, both in the country and abroad, who genuinely love freedom and national sovereignty, have in their hands? How many 'no's' do they have left?"

The women, who had been quietly listening to the men's political discussion, began to feel frustrated. With her experience living under the communist regime for many years, Loan spoke up first:

"Do you all know? A few months after April 30, 1975, they rounded up all the men and sent them to reeducation camps in the mountains. At home, it seemed like only the elderly, women, and children were left. They organized neighborhood committees and forced people to attend constant study sessions. The communist cadres, who were often uneducated like cowherds, always took on a condescending tone, treating everyone like dirt. Everyone just sat there, taking in whatever they said, like water off a duck's back—nobody dared to express an opinion. Once, Uncle Tư from the next village got fed up with their lies and nonsense, so he raised his hand to speak up. After the session, when everyone else had left, they asked him to stay behind for an afternoon session to help him better understand Marxist-Leninist philosophy. After that, even the old women who couldn't read a single letter and the children all claimed they had thoroughly absorbed Marxist-Leninist teachings. No one had any more questions or opinions."

When Loan finished speaking, everyone burst out laughing. Tâm added:

"My wife is right. I don't know how they behaved in Saigon, but in the smaller provinces and rural areas, just the sight of those 'baby' policemen was enough to scare people. They had no laws, and they could do whatever they wanted, following the 'village rules over king's law' principle.

In civilized countries, democracy and freedom are expressed through open and proper debates among opposing parties to perfect the nation and society. But under communism, it seems there's no such thing because they don't accept opposition. In a country or community, if there are no opposing parties or individuals, it's synonymous with military dictatorship.

To better understand the causes of the Vietnam War, as everyone knows, the war in Vietnam originated during the French colonial period, initially led by patriotic Vietnamese who organized to resist the invading power. Then, in the following period, international communism, driven by Russia, emerged.

Hồ Chí Minh, a young man trained in Russia's tactics and strategy, was brought back to Vietnam to organize the Communist Party of Vietnam to fight against the French colonialists. In the subsequent stages, many nationalists were eliminated by the communist party in various ways because these nationalist parties did not accept Marxism-Leninism. After World War II, although the world situation was relatively stable, and hot wars had nearly ended, the Cold War emerged as a fierce competition for international influence between Russia and the United States. Vietnam was influenced and pressured by the major powers to enter that transition orbit, or in other words, the Vietnam War had two forms:

One was a civil war between the Vietnamese nationalists and the Vietnamese communists.

The other was an international proxy war between the two ideologies: Communism and Capitalism, led by Russia and the United States, respectively."

As Tâm was speaking, Phượng interrupted:

"Based on historical records and what you've said, the North

Vietnamese communists systematically eliminated many Vietnamese patriots because they didn't agree with the communist party. Now, they use every political trick to secure the supreme interests of the party. Sooner or later, this will be exposed to the public, and the history of Vietnam will curse them."

Trung laughed and said:

"Oh, Phượng! Those black-toothed machete-wielders don't fear God or Buddha—who do you think will curse them? The most practical thing is to hope that we, the refugees who fled our homeland, stop tearing each other apart and making fools of ourselves for their amusement! If we can't see each other as friends, then it's better to treat each other like two strangers who've just met on the road. That would look better and more civilized."

Although Thảo wasn't particularly interested in political and social issues, she sat quietly, allowing everyone to express their opinions and thoughts. When she noticed that Trung seemed to be wrapping up the conversation, she spoke up:

"You all have been so engrossed in discussing social issues that you've let your drinks go cold. I'll go to the kitchen to boil more water and make tea!" At the same time, Phượng suddenly remembered something important:

"Thảo! We've been so caught up in talking that I forgot today the clothing stores are having a big sale. How about the three of us go check it out? Let the men stay here and talk about whatever they want—we'll have more fun shopping on our own." Loan and Thảo nodded in agreement immediately.

That afternoon, Loan and Tâm said goodbye to their friends in Portland and drove back to Seattle. At this moment, Tâm was still surprised to have reunited with Tuấn, as he never expected that Tuấn had come to the United States and was living only three hours away by car. Back in Vietnam, Tâm had asked around about Tuấn's family, but no one knew where they had moved. As he thought about this, a warm feeling spread through him, making him smile as he glanced at Loan:

"You know, darling, the earth might be vast and grand geographically compared to our human form, but when it comes to the serendipity arranged by fate, it feels so small, doesn't it?"

"Are you talking about the coincidence of running into an old friend? I used to think I was the unluckiest woman, but after hearing about what

Thảo and Trung went through during their escape at sea, with all the crises, the anxiety, the illnesses, and the lack of medicine, it's terrifying!"

"They were indeed more fortunate than many others who were raped by Thai pirates, and sometimes their boats were sunk, or they were killed and thrown into the sea to cover up the crime. Let's leave those tragic stories in the past. Remembering them only brings more pain, with no benefit. We live for the future, not the past. What delighted me the most was the unexpected reunion with Tuấn. After more than twenty years apart, it feels like a dream. Tuấn looks older and more mature now, but he hasn't changed much."

When Tâm mentioned this, Loan imagined the connection between Tâm and Lệ when they were still in Pleiku. She silently gazed out the window, aware that her husband still cherished the memory of his late first wife. This didn't make her jealous but rather left her with a feeling of melancholy.

After a long pause, with Loan not saying a word, only staring outside, Tâm asked:

"What are you looking at?" Loan turned back with a faint smile:

"I'm not looking at anything! The sound of the car engine makes me sleepy."

"Why don't you lean your head back and take a nap? We still have a long way to go to Seattle."

"No, I'll stay awake and keep you company. By the way, I've heard that many people are returning to Vietnam to visit. It seems the communists there are more lenient now, perhaps because they want to attract foreign currency. Airfare is cheaper too. Should we go back and visit your parents to make them happy?"

In a moment of contemplation, he mumbled, "It's been nearly five years already..." Tâm turned to his wife, seeming to agree:

"Our life here has stabilized, so we should go back to visit our parents and see how the country has changed." After a brief hesitation, Loan suggested:

"If you plan to go, I think we should invite Tuấn and his wife to join us. We could also visit Lệ's grave in Pleiku. What do you think?"

Tâm was startled and surprised by the sudden suggestion to visit Lệ's grave in Pleiku. Although Tâm had long harbored this desire, he feared it would upset Loan, so he kept it hidden, waiting for the right moment to express it. Today, unexpectedly, she opened the door for him.

A look of happiness appeared on his face as Tâm silently thanked his wife for her kindness and generosity. He gently stroked her shoulder:

"Pleiku is a poor city. Aside from the occasional sight of Highlanders carrying their children on their backs and balancing bundles of firewood on their heads as they walk along the road, it doesn't have the bustling or beautiful scenery like here!" Loan teased her husband:

"Are you saying Cà Mau is rich? While there may not be Highlanders carrying children and wearing short skirts, there are still peasant girls, farmers with conical hats, scarves wrapped around their necks, and trousers rolled up to their knees, paddling boats or wading through the fields! My purpose in going up there is to visit her grave and to finally see Pleiku since I've only heard about it and never seen it with my own eyes."

Around July of the following summer, Phượng, Tuấn, Loan, and Tâm were in Saigon and on their way back to Pleiku on a regular Vietnam Airlines flight.

Through the plane's window, scattered white clouds drifted aimlessly as if journeying to some far-off horizon. The sky remained high and blue like the ocean, and below, the familiar mountain ranges stood proudly, defying the passage of time. The dense green forests of the "Royal Domain" Ban Mê Thuột were full of mysteries.

The winding streams snaked through the hills, paralleling the provincial road that ran through Cheo Reo and Phú Bổn, the route chosen by General Phú, Commander of the II Corps Tactical Zone, for the withdrawal of Pleiku's forces to Tuy Hòa in March 1975, following President Nguyễn Văn Thiệu's strategic retreat plan. Sadly, the gateway to freedom for the retreating forces became a death trap due to accidents, disease, hunger, and the brutal violence of war.

Now, all that remains here and throughout this land are the faded remnants of a time to remember and also to forget! All killing eventually ends, and every wound heals with time. What belongs to the past should be left in the past.

The steady hum of the jet engine and the sound of the wind continued. Phượng and Loan had closed their eyes, trying to sleep in their seats ahead. Tâm sat in the back, near the window, still gazing outside as if observing something very important. In reality, he was reminiscing about his days as Lieutenant Lê Văn Tâm, a helicopter pilot with the HU-1 squadron stationed in Đà Nẵng. He had flown over this very sky many times during missions for allied units. Today, thirty years later, he found himself back in this sky, looking down at the mountains and the familiar white clouds drifting serenely across the horizon.

The seatbelt sign illuminated, and the flight attendant's announcement for passengers to prepare for landing brought Tâm back to reality. Tuấn, sitting beside him, had just finished reading a fascinating article in the Tuổi Trẻ newspaper he had bought earlier in Saigon, about the "Tamexco" fraud case involving millions of US dollars masterminded by a group of communist cadres. They had squandered the money on luxury goods and personal property, with names like Phạm Huy Phước, director of the state-owned Tamexco company, Lê Minh Hải, Trần Quang Vinh, and Lê Đức Cảnh.

Tuấn shook his head in disbelief and handed the newspaper to Tâm:

"Here, take a look at this case. It seems the communist cadres have succeeded in creating a paradise on earth, just as Uncle Hồ envisioned." Tâm understood and replied with a faint smile:

"You've been away from the country for a while, so this surprises you. But we're not too shocked by these 'red capitalists.' The fraud case involving millions of dollars in lost state funds—money that comes from the hard-earned taxes of the people—is committed by the elite of the party who have the opportunity to grab high-stakes profits. As for the lower-level cadres, like the traffic police, I don't want to generalize because there are still some good and honest ones. But most of them, struggling to make ends meet on meager salaries, take whatever bribes they can, turning a blind eye to let smugglers slip through the law."

"Let me tell you a hard-to-believe but true story about a street sweeper who accepted bribes. In the commercial areas of Saigon and Chợ Lớn, street sweepers usually gather trash into piles to be collected later by garbage trucks. If your house is along the street and you don't bribe the street sweeper with money or gifts, the trash pile will end up right in front of your door. Neighbors will freely add their waste to the pile, attracting flies that swarm like a feast. But if you pay the bribe, the trash pile will be moved to the next house if that homeowner isn't as generous as you. That's the socialist society in our country now! In the old days, during the first and second Republics, there was corruption and bribery among officials, but compared to now, it's like child's play compared to the swindling, cheating, and deceit of these red capitalists."

Tuấn fell silent as if accepting the harsh reality. He gazed out the window aimlessly. The plane banked to the right as it entered its final approach. The barren hills surrounding the airport below appeared in a familiar yellow hue, and the green coffee plantations still stood, divided into square plots as if waiting for someone from the past to return.

A sense of anticipation, almost indescribable, pulsed through their hearts as Tâm and Tuấn, two Vietnamese men now temporarily acting as foreign tourists with no friends or relatives waiting for them at the airport, returned to revisit their homeland. The warm and gentle memories of the past resurfaced, replacing the absent embrace of loved ones.

The plane landed safely and taxied to the gate. Through the window, Tâm, Tuấn, Phượng, and Loan noticed the buzz of excitement among the locals waiting to reunite with their loved ones on this flight. The four of them felt uplifted by the joyful atmosphere of these family reunions, so they allowed others to disembark first before following suit.

As soon as they exited the plane, Loan hurried over to hold her husband's hand:

"It feels cooler here than in Saigon, doesn't it?"

"Yes! It's much better than Saigon, thanks to the highland climate. But sometimes, the mountain winds from lower Laos can get pretty hot too!" Tuấn and Phượng, walking ahead of them, turned back and asked:

"Do you notice anything different, Tâm?" Glancing around briefly to take in the surroundings, he replied succinctly:

"No, it looks the same as before, or maybe even worse." After more than two decades, the Pleiku Cu Hanh Airport still retains its original appearance. The only difference was that the paint had faded and peeled, and wild grass grew wildly, turning yellow with neglect.

Suddenly, Tâm hastened his pace, pulling his wife along. Loan felt her husband was behaving strangely with this sudden urgency, so she complained:

"Why are you rushing? Slow down so I can keep up!" Tâm pretended as if nothing was unusual:

"Hurry up, we need to get our luggage so we can head into town early!" Noticing Tâm's abrupt behavior, Tuấn understood the unspoken reason, as it was at this very place that Tâm first met Lệ through Tuấn's introduction. The two fell deeply in love and married. Now, perhaps Tâm wanted to escape these personal memories.

Tuấn nodded at Tâm as if to convey his understanding:

"That's right, Loan. Seeing the old airport just makes us sadder. It'll be more fun in town!" Phượng, like a country girl visiting the city for the first time, found everything fascinating. She remarked:

"Tuấn! The landscape here, with its bare hills, looks a lot like California, doesn't it?" Tuấn smiled indulgently at his wife:

"You know, we used to call this the 'Thượng' region, also known as the 'dusty sunshine, muddy rain' land. When it's sunny and windy, or when cars speed by, the dust clouds up the sky. And when it rains, well, going barefoot is better than wearing shoes! But you know, one special thing about this place is that the girls here have beautiful, fair skin, just like in Ban Mê Thuột or Đà Lạt." Phượng chimed in:

"Then Lệ must have been very beautiful, right?" Tuấn didn't reply, just nodded, raising a finger to his lips to signal silence. Phượng understood and said no more.

After finding a "hotel" in the main part of town for their stay, the sun was almost directly overhead. To avoid wasting time, the four of them decided to have lunch at a nearby pho shop across the street and also inquire about renting a vehicle for transportation.

The pho shop owner, dressed in shorts and a bare chest, was friendly and eager to assist his out-of-town customers. He even introduced them to a "xe thổ" driver who frequented the shop for coffee.

Once they had arranged everything, Loan suggested:

"It's still early. I suggest we visit Lệ's grave first, and then we can decide what to do in the afternoon!" Phượng quickly agreed:

"Loan is right! But first, we should stop by the market to buy some fruit and incense for the offering."

The "xe thổ" three-wheeler carrying Tuấn and Phượng led the way toward Lệ's old neighborhood, east of Pleiku Airport.

Tâm remained silent, his face reflecting sadness as he clutched a bundle of incense and candles in his hand. Loan, sitting beside him, felt a deep sympathy for her husband and didn't say a word. The xe thổ steadily and unfeelingly carried him back to the old village road, a road filled with memories, bearing the marks of love and sorrow from a bygone time between Tâm and Lệ.

After about fifteen minutes, Loan squeezed her husband's wrist and asked in a somber tone:

"Are you thinking about her?" Startled by his wife's accurate guess, Tâm quickly tried to cover up his emotions with a vague response:

"I'm thinking because we're returning to the old neighborhood. Ah, we're just a few houses away!" Tâm paused here, and Loan continued his unfinished sentence:

"Is that Lệ's house?"

"Yes!"

The xe thổ carrying Tuấn and Phượng slowed down, finally stopping next to a sugarcane juice stall on the left side of the road.

Tâm was taken aback and puzzled as he pointed out to his wife:

"Do you see that brick foundation house over there? That was Lệ's house. Why is it now the local police station?" Tuấn and Phượng had gotten off the xe thổ and approached as well.

"They confiscated Lệ's family home. Let's not bother with those shady people—I can't stand them!" Tâm shook his head in disappointment:

"There's no point in going inside. Let's sit here for a bit, drink some sugarcane juice, and then head to the cemetery."

The elderly lady who owned the sugarcane juice stall noticed her customers and hurried out to greet them warmly, inviting them to sit on the wooden chairs arranged around the stall.

Tâm found the woman's voice familiar. He looked closely and, after a moment of recognition, exclaimed in joy:

"Bác Năm, do you remember me?" After a moment of hesitation, Bác Năm stopped turning the sugarcane press and raised her head to look at the stranger:

"Oh! Is that you, Tâm?"

"Yes, it's me, Bác!" She came out and took his hand, looking him up and down as if examining a relic from more than twenty years ago.

"Oh my goodness! We haven't seen you in decades, and everyone thought you were dead. The house and the coffee plantation on your wife's side were all confiscated after the liberation."

"How is Bác Năm's husband?"

"My husband passed away a year after liberation from starvation and lack of medicine. After they took the coffee plantation, my husband was unemployed, and he fell ill repeatedly. With no medicine and little food, sometimes we had to go hungry, surviving on cassava and sweet potatoes. In the old days, your wife's family treated us so kindly, but those people are just as cruel and wicked."

Bác Năm's husband had been the overseer of the coffee plantation for Lệ's parents. After Lệ's father passed away, Bác Năm's husband continued to loyally manage the estate, earning the trust and affection of Lệ's family, who considered him like one of their own.

"Poor him! He was so kind. Whenever we visited the plantation, he would rush to pick fruit for us to eat, and even pack some for us to take home!"

"Where are you living now? Are you just visiting?"

"Yes, Bác, we live far away. We came back to visit the graves, see the old house, and reconnect with neighbors before we go."

"Let me press some sugarcane juice for you to quench your thirst, and then we can visit the graves of the mother and daughter, which are nearby. Poor them, the graves have been overgrown with weeds for years with no one to care for them!"

After bidding farewell to Bác Năm, the four of them walked across a large empty field behind the stall. Two graves, overgrown with weeds, lay side by side on a small mound, looking desolate like abandoned graves.

Loan gently held her husband's hand, her voice tinged with sadness as she asked softly:

"Which one is Lệ's grave?" Confronted with this bleak scene, Tâm felt choked up, his tears welling up and streaming down his cheeks. He couldn't find the words to answer Loan's question, so he simply pointed in silence.

Seeing Tâm overcome with emotion, Tuấn stepped in to explain quietly to Loan:

"The grave on the left is Lệ's, and the one on the right is her mother's." Everyone stood quietly, as if sharing the same feelings of sorrow and regret for the forgotten dead. Without saying a word, Tuấn, Phượng, and Loan bent down to pull out the weeds from the front of the two graves. They then placed offerings of fruit and lit incense and candles.

Tâm was the first to light incense and pray, followed by Tuấn, then Loan, and Phượng, each offering their respects in solemn silence.

Later, Phượng pulled Loan aside and whispered:

"Maybe we should stay a few more days to hire someone to rebuild the graves for Lệ and her mother. What do you think?"

"I'm very glad you suggested that, and I agree. You should discuss it with Tuấn. I'm sure Tâm will be very happy to see Lệ and her mother's graves properly taken care of."

Today is the third afternoon, the agreed-upon completion time with the construction workers for rebuilding the graves. Tâm and Tuấn had prepared all the necessary incense, candles, and offerings and had also brought a monk to perform the requiem for the souls of the deceased.

Thanks to Bác Năm's oversight and urging, the two graves, which had looked like abandoned plots overgrown with weeds just days ago, were now completely transformed. Surrounding the graves was a solid fence of brick and iron, providing protection. A roof was built over the graves to shield them from the rain, and the inside was tiled with red bricks, creating a proper space for offerings. Bác Năm had quickly placed potted chrysanthemums and marigolds around the graves, making them look warm and inviting.

After the neighbors who had come to pay their respects had left, and the sun was beginning to set in the west, Bác Năm excused herself to go home and prepare dinner. Loan and Phượng tidied up the cups and bowls to return them to Bác Năm.

Tâm approached Tuấn, shook his hand, and squeezed it tightly:

"Thank you and Phượng so much. It's thanks to your help that Lệ and her mother have such a warm and secure resting place today." Tuấn, moved by his friend's words, put his arm around Tâm's shoulder as if they were the same close friends from the old days:

"Don't say that, Tâm. This was our duty to take care of. Now that everything is done, let's go say goodbye to Bác Năm and head back." Tâm bent down to help Loan with the tray she was carrying:

"You go with Phượng and Tuấn to Bác Năm's house first. I'll light some more incense and join you shortly." Loan understood her husband's feelings and simply nodded before following Phượng and Tuấn, who were already walking ahead.

As the last incense stick burned brightly in his hand, Tâm felt as though Lệ's spirit had returned to witness his presence today. Kneeling before the tombstone, he silently prayed as if speaking to his lost love, now resting in eternal sleep beneath the grave: "Lệ, my love! The war has ended, just as you always wished. But sadly, the separation between us is now one of life and death. Though I have another wife now, your image remains deeply embedded in my mind and heart, just as it always was. Do you remember when we used to go out, and I would call you my 'phà ca' of the Highlands? Today, you are still that 'phà ca' from those days, deeply cherished in my soul.

I have been saddened many times by the sight of raindrops dissolving and flowing away on the pavement because you once told me that you feared our love would vanish as quickly as those drops! The national radio station no longer reports the supposed glorious victories counted in the bodies of fallen Vietnamese brothers on either side, on the battlefield, which you always found so repugnant. The people on the other side have conquered the South and, in doing so, indirectly sent you to another world. They also forced me, as if I were their mortal enemy, to flee from my homeland. I suffered imprisonment and then had to leave my country, unable to care for your grave, leaving you to rest in a neglected, lonely place. Today, as the wounds of hatred in people's hearts are slowly fading into the past, I have finally had the chance to return and rebuild your grave so that you can rest in warmth and peace."

After his prayer, Tâm bowed down, placing the incense and gently kissing Lệ's tombstone.

The last rays of the setting sun were obscured by a gray cloud that stretched across the horizon above the row of stilt houses in the nearby Thượng village, casting a faint golden glow. The highland borderlands, with their vast stretches of land and sparse population, grew more desolate as evening approached. Occasionally, the rustling of dry leaves could be heard as the wind stirred through the grass nearby.

Tâm walked around the graves one last time, picking a beautiful yellow chrysanthemum, and laid it on her grave: "Lệ, I bid you farewell…"

The EVA Air Boeing 747 had just taken off from Tân Sơn Nhất Airport. Through the small window, the city of Saigon was left behind. The trees and houses gradually shrank in size.

The vastness of the sky was not enough to sever or erase the bonds of love that had connected people for so long. The memories lingered in their minds as their eyes remained fixed on the small window, delaying the moment of departure.

The plane reached cruising altitude at 37,000 feet, and beneath them, a thick layer of white clouds stretched out like cotton. Tâm reached over to help his wife undo her seatbelt:

"Now that you've returned to visit home for the first time, how do you feel?"

"I'm happy but also a little sad. Happy to have seen my parents in Cà Mau, who are still healthy, and to have gone with you to Pleiku to rebuild Lệ's grave and her mother's. But I'm sad because we have to go back to our current life, far away from those we care about and our homeland. What about you? Are you happy?"

"Of course I am! But I especially want to thank you for your kindness and generosity in helping me rebuild Lệ's grave so meticulously." Loan gently covered her husband's mouth to stop him from speaking further.

"That was my intention all along. After all, she once shared her feelings with you, and now that she's gone, taking care of her is like taking care of you." Tâm felt a deep sense of gratitude as he leaned in close to his wife's ear and whispered:

"I'm so fortunate and happy to have you as my wife!"

It was almost noon, and the sun was shining in from the west. Tâm quickly pulled down the window shade to block out the light streaming in. The steady hum of the jet engines continued, with the same constant tone that seemed to stretch on indefinitely. Loan pulled the personal blanket up to her chin, closing her eyes to find a quick nap.

Cà Mau, Saigon, Pleiku—those cities were now truly behind them, relegated to the past. Tâm sat there, seemingly carefree, but his soul was caught up in the storms of life. The bullet wound on his knee was a historical remnant of the civil war in Vietnam, and it still ached from time to time when the weather changed.

In the past, the forced departure from his homeland did not fill him with pride, and if anything, it brought a sense of humiliation. Even his return visit to his homeland did not make him puff up with pride like a new rich man, jingling with coins in his pocket. Let others change their names and hide their past; he remained a proud Vietnamese, steadfast in his loyalty to the simple name given to him by his parents when he was born.

As for the war, he saw it as a fundamental mistake, a repetition of the divisions that occurred in Vietnam during the Trịnh-Nguyễn conflict. The hatred and bloodshed among Vietnamese brothers were masked and justified by false and misleading slogans, turning it into a proxy war between capitalism and communism. Adding to that, personal ambitions and political factions pushed the country to the brink of ruin.

People live for the future, not the past. Harboring hatred and resentment only makes a person bitter, narrow-minded, and lacking in confidence. That's why he tried to cleanse his mind and forget those things.

He returned to revisit the old scenes and villages, to see the familiar canals behind his house, where the water levels rose and fell, and to mend the unfinished relationships of the past.

Anh LÊ VĂN TÂM, a Vietnamese heart, prayed to God to grant him and all other Vietnamese exiles eternal peace.

CHAPTER 11

Last Leaves of the Season

Original Title: "Những Chiếc Lá Cuối Mùa"

Flowers bloom only to wither.
Water hyacinths gather only to drift apart!

This is the natural law of creation. Just as humans are born from somewhere, grow old, and eventually must return to some realm of oblivion!?

This is a pessimistic view of life, what those with religion might call the "fleeting existence." But those with a positive spirit accept that change must follow such a law, and it cannot be reversed. They accept it but do not recognize it as the final condition that every human life must face.

No one forces us to get married, have children, and then say, "Children are karmic debts, and spouses are destined adversaries." Nor does anyone force us to live, to declare life as "a sea of suffering." If everyone used reason to judge and reflect, then everything must be this way and cannot be otherwise!

But, dear friends, "It's easy to say, but when it comes to practice, it's a different story."

According to philosophers, thoughts and conclusions usually stem from three places:

Desires and wants... often arise from the kidneys (gut feeling).

Love and hate... often arise from the heart.

Right and wrong, wickedness... often arise from the mind.

However, this classification depends on each person's understanding and the specific circumstances. Whether a decision is reasonable or not? I am neither a philosopher nor a psychologist, so I dare not delve further into these areas. The reason I discuss this a little is because it relates to the matters I want to express below.

Author: Nguyễn Văn Ba ("Nam")

After I wrote the piece "38 Years Looking Back" for the 63A class reunion in July 2001, now it's July 2003, and I must write something to commemorate the 40th anniversary of our reunion.

After the end of the Vietnam War, my entire family settled in the Pacific Northwest of the United States—"a good land where birds flock." In the blink of an eye, it has been twenty-eight long years. Many times, I never imagined this day would come—the day when I sit alone, "pulling my knees up to talk for fun," because who has the time to talk about everything under the sun with me all day?

###

Age is creeping up on me, following closely!

Now I understand the mindset of the elderly, who often sit by the window, pulling back the curtain to look outside for hours, something that used to puzzle me. Why don't older people like to go outside, or why do they tend to be more conservative?

I've been retired for nearly two years. People say that retirees love to garden and water plants, but I'm too lazy—"thick with grease," as my wife often says. The backyard is full of hundreds of flowers in all colors; occasionally, I'll prune them just to keep the neighbors happy. I spend my days getting up, sitting down, going to the library to borrow books to read, getting bored, and then sleeping—the days pass by like this, one after another.

But don't think I'm lazy! I just want to live according to the circumstances of each time and place. When I set foot in this city around July 1975, after five days at the sponsor's house, I found a job and worked until I retired in 2001. I never collected unemployment benefits, maybe because I'm like an ant, constantly working from dawn to dusk.

I say this so that everyone understands I am not a parasite, just living off society! My children are grown; the youngest is already twenty-nine years old. We've fulfilled our duties as parents, and that's enough to make me happy! Now, I occasionally meet old classmates or others, go out to eat, chat, listen to music, sing karaoke, and travel—these activities help pass the time and keep my mind occupied.

A few months ago, I received an invitation to the 63A reunion from Lê Văn Bút in Southern California, coinciding with a gathering of the 237th Lôi Thanh Squadron, with the Squadron Leader Nguyễn Phú Chính attending. In the evening, there was singing and dancing, organized annually by the Air Force Association of Southern California on Independence Day. We also planned to visit the new Vietnam War Memorial, which was inaugurated a few months ago.

According to the schedule, on July 4th, 2003, from 5 pm to 11 pm, we were to meet 63A friends and have dinner at Nguyễn Kim Chung's house. But we couldn't attend because we had to meet the 237th Lôi Thanh Squadron and other Chinook squadrons at a restaurant in Southern California.

My wife and I flew to Burbank Airport at 8:30 pm on July 3rd, 2003, where our youngest daughter, who works at Children's Hospital, picked us up. The next day, we were driven by Đại Bàng Đặng Đức Cường to the reunion with the Chinook squadron members, where we met Nguyễn Văn Mai, Nguyễn Phú Chính, Nguyễn Văn Hoa, Đinh Văn Huê, Nguyễn Đức Lợi, and other brothers like Bá Hùng, Cầu, Tôn, Châu, Quế, Thục, Phước, Nguyên, Chánh, Vũ, Ngọc, and many others who attended previous gatherings.

Almost three years later, seeing everyone again, they were much the same, except for one Đại Bàng with long, snow-white hair, who looked like an old sage descending from the heavens. I almost bowed in respect, but upon closer inspection, it turned out to be Đại Úy, the Squadron Treasurer "Lợi Móm."

I heard he had surgery recently, so seeing him healthy again is a relief. Wishing Đại Bàng a speedy recovery. It had been a long time since I last met Đinh Văn Huê, who now lives in Florida; he still looks young and full of energy. A while back, I met him at SeaTac Seattle Airport when he was training Chinese pilots on the H-34 helicopter, and after that, we lost touch.

But as with all parties, the one at the restaurant eventually came to an end. Phước and his wife kindly invited everyone to continue the party at their house, where we drank and sang until late at night. Lê Văn Cầu went solo this time, offering coffee cups with the Lôi Thanh insignia and the 2003 reunion logo as gifts to the participants. The members also contributed money to buy gifts for the remaining Chinook squadron members in Vietnam.

A few days later, we traveled to San Diego to visit Hoa and his wife, a former Operations Officer of the 237th Squadron who later worked in the Helicopter Division at the Air Force Headquarters. Since leaving the refugee camp at Camp Pendleton in July 1975, I've only met them twice. We didn't know their address to visit, but my wife suggested that since we were in the area, we should take the opportunity to visit and invite them to come up to the Pacific Northwest to see the scenery here.

Their home is about a 20-minute drive from San Diego, in a small county nestled in a valley surrounded by low mountains, similar to Camp Pendleton, the training base for the U.S. Marine Corps. The vegetation is sparse, unlike the lush greenery of the Pacific Northwest. During the summer, water conservation is necessary, so most of the grass has dried up.

Only one or two Vietnamese families live here, and there are no major factories; people mainly live off small businesses, so there aren't any large skyscrapers like in other places. Their children are all grown and have their own families. Hoa, who used to be plump and hearty, now appears thinner and moves slowly like an old man. I heard he had two strokes, which have significantly affected his health. His wife jokingly said to my wife,

"Our family was different back then, but now the kids have all moved out, leaving only two old monkeys staring at each other!"

In truth, every family is like that—water runs over stones and eventually wears them down, just as time erodes the body. "Birth, Aging, Sickness, Death"—whether we want it or not, everyone must accept this cycle!

On Saturday, July 5th, 2003, around 9:00 am, Vũ Đăng Hùng and his wife picked us up at our residence to go to the Emerald Bay Restaurant in Santa Ana to meet 63A classmates. There, I saw familiar faces like Bửu Vi, Cửu, Bút, etc. I noticed that Bút still holds on to the saying, "The squadron is paradise..." as he busily handled the responsibilities as the head of the organizing committee. I, on the other hand, was still my usual carefree self—truthfully, even if I wanted to help, I wouldn't know what to do, so I invited Hùng and Bửu Vi's wife to have breakfast, fill our stomachs, and then deal with whatever came our way.

The Southern California sun is as intense as in the tropics; it burns the skin but doesn't leave the neck or armpits sticky, sweaty, and humid like in Vietnam or Texas. I remember once visiting a friend in Cowboy Country; we went for a walk along the Riverwalk. From 11:00 am until the afternoon, the sky suddenly unleashed a torrential downpour like in Saigon, but the rain here wasn't cold like where we live. The warm raindrops felt as if they had been heated by the sun earlier.

I suddenly felt young again, recalling the time when I was five or six years old in the countryside, running out to bathe in the rain or catching fish in the ditches, watching the water flow from the garden down to the pond. But those days are long gone—the days of sitting at the bamboo gate in the morning, waiting for my mother to return from the market with a string of sticky rice cakes around my neck or a piece of bánh bò or bánh còng and feeling so happy that I would jump up and down. Now, those memories are far behind me, over fifty years ago!

Sometimes, sitting alone and reflecting, I realize how illogical human beings can be. Why do people have to chase after what is called "civilization" instead of living according to primitive ways? "The more you strive, the longer the suffering lasts." In the end, it's only humans who make each other suffer, not any god, Buddha, or saint!

Well, life is like a stream, originating from the high mountains and flowing out to the sea; even if we want to stop it, we can't. When it reaches the ocean and mixes with the rest, it evaporates, and the wind carries it back to the mountains. There, it condenses into rain and flows back down to the sea. This cycle repeats endlessly, never changing.

At around 10:00 am, we returned to the Emerald Bay seafood restaurant to meet everyone and see if there were any new faces. Besides those who attended in previous years, there were new participants like Tôn Thất Thuần, Phan Hiền Tính, Nguyễn Quốc Đạt, Nguyễn Quí An, Trần Quốc Bàn, Nguyễn Thành Cứ, Trần Văn Nghiêm, Bửu Vi, and others.

I must honestly say that the 63A organizing committee is very thorough. They had a solemn altar for the Nation on stage, complete with military drums and gongs. After the ceremony, as usually done by national associations and organizations, Chi Tạ Thương Tứ acted as the MC, introducing group songs performed by the grandmothers (pilots' wives) of the 63A class. Though they were older, they still looked fresh in their green áo dài from the past! Then came the solo performances by the elderly men, singing various songs in Vietnamese, French, English, etc.

But it lacked the traditional operatic segment, making us rural folk feel something was missing! In summary, the organization was very thoughtful, and the main content of the reunion was to give classmates the chance to see each other again after many years in this foreign land. My family and I thank the 63A organizing committee.

In the evening, we joined Mr. and Mrs. Vũ Đăng Hùng at the "Space Reunion Night" at the DoubleTree Hotel in Orange County, where Mrs. Hùng was part of the organizing committee. Let me take this opportunity to share a bit about Mrs. Hùng. She is the sister of Lieutenant Đặng Đức Cường, and her real name is Đặng Thị Tuyết Hường. Before 1975, Cường was with the 237th Squadron, flying with me. Cường had another sister, who was the wife of Lieutenant Quang (Quang "Soft Eyes"), who also flew in the same squadron. Quang was very handsome, but he passed away before 1975, leaving behind a young wife and daughter. We met Mrs. Quang again at Hùng's house in 2001. Mrs. Hùng seems to have a special connection with the Air Force, as she married Hùng, who was in the same class as us.

She is very active in social work, and both Mr. and Mrs. Hùng are part of the 63A organizing committee. They have spent a lot of time printing the book "63/SVSQ 40 Years Reunion" and making coffee mugs with the names of the classmates to distribute as souvenirs of the reunion day.

In addition, they are also involved in the organizing committee of the Air Force Association in Southern California, and Mrs. Hùng volunteers to help with the publication of the Lý Tưởng Air Force magazine, encouraging everyone to subscribe. When we arrived, Mr. and Mrs. Hùng were very hospitable, welcoming us warmly, and we are very grateful for their kindness and warmth towards our friends.

On Sunday morning, July 6th, 2003, we met for breakfast one last time before parting ways. Trần Phước Hội, the author of the article "Air Force in the Forgotten Times" in Lý Tưởng magazine, called on everyone to contribute a little to buy gifts for the 63A classmates currently facing hardship in Vietnam.

We also appreciate the spirit of Tạ Thượng Tứ, who volunteered to be the head of the social committee, helping to share news, send condolence wreaths, and post obituaries when any of us takes our final flight to the great beyond so that everyone can say "See you later!" or "Goodbye" to that friend. Mentioning this does bring some sadness, doesn't it, friends? But is there any golden leaf that doesn't fall from the branch when the first winds of winter blow?

In the afternoon, Mr. and Mrs. Hùng invited us back to the clubhouse, where Võ Ý was introducing his first book, "The Vertical Biography of Thảo." I remember back in the day when we were stationed in Nha Trang, the Thần Tượng Squadron was near the 114th Squadron, so I knew what Võ Ý looked like—a fair-skinned, handsome man with a cheerful personality. But back then, he was too young to be as "enlightened" as he is now. After more than 30 years, I met him again, and he has changed a lot; perhaps now he is more "enlightened," looking more solemn like an old sage, speaking less than before. I wish him good luck with the book sales so that he can recoup the printing costs.

Southern California, the land of charming girls with honey-colored skin, where the four seasons are almost alike, is quite different from where we live. In the fall and winter, we spend most of our days indoors, looking out at the yard, watching the yellow leaves fall and flutter in the drizzle, feeling a deep melancholy.

If you are a heavy-hearted poet, I'm sure it will inspire you to write some really good verses, perhaps even better than the great poets of the Tang Dynasty. Speaking of continuous rain, it reminds me of the time when I flew the H-34 on missions in the Central Highlands, where it was "Dust in the Sun, Mud in the Rain." After flying operations, we would pile into jeeps, go out for dinner, and then split up to visit familiar coffee shops, listening to Trịnh Công Sơn's music and watching the rain drip onto the street. Looking back, those were precious times—youthful days with little contemplation, lots of romantic love, living fast, and not knowing if there would be a tomorrow while flying missions over enemy territory.

Well, bringing up old stories only adds to the sadness, with no benefit at this point! All of us now have gray hair, and in a few years, when we meet again, a warm smile and a firm handshake are enough; there's no need to bond as deeply as Bá Nha and Tử Kỳ.

On the night of July 4th, 2003, I deeply regretted not being able to make it to Nguyễn Kim Chung's house. But thanks to the DVD that Bút sent, I could follow what happened that evening. I heard Bút announce in front of everyone that he would resign as President of the 63A SVSQ Association, reasoning that it would give others a chance to organize future events. Some, like Bàn, Tứ, and Nghiêm, raised different opinions.

But I think (speaking like a soldier), don't "abandon the group." Even though we weren't present that night, I have some thoughts: "Organizing is not difficult, but gaining the goodwill of people's hearts is incredibly challenging..." It doesn't matter where it's held; it's easy for those with abundant resources but for those who lack the means, traveling far, renting hotels and cars... it's costly, isn't it?

But money isn't the deciding factor in whether to go or not?! But I value the leadership and warm camaraderie of our classmates, so whether by foot or facing difficulties, I will still make it. This is no longer about the military or factions; it's about the leadership and warm friendship of classmates. I won't discuss this further, but I suggest keeping the core leadership team in Southern California as it is. The "Organizer" can be based on the decision of the majority.

For example, if the event is held in Texas, Nghiêm can be the local organizing chairman, responsible for coordinating logistics, transportation, and accommodations, in collaboration with the executive committee in California, sharing tasks like sending invitations to 63A members to participate (I think Nghiêm has the capability to handle that. But what about when it's held elsewhere?).

In summary, if you want the 63A class to stay together, should the executive committee in Southern California reflect on the ideals they initially set forth and reconsider? I know that while organizing is hard, criticizing and judging are as easy as flipping a hand. Even the weather gets criticized, let alone people!

But as long as you take pride in your conscience, that's what matters. Ladies, if you read this, please discuss it with your husbands—continue with joy and ignore any unkind words. The 63A SVSQ class will never abandon you!

Please, don't be like the last leaves of the season!

Redmond SVSQ/63A
Nguyễn Văn Ba [Nam]

Nguyễn Văn Ba
01/10/1940 – 06/17/2013

ABOUT THE AUTHOR

Mr. Ba Van Nguyen (formal Vietnamese name/spelling: Nguyễn Văn Ba), a remarkable figure, was born on January 10th, 1940, in Long An, Vietnam. He was the second oldest of four brothers and five sisters in a family that deeply valued resilience and unity. At the age of 23, Mr. Nguyen joined the South Vietnamese Air Force, where he trained as a pilot and demonstrated exceptional skill and bravery. His military career, marked by rigorous training in both Vietnam and the United States, led him to eventually become a Lieutenant Colonel in the Air Force, flying the formidable Chinook (CH-47) among many other helicopters.

On April 29th, 1975, in a heroic act of love and courage, Mr. Nguyen piloted his Chinook helicopter to rescue his family from the turmoil of the Vietnam War. This extraordinary feat of flying and quick thinking brought his loved ones to safety aboard the USS Kirk, a moment immortalized in military history and remembered as an act of profound bravery. His heroic story was a key highlight in the 2015 Oscar-nominated film "Last Days in Vietnam," written and produced by Rory Kennedy, the youngest daughter of Robert F. Kennedy.

Following his escape to the United States, Mr. Nguyen, along with his wife, Nho Tran Nguyen, rebuilt their lives from the ground up, showcasing his relentless determination. He pursued education, worked multiple jobs, and eventually secured a position at Boeing, where he contributed to military electronics for 18 years until his retirement in 2002. Despite the challenges of starting anew in a foreign land, Mr. Nguyen's dedication to his family and his unwavering spirit never faltered.

Mr. Nguyen was also a passionate writer, contributing to Vietnamese newspapers, and he enjoyed various hobbies, including playing the keyboard, singing karaoke, and spending time with his grandchildren – Lexi, Liam, Lucas, Miles, and Lila. His life was a testament to the power of hope, resilience, and love.

Diagnosed with Frontal Temporal Dementia in 2006, Mr. Nguyen faced his illness with the same courage that defined his life. He passed away peacefully on June 17th, 2013, leaving behind a legacy of bravery, love, and inspiration. He is survived by his devoted wife, Mrs. Nho Nguyen, and his children and grandchildren, who continue to honor his memory. Mr. Ba Van Nguyen's story is one of a true hero, a loving father, and an enduring symbol of the human spirit.

www.nguyenvanba.com

www.ingramcontent.com/pod-product-compliance
Lightning Source LLC
Chambersburg PA
CBHW020428130626
46549CB00001B/33